Universitext

Universitext

Universitext is a series of textbooks that presents material from a wide variety of mathematical disciplines at master's level and beyond. The books, often well class-tested by their author, may have an informal, personal even experimental approach to their subject matter. Some of the most successful and established books in the series have evolved through several editions, always following the evolution of teaching curricula, to very polished texts.

Thus as research topics trickle down into graduate-level teaching, first textbooks written for new, cutting-edge courses may make their way into *Universitext.*

For further volumes:
http://www.springer.com/series/223

Donu Arapura

Algebraic Geometry over the Complex Numbers

Donu Arapura
Department of Mathematics
Purdue University
150 N. University Street
West Lafayette, IN 47907
USA
dvb@math.purdue.edu

ISSN 0172-5939 e-ISSN 2191-6675
ISBN 978-1-4614-1808-5 e-ISBN 978-1-4614-1809-2
DOI 10.1007/978-1-4614-1809-2
Springer New York Dordrecht Heidelberg London

Library of Congress Control Number: 2012930383

Mathematics Subject Classification (2010): 14-XX, 14C30

Printed on acid-free paper

Springer is part of Springer Science+Business Media (www.springer.com)

To my parents, who taught me that knowledge is something to be valued

Preface

Algebraic geometry is the geometric study of sets of solutions to polynomial equations over a field (or ring). These objects, called algebraic varieties (or schemes or ...), can be studied using tools from commutative and homological algebra. When the field is the field of complex numbers, these methods can be supplemented with transcendental ones, that is, by methods from complex analysis, differential geometry, and topology. Much of the beauty of the subject stems from the rich interplay of these various techniques and viewpoints. Unfortunately, this also makes it a hard subject to learn. This book evolved from various courses in algebraic geometry that I taught at Purdue. In these courses, I felt my job was to act as a guide to the vast terrain. I did not feel obligated to cover everything or to prove everything, because the standard accounts of the algebraic and transcendental sides of the subject by Hartshorne [60] and Griffiths and Harris [49] are remarkably complete, and perhaps a little daunting as a consequence. In this book I have tried to maintain a reasonable balance between rigor, intuition, and completeness. As for prerequisites, I have tried not to assume too much more than a mastery of standard graduate courses in algebra, analysis, and topology. Consequently, I have included discussions of a number of topics that are technically not part of algebraic geometry. On the other hand, since the basics are covered quickly, some prior exposure to elementary algebraic geometry (at the level of say Fulton [40], Harris [58, Chapters 1–5] or Reid [97]) and calculus with manifolds (as in Guillemin and Pollack [56, Chapters 1 & 4] or Spivak [109]) would certainly be desirable, although not absolutely essential.

This book is divided into a number of somewhat independent parts with slightly different goals. The starred sections can be skipped without losing much continuity. The first part, consisting of a single chapter, is an extended informal introduction illustrated with concrete examples. It is really meant to build intuition without a lot of technical baggage. Things really get going only in the second part. This is where sheaves are introduced and used to define manifolds and algebraic varieties in a unified way. A watered-down notion of scheme—sufficient for our needs—is also presented shortly thereafter. Sheaf cohomology is developed quickly from scratch in Chapter 4, and applied to de Rham theory and Riemann surfaces in the next few chapters. By Part III, we move into Hodge theory, which is really the heart of transcendental algebraic geometry. This is where algebraic geometry meets differential geometry on the one hand, and some serious homological algebra on the other. Although I have skirted around some of the analysis, I did not want to treat this entirely as a black box. I have included a sketch of the heat equation proof of the Hodge theorem, which I think is reasonably accessible and quite pretty. This theorem along with the weak and hard Lefschetz theorems have some remarkable consequences for the geometry and topology of algebraic varietes. I discuss some of these applications in the remaining chapters. From Hodge theory, one extracts a set of useful invariants called Hodge numbers, which refine the Betti numbers. In the fourth part, I consider some methods for actually computing these numbers for various examples, such as hypersurfaces. The task of computing Hodge numbers can be converted to an essentially algebraic problem, thanks to the GAGA theorem, which is explained here as well. This theorem gives an equivalence between certain algebraic and analytic objects called coherent sheaves. In the fifth part, I end the book by touching on some of the deeper mysteries of the subject, for example, that the seemingly separate worlds of complex geometry and characteristic p geometry are related. I will also explain some of the conjectures of Grothendieck, Hodge, and others along with a context to put them in.

I would like to thank Bill Butske, Harold Donnelly, Ed Dunne, Georges Elencwajg, Anton Fonarev, Fan Honglu, Su-Jeong Kang, Mohan Ramachandran, Peter Scheiblechner, Darren Tapp, and Razvan Veliche for their suggestions and clarifications. My thanks also to the NSF for their support over the years.

Donu Arapura
Purdue University
November, 2011

Contents

Part I
Introduction through Examples

Chapter 1
Plane Curves

Algebraic geometry *is* geometry. This sounds like a tautology, but it will be easy to forget once we start learning about sheaves, cohomology, Hodge structures, and so on. So perhaps it is a good idea to keep ourselves grounded by taking a very quick tour of the classical theory of complex algebraic curves in the plane, using only primitive, and occasionally nonrigorous, tools. This will hopefully provide a better sense of where the subject comes from and where we want to go. Once we have laid the proper foundations in later chapters, we will revisit these topics and supply some of the missing details.

The treatment here is very much inspired by Clemens's wonderful book [20] as well as the first chapter of Arbarello, Cornalba, Griffiths, and Harris's treatise [5].

1.1 Conics

A complex *affine algebraic plane curve* is the set of zeros

$$X = V(f) = \{(x,y) \in \mathbb{C}^2 \mid f(x,y) = 0\} \tag{1.1.1}$$

of a nonconstant polynomial $f(x,y) \in \mathbb{C}[x,y]$. Notice that we call this a curve because it has one complex dimension. However, we will be slightly inconsistent and refer to this occasionally as a surface, especially when we want to emphasize its topological aspects. The curve X is called a conic if f is a quadratic polynomial. The study of conics over $\mathbb{R}$ is something one learns in school. The complex case is actually easier, since distinctions between ellipses and hyperbolas disappear. The group of affine transformations

$$\begin{pmatrix} x \\ y \end{pmatrix} \mapsto \begin{pmatrix} a_{11}x + a_{12}y + b_1 \\ a_{21}x + a_{22}y + b_2 \end{pmatrix}$$

with $\det(a_{ij}) \neq 0$ acts on $\mathbb{C}^2$. High-school methods can be used to show that after making a suitable affine transformation, there are three possibilities along with subcases:

1. A union of two (possibly identical, parallel, or incident) lines.
2. A circle $x^2 + y^2 = 1$.
3. A parabola $y = x^2$.

Things become simpler if we add a line at infinity. This can be achieved by passing to the projective plane $\mathbb{P}^2 = \mathbb{P}^2_{\mathbb{C}}$, which is the set of lines in $\mathbb{C}^3$ containing the origin. To any $(x_0, x_1, x_2) \in \mathbb{C}^3 - \{0\}$, there corresponds a unique point $[x_0, x_1, x_2] = \mathrm{span}\{(x_0, x_1, x_2)\} \in \mathbb{P}^2$. We embed $\mathbb{C}^2 \subset \mathbb{P}^2$ as an open set by sending $(x, y) \mapsto [x, y, 1]$. The line at infinity is the complement given by $x_2 = 0$. The x_i are called homogeneous coordinates, although these are not coordinates in the technical sense of the word. The true coordinates are given by the ratios $x_0/x_2, x_1/x_2$ on the chart $\{x_2 \neq 0\}$, $x_0/x_1, x_2/x_1$ on $\{x_1 \neq 0\}$, and $x_1/x_0, x_2/x_0$ on $\{x_0 \neq 0\}$. We identify $x = x_0/x_2$, $y = x_1/x_2$. The closure of an affine plane curve $X = V(f)$ in $\mathbb{P}^2$ is the *projective algebraic plane curve*

$$\overline{X} = \{[x_0, x_1, x_2] \in \mathbb{P}^2 \mid F(x_0, x_1, x_2) = 0\}, \tag{1.1.2}$$

where

$$F(x_0, x_1, x_2) = x_2^{\deg f} f(x_0/x_2, x_1/x_2)$$

is the homogenization of f.

The projective linear group $\mathrm{PGL}_3(\mathbb{C}) = \mathrm{GL}_3(\mathbb{C})/\mathbb{C}^*$ acts on $\mathbb{P}^2$ via the standard $\mathrm{GL}_3(\mathbb{C})$ action on $\mathbb{C}^3$. The game is now to classify the projective conics up to a projective linear transformation. The list simplifies to three cases including all the degenerate cases: a single line, two distinct lines that meet, and the projectivized parabola C given by

$$x_0^2 - x_1 x_2 = 0. \tag{1.1.3}$$

If we allow nonlinear transformations, then things simplify further. The map from the complex projective line to the plane given by $[s,t] \mapsto [st, s^2, t^2]$ gives a bijection of $\mathbb{P}^1$ to C. The inverse can be expressed as

$$[x_0, x_1, x_2] \mapsto \begin{cases} [x_1, x_0] & \text{if } (x_1, x_0) \neq 0, \\ [x_0, x_2] & \text{if } (x_0, x_2) \neq 0. \end{cases}$$

Note that these expressions are consistent by (1.1.3). These formulas show that C is homeomorphic, and in fact isomorphic in a sense to be explained in the next chapter, to $\mathbb{P}^1$. Topologically, this is just the two-sphere S^2.

Exercises

1.1.1. Show that the subgroup of $\mathrm{PGL}_3(\mathbb{C})$ fixing the line at infinity is the group of affine transformations.

1.1.2. Deduce the classification of projective conics from the classification of quadratic forms over $\mathbb{C}$.

1.1.3. Deduce the classification of affine conics from Exercise 1.1.1.

1.2 Singularities

We recall a version of the implicit function theorem:

Theorem 1.2.1. *If $f(x,y)$ is a polynomial such that $f_y(0,0) = \frac{\partial f}{\partial y}(0,0) \neq 0$, then in a neighborhood of $(0,0)$, $V(f)$ is given by the graph of an analytic function $y = \phi(x)$ with $\phi'(0) \neq 0$.*

In outline, we can use Newton's method. Set $\phi_0(x) = 0$, and

$$\phi_{n+1}(x) = \phi_n(x) - \frac{f(x,\phi_n(x))}{f_y(x,\phi_n(x))}.$$

Then ϕ_n will converge to ϕ. Proving this requires some care, of course.

A point (a,b) on an affine curve $X = V(f)$ is a *singular* point if

$$\frac{\partial f}{\partial x}(a,b) = \frac{\partial f}{\partial y}(a,b) = 0;$$

otherwise, it is *nonsingular*. In a neighborhood of a nonsingular point, we can use the implicit function theorem to write x or y as an analytic function of the other variable. So locally at such a point, X looks like a disk. By contrast, the nodal curve $y^2 = x^2(x+1)$ looks like a union of two disks touching at $(0,0)$ in a small neighborhood of this point given by $|t \pm 1| < \varepsilon$ in the parameterization

$$x = t^2 - 1,$$
$$y = xt.$$

See Figure 1.1 for the real picture.

The two disks are called branches of the singularity. Singularities may have only one branch, as in the case of the cusp $y^2 = x^3$ (Figure 1.2).

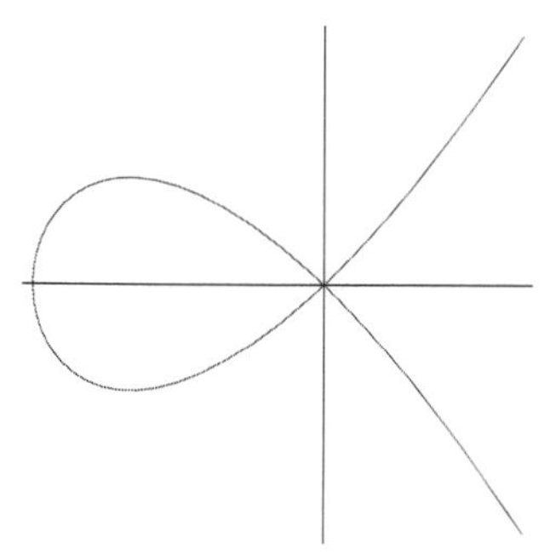

Fig. 1.1 Nodal curve.

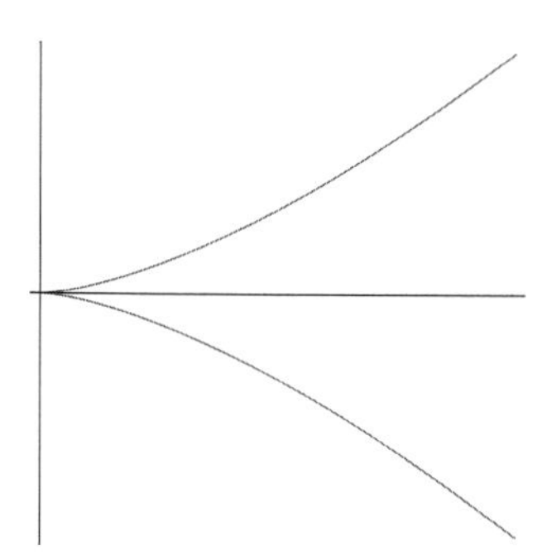

Fig. 1.2 Cuspidal curve.

In order to get a better sense of the topology of a complex singularity, we can intersect the singularity $f(x,y) = 0$ with a small 3-sphere,

$$S^3 = \{(x,y) \in \mathbb{C}^2 \mid |x|^2 + |y|^2 = \varepsilon^2\},$$

to get a circle S^1 embedded in S^3 in the case of one branch. The embedded circle is unknotted when this is nonsingular, but it would be knotted otherwise. For the cusp, we would get a trefoil or $(2,3)$ torus knot [87].

The affine plane curve X (1.1.1) is called nonsingular if all its points are nonsingular. The projective curve $\overline{X}$ (1.1.2) is nonsingular if all of its points including points at infinity are nonsingular. In explicit terms, this means that the affine curves $f(x,y) = F(x,y,1) = 0$, $F(1,y,z) = 0$, and $F(x,1,z) = 0$ are all nonsingular. A nonsingular curve is an example of a Riemann surface or a one-dimensional complex manifold.

Exercises

1.2.2. Prove the convergence of Newton's method in the ring of formal power series $\mathbb{C}[[x]]$, where $\phi_n \to 0$ if and only if the degree of its leading term $\to \infty$. Note that this ring is equipped with the x-adic topology, where the ideals (x^N) form a fundamental system of neighborhoods of 0.

1.2.3. Prove that Fermat's curve $x_0^n + x_1^n + x_2^n = 0$ in $\mathbb{P}^2$ is nonsingular.

1.3 Bézout's Theorem

An important feature of the projective plane is that any two lines meet. In fact, it has a much stronger property:

Theorem 1.3.1 (Weak Bézout's theorem). *Any two algebraic curves in* $\mathbb{P}^2$ *intersect.*

We give an elementary classical proof here using resultants. Given two monic polynomials

$$f(y) = y^n + a_{n-1}y^{n-1} + \cdots + a_0 = \prod_{i=1}^{n}(y - r_i),$$

$$g(y) = y^m + b_{m-1}y^{m-1} + \cdots + b_0 = \prod_{j=1}^{m}(y - s_j),$$

their resultant is the expression

$$\operatorname{Res}(f,g) = \prod_{ij}(r_i - s_j).$$

It is obvious that $\operatorname{Res}(f,g) = 0$ if and only if f and g have a common root. From the way we have written it, it is also clear that $\operatorname{Res}(f,g)$ is a polynomial of degree mn in $r_1,\ldots,r_n,s_1,\ldots,s_m$ that is symmetric separately in the r's and s's. So it can be rewritten as a polynomial in the elementary symmetric polynomials in the r's and s's. In other words, $\operatorname{Res}(f,g)$ is a polynomial in the coefficients a_i and b_j. Standard formulas for it can be found, for example, in [76].

Proof. Assume that the curves are given by homogeneous polynomials $F(x,y,z)$ and $G(x,y,z)$ respectively. After translating the line at infinity if necessary, we can assume that the polynomials $f(x,y) = F(x,y,1)$ and $g(x,y) = G(x,y,1)$ are both nonconstant in x and y. Treating these as polynomials in y with coefficients in $\mathbb{C}[x]$, the resultant $\operatorname{Res}(f,g)(x)$ can be regarded as a nonconstant polynomial in x. Since $\mathbb{C}$ is algebraically closed, $\operatorname{Res}(f,g)(x)$ must have a root, say a. Then $f(a,y) = 0$ and $g(a,y) = 0$ have a common solution. □

It is worth noting that this argument is entirely algebraic, and therefore applies to any algebraically closed field, such as the field of algebraic numbers $\overline{\mathbb{Q}}$. So as a bonus, we get the following arithmetic consequence.

Corollary 1.3.2. *If the curves are defined by equations with coefficients in $\overline{\mathbb{Q}}$, then the there is a point of intersection with coordinates in $\overline{\mathbb{Q}}$.*

Suppose that the curves $C, D \subset \mathbb{P}^2$ are irreducible and distinct. Then it is not difficult to see that $C \cap D$ is finite. We can ask how many points are in the intersection. To get a more refined answer, we can assign a multiplicity to the points of intersection. If the curves are defined by polynomials $f(x,y)$ and $g(x,y)$ with a common isolated zero at the origin $O = (0,0)$, then define the *intersection multiplicity* at O by

$$i_O(C,D) = \dim \mathbb{C}[[x,y]]/(f,g),$$

where $\mathbb{C}[[x,y]]$ is the ring of formal power series in x and y. The ring of convergent power series can be used instead, and it would lead to the same result. The multiplicities can be defined at other points by a similar procedure. While this definition is concise, it does not give us much geometric insight. Here is an alternative: $i_p(D,E)$ is the number of points close to p in the intersection of small perturbations of these curves. More precisely, we have the following:

Lemma 1.3.3. *$i_p(D,E)$ is the number of points in $\{f(x,y) = \varepsilon\} \cap \{g(x,y) = \eta\} \cap B_\delta(p)$ for small positive $|\varepsilon|, |\eta|, \delta$, where $B_\delta(p)$ is a δ-ball around p.*

Proof. This follows from [42, 1.2.5e]. □

There is another nice interpretation of this number worth mentioning. If $K_1, K_2 \subset S^3$ are disjoint knots, perhaps with several components, their linking number is roughly the number of times one of them passes through the other. A precise definition can be found in any basic book on knot theory (e.g., [98]).

Theorem 1.3.4. *Given a small sphere S^3 about p, $i_p(D,E)$ is the linking number of $D \cap S^3$ and $E \cap S^3$.*

Proof. See [42, 19.2.4]. □

We can now state the strong form of Bézout's theorem. We will revisit this in Corollary 11.2.7.

Theorem 1.3.5 (Bézout's theorem). *Suppose that C and D are algebraic curves with no common components. Then the sum of intersection multiplicities at points of $C \cap D$ equals the product of degrees $\deg C \cdot \deg D$, where $\deg C$ and $\deg D$ are the degrees of the defining polynomials.*

Corollary 1.3.6. *The cardinality $\#C \cap D$ is at most $\deg C \cdot \deg D$.*

Exercises

1.3.7. Show that the vector space $\mathbb{C}[[x,y]]/(f,g)$ considered above is finite-dimensional if $f=0$ and $g=0$ have an isolated zero at $(0,0)$.

1.3.8. Suppose that $f=y$. Using the original definition show that $i_O(C,D)$ equals the multiplicity of the root $x=0$ of $g(x,0)$. Now prove Bézout's theorem when C is a line.

1.4 Cubics

We now turn our attention to the very rich subject of cubic curves. In the degenerate case, the polynomial factors into a product of a linear and quadratic polynomial or three linear polynomials. Then the curve is a union of a line with a conic or three lines. So now assume that X (1.1.2) is defined by an irreducible cubic polynomial. It is called an *elliptic curve* because of its relationship to elliptic functions and integrals.

Lemma 1.4.1. *After a projective linear transformation, an irreducible cubic can be transformed into the projective closure of an affine curve of the form $y^2 = p(x)$, where $p(x)$ is a cubic polynomial. This is nonsingular if and only if $p(x)$ has no multiple roots.*

Proof. See [105, III §1]. □

We note that nonsingular cubics are very different from conics, even topologically.

Proposition 1.4.2. *A nonsingular cubic X is homeomorphic to a torus $S^1\times S^1$.*

There is a standard way to visualize this (see Figure 1.3). Mark four points $a,b,c,d=\infty$ on $\mathbb{P}^1$, where the first three are the roots of $p(x)$. Join a to b and c to d by nonintersecting arcs α and β. The preimage of the complement $Y=\mathbb{P}^1-(\alpha\cup\beta)$ in X should fall into two pieces both of which are homeomorphic to Y. So in other words, we can obtain X by first taking two copies of the sphere, slitting them along α and β, and then gluing them along the slits to obtain a torus.

Perhaps that was not very convincing. Instead, we will use a parameterization by elliptic functions to verify Proposition 1.4.2 and more. By applying a further projective linear transformation, we can put our equation for X into Weierstrass form

$$y^2 = 4x^3 - a_2x - a_3 \tag{1.4.1}$$

with discriminant $a_2^3 - 27a_3^2 \neq 0$. The idea is to parameterize the cubic by the elliptic integral

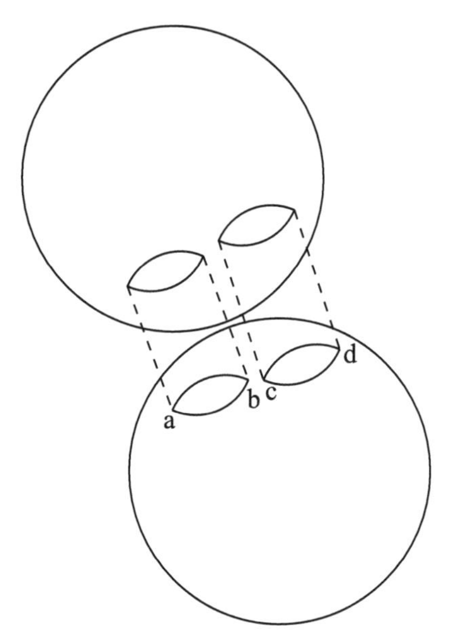

Fig. 1.3 Visualizing the cubic.

$$E(z) = \int_{z_0}^{z} \frac{dx}{y} = \int_{z_0}^{z} \frac{dx}{\sqrt{4x^3 - a_2x - a_3}}. \tag{1.4.2}$$

While the integrand appears to have singularities at the zeros of $p(x) = 4x^3 - a_2x - a_3$, by differentiating $y^2 = p(x)$ and substituting, we see that

$$\frac{dx}{y} = \frac{2dy}{p'(x)}$$

has no singularities at these points. Thus the integral (1.4.2) should determine a holomorphic function E, but it would be "multivalued" because it depends on the path of integration. We should understand this to mean that E is really a holomorphic function on the universal cover $\tilde{X}$ of X. To understand the multivaluedness more precisely, let us introduce the set of periods $L \subset \mathbb{C}$ as the set of integrals of dx/y around closed loops on X. The set L is actually a subgroup. To see this, let Loop(X) be the free abelian group consisting of finite formal integer linear combinations of $\sum n_i \gamma_i$ of closed loops on X. The map $\gamma \mapsto \int_\gamma dx/y$ gives a homomorphism of Loop(X) $\to \mathbb{C}$. The image is exactly L, and it is isomorphic to the first homology group $H_1(X, \mathbb{Z})$, which will discussed in more detail later on. We can see that E descends to a map $X \to \mathbb{C}/L$, which is in fact the homeomorphism alluded to in Proposition 1.4.2.

The above story can be made more explicit by working backward in some sense. First, we characterize the group L in a different way.

Theorem 1.4.3. *There exists a unique lattice $L \subset \mathbb{C}$, i.e., an abelian subgroup generated by two $\mathbb{R}$-linearly independent numbers, such that*

$$a_2 = g_2(L) = 60 \sum_{\lambda \in L, \lambda \neq 0} \lambda^{-4},$$

$$a_4 = g_3(L) = 140 \sum_{\lambda \in L, \lambda \neq 0} \lambda^{-6}.$$

Proof. [106, I 4.3]. □

Fix the period lattice L as above. The Weierstrass $\wp$-function is given by

$$\wp(z) = \frac{1}{z^2} + \sum_{\lambda \in L, \lambda \neq 0} \left(\frac{1}{(z-\lambda)^2} - \frac{1}{\lambda^2} \right).$$

This converges to an elliptic function, which means that it is meromorphic on $\mathbb{C}$ and doubly periodic: $\wp(z+\lambda) = \wp(z)$ for $\lambda \in L$ [105]. This function satisfies the Weierstrass differential equation

$$(\wp')^2 = 4\wp^3 - g_2(L)\wp^2 - g_3(L).$$

Thus $\wp$ gives exactly the inverse to the integral E. We get an embedding $\mathbb{C}/L \to \mathbb{P}^2$ given by

$$z \mapsto \begin{cases} [\wp(z), \wp'(z), 1] & \text{if } z \notin L, \\ [0,1,0] & \text{otherwise.} \end{cases}$$

The image is the cubic curve X defined by (1.4.1). This shows that X is a torus topologically as well as analytically. See [105, 106] for further details.

Exercises

1.4.4. Prove that the projective curve defined by $y^2 = p(x)$ is nonsingular if and only if $p(x)$ has no repeated roots.

1.4.5. Prove that the singular projective curve $y^2 = x^3$ is homeomorphic to the sphere.

1.5 Genus 2 and 3

A compact orientable surface is classified up to homeomorphism by a single number called the *genus*. The genus is 0 for a sphere, 1 for a torus, and 2 for the surface depicted in Figure 1.4.

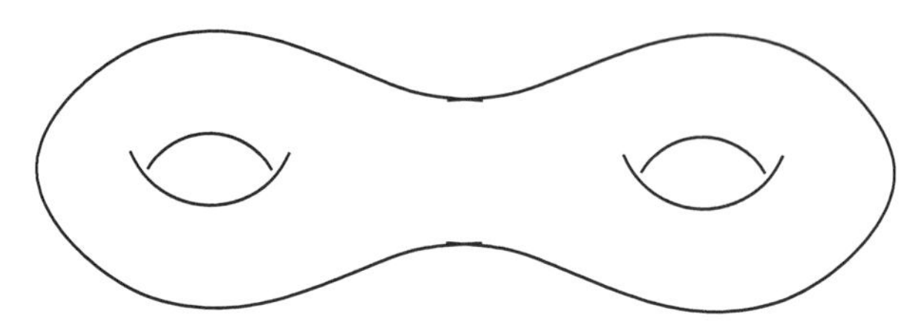

Fig. 1.4 Genus-2 surface.

We claim that a nonsingular quartic in $\mathbb{P}^2$ is a three-holed or genus-3 surface. A heuristic argument is as follows. Let $f \in \mathbb{C}[x,y,z]$ be the defining equation of our nonsingular quartic, and let $g = (x^3+y^3+z^3)x$. The degenerate quartic $g = 0$ is the union of a nonsingular cubic and a line. Topologically this is a torus meeting a sphere in three points (Figure 1.5). Consider the pencil $f_t = tf + (1-t)g$. As t evolves from $t = 0$ to 1, the three points of intersection in $f_t = 0$ open up into circles, resulting in a genus-3 surface (Figure 1.6).

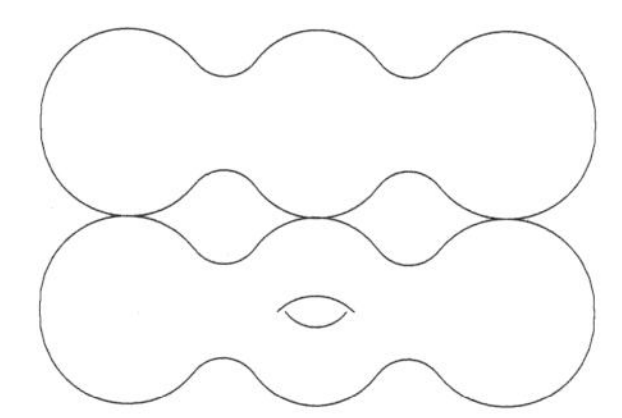

Fig. 1.5 Degenerate quartic.

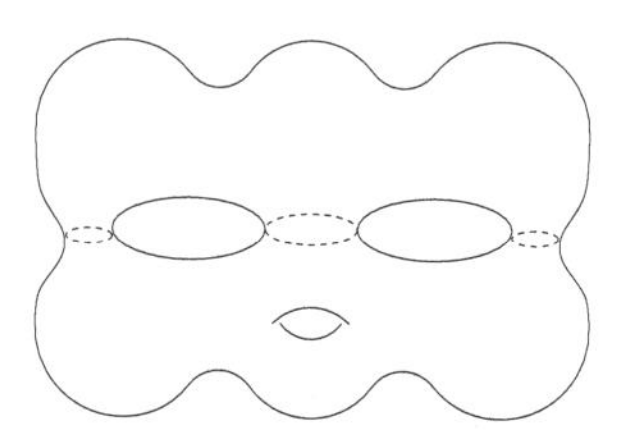

Fig. 1.6 Nonsingular quartic.

In going from degree 3 to 4, we seem to have skipped over genus 2. It is possible to realize such a surface in the plane, but only by allowing singularities. Consider the curve $X \subset \mathbb{P}^2$ given by

$$x_0^2x_2^2 - x_1^2x_2^2 + x_0^2x_1^2 = 0.$$

This has a single singularity at the origin $[0,0,1]$. To analyze this, switch to affine coordinates $x = x_0/x_2, y = x_1/x_2$. Then the polynomial $x^2 - y^2 + x^2y^2$ is irreducible, so it cannot be factored into polynomials, but it can be factored into convergent power series

$$x^2 - y^2 + x^2y^2 = \underbrace{(x + y + \sum a_{ij}x^iy^j)}_{f}\underbrace{(x - y + \sum b_{ij}x^iy^j)}_{g}.$$

By the implicit function theorem, the branches $f = 0$ and $g = 0$ are local analytically equivalent to disks. It follows that in a neighborhood of the origin, the curve looks like two disks touching at a point. We get a genus-2 surface by pulling these apart (Figure 1.7).

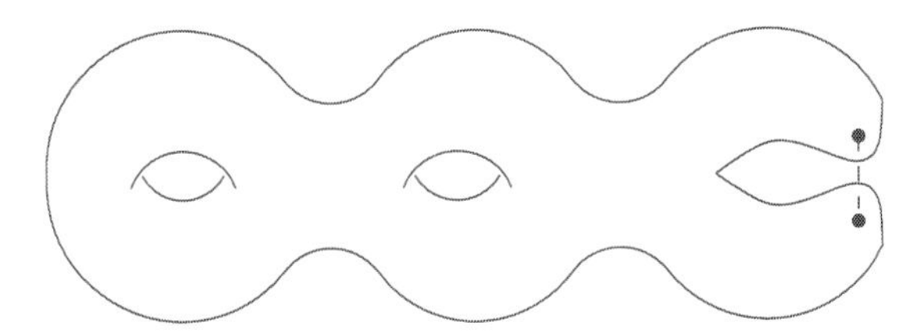

Fig. 1.7 Normalization of singular quartic.

The procedure of pulling apart the points described above can be carried out within algebraic geometry. It is called *normalization*:

Theorem 1.5.1. *Given a curve X, there exist a nonsingular curve $\tilde{X}$ (called the normalization of X) and a proper surjective morphism $H : \tilde{X} \to X$ that is finite-to-one everywhere and one-to-one over all but finitely many points. This is uniquely characterized by these properties.*

The word "morphism" will not be defined precisely until the next chapter. For the present, we should understand it to be a map definable by algebraic expressions such as polynomials. We sketch the construction, which is entirely algebraic. Further details will be given later on. Given an integral domain R with field of fractions K, the *integral closure* of R is the set of elements $x \in K$ such that $x^n + a_{n-1}x^{n-1} + \cdots + a_0 = 0$ for some $a_i \in R$. This is closed under addition and multiplication. Therefore it forms a ring [8, Chapter 5]. The basic facts can be summarized as follows:

Theorem 1.5.2. *If $f \in \mathbb{C}[x,y]$ is an irreducible polynomial, then the integral closure $\tilde{R}$ of the domain $R = \mathbb{C}[x,y]/(f)$ is finitely generated as an algebra. If $\mathbb{C}[x_1,\ldots,x_n] \to \tilde{R}$ is a surjection, and $f_1,\ldots,f_N$ generators for the kernel, then*

$$V(f_1,\ldots,f_N) = \{(a_1,\ldots,a_n) \mid f_i(a_1,\ldots,a_n) = 0\}$$

is nonsingular in the sense that the Jacobian matrix has expected rank (§2.5).

Proof. See [8, Proposition 9.2] and [33, Corollary 13.13]. □

Suppose that $X = V(f)$. Then we set $\tilde{X} = V(f_1,\dots,f_N) \subset \mathbb{C}^n$. We can lift the inclusion $R \subset \tilde{R}$ to a homomorphism of polynomial rings by completing the diagram

$$\begin{array}{ccc} \mathbb{C}[x_1,\dots,x_n] & \longrightarrow & \tilde{R} \\ \uparrow h & & \uparrow \\ \mathbb{C}[x,y] & \longrightarrow & R \end{array}$$

This determines a pair of polynomials $h(x), h(y) \in \mathbb{C}[x_1,\dots,x_n]$, which gives a polynomial map $H : \mathbb{C}^n \to \mathbb{C}^2$. By restriction, we get our desired map $H : \tilde{X} \to X$. This is the construction in the affine case. In general, we proceed by gluing these affine normalizations together. The precise construction will be given in §3.4.

Exercises

1.5.3. Verify that $x_0^2x_2^2 - x_1^2x_2^2 + x_0^2x_1^2 = 0$ is irreducible and has exactly one singular point.

1.5.4. Verify that $x^2 - y^2 + x^2y^2$ can be factored as above using formal power series.

1.5.5. Show that $t = y/x$ lies in the integral closure $\tilde{R}$ of $\mathbb{C}[x,y]/(y^2 - x^3)$. Show that $\tilde{R} \cong \mathbb{C}[t]$.

1.5.6. Show that $t = y/x$ lies in the integral closure $\tilde{R}$ of $\mathbb{C}[x,y]/(x^2 - y^2 + x^2y^2)$. Show that $\tilde{R} \cong \mathbb{C}[x,t]/(1 - t^2 - x^2t^2)$.

1.6 Hyperelliptic Curves

An affine hyperelliptic curve is a curve of the form $y^2 = p(x)$, where $p(x)$ has distinct roots. The associated *hyperelliptic* curve X is gotten by taking the closure in $\mathbb{P}^2$ and then normalizing to obtain a nonsingular curve. (We are bending the rules a bit here; usually the term hyperelliptic is applied only when the degree of $p(x)$ is at least 5.) Once again we start by describing the topology.

Proposition 1.6.1. *X is a genus-$g = \lceil \deg p(x)/2 \rceil - 1$ surface, where $\lceil\,\rceil$ means round up to the nearest integer.*

We postpone a rigorous proof. For now, we can see this by using a cut-and-paste construction generalizing what we did for cubics. Let $a_1,\dots,a_n$ denote the roots of $p(x)$ if $\deg p(x)$ is even, or the roots together with ∞ otherwise. These points, called branch points, are exactly where the map $X \to \mathbb{P}^1$ is one-to-one. Take two copies of

$\mathbb{P}^1$ and make slits from a_1 to a_2, a_3 to a_4, and so on, and then join them along the slits. The genus of the result is $n/2-1$.

Corollary 1.6.2. *Every natural number is the genus of some algebraic curve.*

The original interest in hyperelliptic curves stemmed from the study of integrals of the form $\int q(x)dx/\sqrt{p(x)}$. As with cubics, these are well defined only modulo their periods $L_q = \{\int_\gamma q(x)dx/\sqrt{p(x)} \mid \gamma \text{ closed}\}$. However, this is usually no longer a discrete subgroup, so $\mathbb{C}/L_q$ would be a very strange object. What turns out to be better is to consider all these integrals simultaneously.

Theorem 1.6.3. *The differentials* $\frac{x^i dx}{\sqrt{p(x)}}$, $i=0,\dots,g-1$, *span the space of holomorphic differentials on* X, *and the set*

$$L = \left\{ \left(\int_\gamma \frac{x^i dx}{\sqrt{p(x)}} \right)_{0\le i<g} \mid \gamma \text{ closed} \right\}$$

is a lattice in $\mathbb{C}^g$, *i.e.,* L *is a discrete subgroup of maximal rank* $2g$.

So it appears that the genus plays a deeper role than one might have initially suspected. That these differentials are holomorphic can be seen by the same sort of calculation we did in Section 1.4. The remaining assertions will follow almost immediately from the Hodge decomposition (see Section 10.3). We thus get a well-defined map from X to the torus $J(X) = \mathbb{C}^g/L$ given by

$$\alpha(x) = \left(\int_{x_0}^{x} \frac{x^i dx}{\sqrt{p(x)}} \right) \mod L. \tag{1.6.1}$$

The torus $J(X)$ is called the *Jacobian* of X, and α is called the *Abel–Jacobi* map. Together these form one of the cornerstones of algebraic curve theory.

We can make this more explicit in an example.

Example 1.6.4. Consider the curve X defined by $y^2 = x^6 - 1$. This has genus two, so that $J(X)$ is a two-dimensional torus. Let E be the elliptic curve given by $v^2 = u^3 - 1$. We have two morphisms $\pi_i : X \to E$ defined by

$$\pi_1 : u = x^2, \quad v = y,$$
$$\pi_2 : u = x^{-2}, \quad v = \sqrt{-1}yx^{-3}.$$

The second map appears to have singularities, but one can appeal to either general theory or explicit calculation to show that it is defined everywhere. We can see that the differential du/v on E pulls back to $2xdx/y$ and $2\sqrt{-1}dx/y$ under π_1 and π_2 respectively. Combining these yields a map $\pi_1 \times \pi_2 : X \to E \times E$ under which the lattice defining $E \times E$ corresponds to a sublattice $L' \subseteq L$. Therefore

$$J(X) = \mathbb{C}^2/L = (\mathbb{C}^2/L')/(L/L') = (E \times E)/(L/L').$$

We express this relation by saying that $J(X)$ is *isogenous* to $E \times E$, which means that it is a quotient of the second by a finite abelian group.

It is worthwhile understanding what is going on at a more abstract level, so as to identify some important players later on in the story. The lattice L can be identified with either the first homology group $H_1(X,\mathbb{Z})$ via the homomorphism $\mathrm{Loop}(X) \to \mathbb{C}^g$ as before, or with the first cohomology group $H^1(X,\mathbb{Z}) = \mathrm{Hom}(H_1(X,\mathbb{Z}),\mathbb{Z})$ using Poincaré duality. Using the second description, we have a map

$$H^1(E,\mathbb{Z}) \oplus H^1(E,\mathbb{Z}) \cong H^1(E \times E,\mathbb{Z}) \to H^1(X,\mathbb{Z}),$$

and L' is the image.

A hyperelliptic curve with eight branch points has genus 3. We have also encountered genus-3 curves as quartics in $\mathbb{P}^2$. These constructions yield distinct classes of examples:

Proposition 1.6.5. *A genus-3 curve is either a quartic in $\mathbb{P}^2$ or hyperelliptic, and these cases are mutually exclusive.*

We give just the broad outline of the last part. First, note that $\alpha : X \to J(X)$ can be defined for any curve X hyperelliptic or otherwise, by a similar recipe replacing $x^i dx/\sqrt{p(x)}$ with a basis of holomorphic differentials (see §10.3). Let us see how it can used to distinguish these cases. As we shall see later, $\dim J(X) = 3$ because X has genus 3. The set of tangent spaces to any manifold can be assembled into an object called the tangent bundle, and for $J(X)$ it turns out to be trivial. Thus we may sensibly identify all the tangent spaces of $J(X)$ with a fixed $\mathbb{C}^3$. So now to every $x \in X$, we have a line $T_x \subset \mathbb{C}^3$ given by the image of the derivative of the Abel–Jacobi map. In this way, we get a map $\alpha' : X \to \mathbb{P}^2$, sometimes called the Gauss map, which is the key to the proposition. In the hyperelliptic case $y^2 = p(x)$, we can calculate the Gauss map by formally differentiating (1.6.1) and projecting to $\mathbb{P}^2$ to obtain the map

$$\alpha'(x) = \left[\frac{1}{y}, \frac{x}{y}, \frac{x^2}{y}\right] = \left[1, x, x^2\right],$$

which is nothing but the original map $X \to \mathbb{P}^1$, defined by $(x,y) \mapsto x$, followed by the embedding of $\mathbb{P}^1 \to \mathbb{P}^2$ as a conic. In particular, α' would be two-to-one in this case.

In the nonhyperelliptic case, X is defined by a homogeneous quartic polynomial $F(x_0,x_1,x_2)$ in homogeneous coordinates. Setting $x = x_0/x_2, y = x_1/x_2$ as usual, a holomorphic differential on $X \cap \mathbb{C}^2$ is simply given as a restriction of $\omega = g(x,y)dx + h(x,y)dy$ with g and h holomorphic. It can be shown that a nonzero ω would become singular at infinity, and this makes it hard to find forms such that $\omega|_X$ is holomorphic everywhere. Instead, we do an indirect construction using residues. We recall that the wedge product is determined by

$$(gdx + hdy) \wedge (pdx + qdy) = (gq - hp)dx \wedge dy,$$

where $dx \wedge dy \neq 0$ is a symbol. Let $f(x,y) = F(x,y,1)$. We can wedge ω and df/f together to obtain a 2-form

$$\omega' = \omega \wedge \frac{df}{f} = \omega \wedge \left(\frac{f_x}{f}dx + \frac{f_y}{f}dy\right).$$

The inverse $\omega' \mapsto \omega$ is well defined modulo df. So the restriction $\omega|_{X \cap \mathbb{C}^2}$ depends only on ω' and is called the Poincaré residue of ω'. (A more complete abstract treatment of residues will be given in Section 12.6.) Now consider the forms

$$\omega_i = \frac{p_i(x,y)dx \wedge dy}{f(x,y)}, \quad p_i = x, y, 1. \tag{1.6.2}$$

At infinity, we can switch to new coordinates $v = x_2/x_1 = y^{-1}, u = x_0/x_1 = xy^{-1}$ and $t = x_2/x_0 = x^{-1}, s = x_1/x_0 = yx^{-1}$. A direct calculation in these coordinates shows that these forms have poles of order 1 along X and no other singularities (Exercise 1.6.8). Therefore their residues will be holomorphic everywhere along X, and in fact, they give a basis for the space of such differentials. Thus

$$\alpha' = \left[\frac{x}{f}, \frac{y}{f}, \frac{1}{f}\right] = [x, y, 1],$$

and this coincides with the given embedding $X \subset \mathbb{P}^2$.

So we see that the geometry of α' separates the hyperelliptic and nonhyperelliptic cases.

Exercises

1.6.6. The curve X above can be constructed in Example 1.6.4 explicitly by gluing charts defined by $y^2 = x^6 - 1$ and $y_2^2 = 1 - x_2^6$ via $x_2 = x^{-1}$, $y_2 = yx^{-3}$. Check that X is nonsingular and that it maps onto the projective closure of $y^2 = x^6 - 1$.

1.6.7. Using the coordinates from the previous exercise, show that the above formulas for π_i define maps from X to the projective closure of $v^2 = u^3 - 1$.

1.6.8. Set $v = x_2/x_1 = y^{-1}, u = x_0/x_1 = xy^{-1}$. Rewrite the 2-forms given in (1.6.2) in these new coordinates. Verify that these are holomorphic multiples of $du \wedge dv/F(u,1,v)$. Ditto for the coordinates s and t.

Part II
Sheaves and Geometry

Chapter 2
Manifolds and Varieties via Sheaves

In rough terms, a manifold is a "space" that looks locally like Euclidean space. An algebraic variety can be defined similarly as a "space" that looks locally like the zero set of a collection of polynomials. Point set topology alone would not be sufficient to capture this notion of space. These examples come with distinguished classes of functions (C^∞ functions in the first case, and polynomials in the second), and we want these classes to be preserved under the above local identifications. Sheaf theory provides a natural language in which to make these ideas precise. A sheaf on a topological space X is essentially a distinguished class of functions, or things that behave like functions, on open subsets of X. The main requirement is that the condition to be distinguished be *local*, which means that it can be checked in a neighborhood of every point of X. For a sheaf of rings, we have an additional requirement, that the distinguished functions on $U \subseteq X$ should form a commutative ring. With these definitions, the somewhat vague idea of a space can be replaced by the precise notion of a *concrete ringed space*, which consists of a topological space together with a sheaf of rings of functions. Both manifolds and varieties are concrete ringed spaces.

Sheaves were first defined by Leray in the late 1940s. They played a key role in the development of algebraic and complex analytic geometry, in the pioneering works of Cartan, Grothendieck, Kodaira, Serre, Spencer, and others in the following decade. Although it is rarely presented in this way in introductory texts (e.g., [110, 111, 117]), basic manifold theory can also be developed quite naturally in this framework. In this chapter we want to lay the basic foundation for the rest of the book. The goal here is to introduce the language of sheaves, and then to carry out a uniform treatment of real and complex manifolds and algebraic varieties from this point of view. This approach allows us to highlight the similarities, as well as the differences, among these spaces.

2.1 Sheaves of Functions

As we said above, we need to define sheaves in order eventually to define manifolds and varieties. We start with a more primitive notion. In many parts of mathematics, we encounter topological spaces with distinguished classes of functions on them: continuous functions on topological spaces, C^∞-functions on $\mathbb{R}^n$, holomorphic functions on $\mathbb{C}^n$, and so on. These functions may have singularities, so they may be defined only over subsets of the space; we will be interested primarily in the case that these subsets are open. We say that such a collection of functions is a presheaf if it is closed under restriction. Given sets X and T, let $\mathrm{Map}_T(X)$ denote the set of maps from X to T. Here is the precise definition of a presheaf, or rather of the kind of presheaf we need at the moment.

Definition 2.1.1. Suppose that X is a topological space and T a nonempty set. A presheaf of T-valued functions on X is a collection of subsets $\mathscr{P}(U) \subseteq \mathrm{Map}_T(U)$, for each open $U \subseteq X$, such that the restriction $f|_V$ belongs to $\mathscr{P}(V)$ whenever $f \in \mathscr{P}(U)$ and $V \subset U$.

The collection of all functions $\mathrm{Map}_T(U)$ is of course a presheaf. Less trivially:

Example 2.1.2. Let T be a topological space. Then the set of continuous functions $\mathrm{Cont}_{X,T}(U)$ from $U \subseteq X$ to T is a presheaf.

Example 2.1.3. Let X be a topological space and let T be a set. The set $T^P(U)$ of constant functions from U to T is a presheaf called the constant presheaf.

Example 2.1.4. Let $X = \mathbb{R}^n$. The sets $C^\infty(U)$ of C^∞ real-valued functions form a presheaf.

Example 2.1.5. Let $X = \mathbb{C}^n$. The sets $\mathscr{O}(U)$ of holomorphic functions on U form a presheaf. (A function of several variables is holomorphic if it is C^∞ and holomorphic in each variable.)

Example 2.1.6. Let L be a linear differential operator on $\mathbb{R}^n$ with C^∞ coefficients (e.g., $\sum \partial^2/\partial x_i^2$). Let $S(U)$ denote the space of C^∞ solutions to $Lf = 0$ in U. This is a presheaf with values in $\mathbb{R}$.

Example 2.1.7. Let $X = \mathbb{R}^n$. The sets $L^p(U)$ of measurable functions $f : U \to \mathbb{R}$ satisfying $\int_U |f|^p < \infty$ form a presheaf.

Upon comparing these examples, we see some qualitative differences. The continuity, C^∞, and holomorphic conditions are *local* conditions, which means that they can be checked in a neighborhood of a point. The other conditions such as constancy or L^p-ness, by contrast, are not. A presheaf is called a sheaf if the defining property is local. More precisely:

Definition 2.1.8. A presheaf of functions $\mathscr{P}$ is called a sheaf if given any open set U with an open cover $\{U_i\}$, a function f on U lies in $\mathscr{P}(U)$ if $f|_{U_i} \in \mathscr{P}(U_i)$ for all i.

Examples 2.1.2, 2.1.4, 2.1.5, and 2.1.6 are sheaves, while the other examples are not, except in trivial cases. More explicitly, suppose that T has at least two elements t_1, t_2, and that X contains a disconnected open set U. Then we can write $U = U_1 \cup U_2$ as a union of two disjoint open sets. The function τ taking the value of t_i on U_i is not in $T^P(U)$, but $\tau|_{U_i} \in T^P(U_i)$. Therefore T^P is not a sheaf. Similarly, L^p is not a sheaf for $0 < p < \infty$ because the constant function 1 is not in $L^p(\mathbb{R}^n)$, even though $1 \in L^p(B)$ for any ball B of finite radius.

However, there is a simple remedy.

Example 2.1.9. A function is locally constant if it is constant in a neighborhood of a point. For instance, the function τ constructed above is locally constant but not constant. The set of locally constant functions, denoted by $T(U)$ or $T_X(U)$, is now a sheaf, precisely because the condition can be checked locally. A sheaf of this form is called a constant sheaf.

We can always create a sheaf from a presheaf by the following construction.

Example 2.1.10. Given a presheaf $\mathscr{P}$ of functions from X to T. Define

$$\mathscr{P}^s(U) = \{f : U \to T \mid \forall x \in U, \exists \text{ a neighborhood } U_x \text{ of } x \text{ such that } f|_{U_x} \in \mathscr{P}(U_x)\}.$$

This is a sheaf called the sheafification of $\mathscr{P}$.

When $\mathscr{P}$ is a presheaf of constant functions, $\mathscr{P}^s$ is exactly the sheaf of locally constant functions. When this construction is applied to the presheaf L^p, we obtain the sheaf of locally L^p functions.

Exercises

2.1.11. Check that $\mathscr{P}^s$ is a sheaf.

2.1.12. Let $\mathscr{B}$ be the presheaf of bounded continuous real-valued functions on $\mathbb{R}$. Describe $\mathscr{B}^s$ in explicit terms.

2.1.13. Let $\pi : B \to X$ be a surjective continuous map of topological spaces. Prove that the presheaf of sections

$$B(U) = \{\sigma : U \to B \mid \sigma \text{ continuous}, \forall x \in U, \pi \circ \sigma(x) = x\}$$

is a sheaf.

2.1.14. Given a sheaf $\mathscr{P}$ on X and an open set $U \subset X$, let $\mathscr{P}|_U$ denote the presheaf on U defined by $V \mapsto \mathscr{P}(V)$ for each $V \subseteq U$. Check that $\mathscr{P}_U$ is a sheaf when $\mathscr{P}$ is.

2.1.15. Let $Y \subset X$ be a closed subset of a topological space. Let $\mathscr{P}$ be a sheaf of T-valued functions on X. For each open $U \subset Y$, let $\mathscr{P}_Y(U)$ be the set of functions

$f : U \to T$ locally extendible to an element of $\mathscr{P}$, i.e., $f \in \mathscr{P}_Y(U)$ if and only if for each $y \in U$, there exist a neighborhood $V \subset X$ and an element of $\mathscr{P}(V)$ restricting to $f|_{V\cap U}$. Show that $\mathscr{P}_Y$ is a sheaf.

2.1.16. Let $F : X \to Y$ be surjective continuous map. Suppose that $\mathscr{P}$ is a sheaf of T-valued functions on X. Define $f \in \mathscr{Q}(U) \subset \mathrm{Map}_T(U)$ if and only if its *pullback* $F^*f = f \circ F|_{f^{-1}U}$ belongs to $\mathscr{P}(F^{-1}(U))$. Show that $\mathscr{Q}$ is a sheaf on Y.

2.2 Manifolds

As explained in the introduction, a manifold consists of a topological space with a distinguished class of functions that looks locally like $\mathbb{R}^n$. We now set up the language necessary to give a precise definition. Let k be a field. Then $\mathrm{Map}_k(X)$ is a commutative k-algebra with pointwise addition and multplication.

Definition 2.2.1. Let $\mathscr{R}$ be a sheaf of k-valued functions on X. We say that $\mathscr{R}$ is a sheaf of algebras if each $R(U) \subseteq \mathrm{Map}_k(U)$ is a subalgebra when U is nonempty. We call the pair $(X, \mathscr{R})$ a concrete ringed space over k or simply a concrete k-space. We will sometimes refer to elements of $\mathscr{R}(U)$ as distinguished functions.

The sheaf $\mathscr{R}$ is called the structure sheaf of X. In this chapter, we usually omit the modifier "concrete," but we will use it later on after we introduce a more general notion. Basic examples of $\mathbb{R}$-spaces are $(\mathbb{R}^n, \mathrm{Cont}_{\mathbb{R}^n,\mathbb{R}})$ and $(\mathbb{R}^n, C^\infty)$, while $(\mathbb{C}^n, \mathscr{O})$ is an example of a $\mathbb{C}$-space.

We also need to consider maps $F : X \to Y$ between such spaces. We will certainly insist on continuity, but in addition we require that when a distinguished function is precomposed with F, or "pulled back" along F, it remain distinguished.

Definition 2.2.2. A morphism of k-spaces $(X, \mathscr{R}) \to (Y, \mathscr{S})$ is a continuous map $F : X \to Y$ such that if $f \in \mathscr{S}(U)$, then $F^*f \in \mathscr{R}(F^{-1}U)$, where $F^*f = f \circ F|_{f^{-1}U}$.

It is worthwhile noting that this completely captures the notion of a C^∞, or holomorphic, map between Euclidean spaces.

Example 2.2.3. A C^∞ map $F : \mathbb{R}^n \to \mathbb{R}^m$ induces a morphism $(\mathbb{R}^n, C^\infty) \to (\mathbb{R}^m, C^\infty)$ of $\mathbb{R}$-spaces, since C^∞ functions are closed under composition. Conversely, if F is a morphism, then the coordinate functions on $\mathbb{R}^m$ are expressible as C^∞ functions of the coordinates of $\mathbb{R}^n$, which implies that F is C^∞.

Example 2.2.4. Similarly, a continuous map $F : \mathbb{C}^n \to \mathbb{C}^m$ induces a morphism of $\mathbb{C}$-spaces if and only if it is holomorphic.

This is a good place to introduce, or perhaps remind the reader of, the notion of a *category* [82]. A category $\mathscr{C}$ consists of a set (or class) of objects $\mathrm{Obj}\mathscr{C}$ and for each pair $A, B \in \mathscr{C}$, a set $\mathrm{Hom}_{\mathscr{C}}(A,B)$ of morphisms from A to B. There is a composition law

$$\circ : \mathrm{Hom}_{\mathscr{C}}(B,C) \times \mathrm{Hom}_{\mathscr{C}}(A,B) \to \mathrm{Hom}_{\mathscr{C}}(A,C),$$

and distinguished elements $id_A \in \mathrm{Hom}_{\mathscr{C}}(A,A)$ that satisfy

(C1) associativity: $f \circ (g \circ h) = (f \circ g) \circ h$,
(C2) identity: $f \circ \mathrm{id}_A = f$ and $\mathrm{id}_A \circ g = g$,

whenever these are defined. Categories abound in mathematics. A basic example is the category of Sets. The objects are sets, $\mathrm{Hom}_{\mathrm{Sets}}(A,B)$ is just the set of maps from A to B, and composition and id_A have the usual meanings. Similarly, we can form the category of groups and group homomorphisms, the category of rings and ring homomorphisms, and the category of topological spaces and continuous maps. We have essentially constructed another example. We can take the class of objects to be k-spaces, and morphisms as above. These can be seen to constitute a category once we observe that the identity is a morphism and the composition of morphisms is a morphism.

The notion of an isomorphism makes sense in any category. We will spell this out for k-spaces.

Definition 2.2.5. An isomorphism of k-spaces $(X,\mathscr{R}) \cong (Y,\mathscr{S})$ is a homeomorphism $F : X \to Y$ such that $f \in \mathscr{S}(U)$ if and only if $F^*f \in \mathscr{R}(F^{-1}U)$.

Given a sheaf $\mathscr{S}$ on X and an open set $U \subset X$, let $\mathscr{S}|_U$ denote the sheaf on U defined by $V \mapsto \mathscr{S}(V)$ for each $V \subseteq U$.

Definition 2.2.6. An n-dimensional C^∞ manifold is an $\mathbb{R}$-space (X, C_X^∞) such that

1. The topology of X is given by a metric.
2. X admits an open cover $\{U_i\}$ such that each $(U_i, C_X^\infty|_{U_i})$ is isomorphic to $(B_i, C^\infty|_{B_i})$ for some open balls $B_i \subset \mathbb{R}^n$.

Remark 2.2.7. It is equivalent and perhaps more standard to require that the topology be Hausdorff and paracompact rather than metrizable. The equivalence can be seen as follows. The paracompactness of metric spaces is a theorem of Stone [112, 70]. In the opposite direction, a Riemannian metric can be constructed using a partition of unity [110]. The associated Riemannian distance function, which is the infimum of the lengths of curves joining two points, then provides a metric in the sense of point set topology.

The isomorphisms $(U_i, C^\infty|_{U_i}) \cong (B_i, C^\infty|_{B_i})$ correspond to coordinate charts in more conventional treatments. The collection of all such charts is called an atlas. Given a coordinate chart, we can pull back the standard coordinates from the ball to U_i. So we always have the option of writing expressions locally in these coordinates.

There are a number of variations on this idea:

Definition 2.2.8.
1. An n-dimensional topological manifold is defined as above but with $(\mathbb{R}^n, C^\infty)$ replaced by $(\mathbb{R}^n, \mathrm{Cont}_{\mathbb{R}^n,\mathbb{R}})$.
2. An n-dimensional complex manifold can be defined by replacing $(\mathbb{R}^n, C^\infty)$ by $(\mathbb{C}^n, \mathscr{O})$.

The one-dimensional complex manifolds are usually called *Riemann surfaces*.

Definition 2.2.9. A C^∞ map from one C^∞ manifold to another is just a morphism of $\mathbb{R}$-spaces. A holomorphic map between complex manifolds is defined as a morphism of $\mathbb{C}$-spaces.

The class of C^∞ manifolds and maps form a category; an isomorphism in this category is called a *diffeomorphism*. Likewise, the class of complex manifolds and holomorphic maps forms a category, with isomorphisms called *biholomorphisms*. By definition, any point of a manifold has a neighborhood, called a coordinate neighborhood, diffeomorphic or biholomorphic to a ball. Given a complex manifold $(X, \mathcal{O}_X)$, we say that $f : X \to \mathbb{R}$ is C^∞ if and only if $f \circ g$ is C^∞ for each holomorphic map $g : B \to X$ from a coordinate ball $B \subset \mathbb{C}^n$. We state for the record the following:

Lemma 2.2.10. *An n-dimensional complex manifold together with its sheaf of C^∞ functions is a 2n-dimensional C^∞ manifold.*

Proof. An n-dimensional complex manifold $(X, \mathcal{O}_X)$ is locally biholomorphic to a ball in $\mathbb{C}^n$, and hence (X, C^∞_X) is locally diffeomorphic to the same ball regarded as a subset of $\mathbb{R}^{2n}$. □

Later on, we will need to write things in coordinates. The pullbacks of the standard coordinates on a ball $B \subset \mathbb{C}^n$ under local biholomorphism from $X \supset B' \cong B$, are referred to as local *analytic coordinates* on X. We typically denote these by $z_1, \ldots, z_n$. Then the real and imaginary parts $x_1 = \mathrm{Re}(z_1), y_1 = \mathrm{Im}(z_1), \ldots$ give local coordinates for the underlying C^∞-manifold.

Let us consider some examples of manifolds. Certainly any open subset of $\mathbb{R}^n$ (or $\mathbb{C}^n$) is a (complex) manifold in an obvious fashion. To get less trivial examples, we need one more definition.

Definition 2.2.11. Given an n-dimensional C^∞ manifold X, a closed subset $Y \subset X$ is called a closed m-dimensional submanifold if for any point $x \in Y$, there exist a neighborhood U of x in X and a diffeomorphism to a ball $B \subset \mathbb{R}^n$ containing 0 such that $Y \cap U$ maps to the intersection of B with an m-dimensional linear subspace. A similar definition holds for complex manifolds.

When we use the word "submanifold" without qualification, we will always mean "closed submanifold." Given a closed submanifold $Y \subset X$, define C^∞_Y to be the sheaf of continuous functions that are locally extendible to C^∞ functions on X (see Exercise 2.1.15). This means that $f \in C^\infty_Y(U)$ if every point of U possesses a neighborhood $V \subset X$ such $f|_{V \cap Y} = \tilde{f}|_{V \cap Y}$ for some $\tilde{f} \in C^\infty(V)$. For a complex submanifold $Y \subset X$, we define $\mathcal{O}_Y$ to be the sheaf of functions that locally extend to holomorphic functions.

Lemma 2.2.12. *If $Y \subset X$ is a closed submanifold of a C^∞ (respectively complex) manifold, then (Y, C^∞_Y) (respectively $(Y, \mathcal{O}_Y)$) is also a C^∞ (respectively complex) manifold.*

Proof. We treat the C^∞ case; the holomorphic case is similar. Choose a local diffeomorphism (X, C_X^∞) to a ball $B \subset \mathbb{R}^n$ such that Y correponds to $B \cap \mathbb{R}^m$. Then any C^∞ function $f(x_1, \dots, x_m)$ on $B \cap \mathbb{R}^m$ extends trivially to a C^∞ function on B and conversely. Thus (Y, C_Y^∞) is locally diffeomorphic to a ball in $\mathbb{R}^m$. □

With this lemma in hand, it is possible to produce many interesting examples of manifolds starting from $\mathbb{R}^n$. For example, the unit sphere $S^{n-1} \subset \mathbb{R}^n$, which is the set of solutions to $\sum x_i^2 = 1$, is an $(n-1)$-dimensional manifold. Further examples are given in the exercises.

The following example, which was touched upon earlier, is of fundamental importance in algebraic geometry.

Example 2.2.13. Complex projective space $\mathbb{P}^n_{\mathbb{C}} = \mathbb{CP}^n$ is the set of one-dimensional subspaces of $\mathbb{C}^{n+1}$. (We will usually drop the $\mathbb{C}$ and simply write $\mathbb{P}^n$ unless there is danger of confusion.) Let $\pi : \mathbb{C}^{n+1} - \{0\} \to \mathbb{P}^n$ be the natural projection that sends a vector to its span. In the sequel, we usually denote $\pi(x_0, \dots, x_n)$ by $[x_0, \dots, x_n]$. Then $\mathbb{P}^n$ is given the quotient topology, which is defined so that $U \subset \mathbb{P}^n$ is open if and only if $\pi^{-1}U$ is open. Define a function $f : U \to \mathbb{C}$ to be holomorphic exactly when $f \circ \pi$ is holomorphic. Then the presheaf of holomorphic functions $\mathscr{O}_{\mathbb{P}^n}$ is a sheaf (Exercise 2.1.16), and the pair $(\mathbb{P}^n, \mathscr{O}_{\mathbb{P}^n})$ is a complex manifold. In fact, if we set

$$U_i = \{[x_0, \dots, x_n] \mid x_i \neq 0\},$$

then the map

$$[x_0, \dots, x_n] \mapsto (x_0/x_i, \dots, \widehat{x_i/x_i}, \dots, x_n/x_i)$$

induces an isomorphism $U_i \cong \mathbb{C}^n$. The notation $\dots, \widehat{x}, \dots$ means skip x in the list.

Exercises

2.2.14. Given k-spaces X, Y, prove that morphisms from X to Y can be patched, i.e., that the set of morphisms from open subsets of X to Y is a sheaf.

2.2.15. Show that the map $f : \mathbb{R} \to \mathbb{R}$ given by $f(x) = x^3$ is a C^∞ morphism and a homeomorphism, but that it is not a diffeomorphism.

2.2.16. Let $f_1, \dots, f_r$ be C^∞ functions on $\mathbb{R}^n$, and let X be the set of common zeros of these functions. Suppose that the rank of the Jacobian $(\partial f_i / \partial x_j)$ is $n - m$ at every point of X. Then show that X is an m-dimensional submanifold using the implicit function theorem [109, p. 41]. In particular, show that the sphere $x_1^2 + \cdots + x_n^2 = 1$ is a closed $(n-1)$-dimensional submanifold of $\mathbb{R}^n$.

2.2.17. Apply the previous exercise to show that the set $\mathrm{O}(n)$ (respectively $\mathrm{U}(n)$) of orthogonal (respectively unitary) $n \times n$ matrices is a submanifold of $\mathbb{R}^{n^2}$ (respectively $\mathbb{C}^{n^2}$).

2.2.18. A manifold that is also a group with C^∞ group operations is called a *Lie group*. Show that $\mathrm{GL}_n(\mathbb{C})$, $\mathrm{GL}_n(\mathbb{R})$, $\mathrm{O}(n)$, and $\mathrm{U}(n)$ are examples.

2.2.19. Suppose that Γ is a group of diffeomorphisms of a manifold X. Suppose that the action of Γ is fixed-point-free and properly discontinuous in the sense that every point possesses a neighborhood N such that $\gamma(N) \cap N = \emptyset$ unless $\gamma = \mathrm{id}$. Give $Y = X/\Gamma$ the quotient topology and let $\pi : X \to Y$ denote the projection. Define $f \in C^\infty_Y(U)$ if and only if the pullback $f \circ \pi$ is C^∞ in the usual sense. Show that (Y, C^∞_Y) is a C^∞ manifold. Deduce that the torus $T = \mathbb{R}^n/\mathbb{Z}^n$ is a manifold, and in fact a Lie group.

2.2.20. Check that the previous exercise applies to complex manifolds, with the appropriate modifications. In particular, show that $E = \mathbb{C}/\mathbb{Z} + \mathbb{Z}\tau$, $\mathrm{Im}(\tau) > 0$, is Riemann surface (called an elliptic curve).

2.2.21. The complex Grassmannian $G = \mathbb{G}(2,n)$ is the set of 2-dimensional subspaces of $\mathbb{C}^n$. Let $M \subset \mathbb{C}^{2n}$ be the open set of $2 \times n$ matrices of rank 2. Let $\pi : M \to G$ be the surjective map that sends a matrix to the span of its rows. Give G the quotient topology induced from M, and define $f \in \mathscr{O}_G(U)$ if and only if $\pi \circ f \in \mathscr{O}_M(\pi^{-1}U)$. For $i \neq j$, let $U_{ij} \subset M$ be the set of matrices with $(1,0)^t$ and $(0,1)^t$ for the ith and jth columns. Show that

$$\mathbb{C}^{2n-4} \cong U_{ij} \cong \pi(U_{ij})$$

and conclude that G is a $(2n-4)$-dimensional complex manifold.

2.2.22. Generalize the previous exercise to the Grassmannian $\mathbb{G}(r,n)$ of r-dimensional subspaces of $\mathbb{C}^n$.

2.3 Affine Varieties

Algebraic varieties are spaces that are defined by polynomial equations. Unlike the case of manifolds, algebraic varieties can be quite complicated even locally. We first study the local building blocks in this section, before turning to arbitrary algebraic varieties. Standard references for the material of this section and the next are Eisenbud–Harris [34], Harris [58], Hartshorne [60], Mumford [92], and Shafarevich [104].

Let k be an algebraically closed field. Affine space of dimension n over k is defined as $\mathbb{A}^n_k = k^n$. When $k = \mathbb{C}$, we can endow this space with the standard topology induced by the Euclidean metric, and we will refer to this as the *classical topology*. At the other extreme is the *Zariski topology*, which will be defined below. It makes sense for any k, and it is useful even for $k = \mathbb{C}$. Unless stated otherwise, topological notions are with respect to the Zariski topology for the remainder of this section. On $\mathbb{A}^1_k = k$, the open sets consist of complements of finite sets together with the empty set. In general, this topology can be defined to be the weakest topology for which the polynomials $\mathbb{A}_k^n \to k$ are continuous functions. The closed sets of $\mathbb{A}_k^n$ are precisely the sets of zeros

$$Z(S) = \{a \in \mathbb{A}^n \mid f(a) = 0, \forall f \in S\}$$

of sets of polynomials $S \subset R = k[x_1,\ldots,x_n]$. Sets of this form are also called *algebraic*. The Zariski topology has a basis given by open sets of the form $D(g) = X - Z(g)$, $g \in R$. Given a subset $X \subset \mathbb{A}_k^n$, the set of polynomials

$$I(X) = \{f \in R \mid f(a) = 0, \forall a \in X\}$$

is an ideal that is radical in the sense that $f \in (X)$ whenever a power of it lies in $I(X)$. Since k is algebraically closed, Hilbert's Nullstellensatz [8, 33] gives a correspondence:

Theorem 2.3.1 (Hilbert). *Let $R = k[x_1,\ldots,x_n]$ with k algebraically closed. There is a bijection between the collection of algebraic subsets of $\mathbb{A}_k^n$ and radical ideals of R given by $X \mapsto I(X)$ with inverse $I \mapsto Z(I)$.*

This allows us to translate geometry into algebra and back. For example, an algebraic subset X is called *irreducible* if it cannot be written as a union of two proper algebraic sets. This implies that $I(X)$ is *prime* or equivalently that $R/I(X)$ has no zero divisors. We summarize the correspondence below:

Theorem 2.3.2 (Hilbert). *With R as above, the map $I \mapsto Z(I)$ gives a one-to-one correspondence between the objects in the left- and right-hand columns below:*

Algebra	*Geometry*
maximal ideals of R	*points of $\mathbb{A}^n$*
maximal ideals of R/J	*points of $Z(J)$*
prime ideals in R	*irreducible algebraic subsets of $\mathbb{A}^n$*
radical ideals in R	*algebraic subsets of $\mathbb{A}^n$*

If $U \subseteq \mathbb{A}_k^n$ is open, a function $F : U \to k$ is called *regular* if it can be expressed as a ratio of polynomials $F(x) = f(x)/g(x)$ such that g has no zeros on U.

Lemma 2.3.3. *Let $\mathscr{O}_{\mathbb{A}^n}(U)$ denote the set of regular functions on U. Then $U \mapsto \mathscr{O}_{\mathbb{A}^n}(U)$ is a sheaf of k-algebras. Thus $(\mathbb{A}_k^n, \mathscr{O}_{\mathbb{A}^n})$ is a k-space.*

Proof. It is clearly a presheaf. Suppose that $F : U \to k$ is represented by a ratio of polynomials f_i/g_i on $U_i \subseteq U$, where $\bigcup U_i = U$. Since $k[x_1,\ldots,x_n]$ has unique factorization, we can assume that these are reduced fractions. Since $f_i(a)/g_i(a) = f_j(a)/g_j(a)$ for all $a \in U_i \cap U_j$, equality holds as elements of $k(x_1,\ldots,x_n)$. Therefore, we can assume that $f_i = f_j$ and $g_i = g_j$. Thus $F \in \mathscr{O}_X(U)$. □

An *affine algebraic variety* is an irreducible subset of some $\mathbb{A}_k^n$. We give X the topology induced from the Zariski topology of affine space. This is called the Zariski topology of X. Suppose that $X \subset \mathbb{A}_k^n$ is an algebraic variety. Given an open set

$U \subset X$, a function $F : U \to k$ is regular if it is locally extendible to a regular function on an open set of $\mathbb{A}^n$ as defined above, that is, if every point of U has an open neighborhood $V \subset \mathbb{A}^n_k$ with a regular function $G : V \to k$ for which $F = G|_{V \cap U}$.

Lemma 2.3.4. *Let X be an affine variety, and let $\mathscr{O}_X(U)$ denote the set of regular functions on U. Then $U \to \mathscr{O}_X(U)$ is a sheaf of k-algebras and $\mathscr{O}_X(X) \cong k[x_0,\dots,x_n]/I(X)$.*

Proof. The sheaf property of $\mathscr{O}_X$ is clear from Exercise 2.1.15 and the previous lemma. So it is enough to prove the last statement. Let $S = k[x_0,\dots,x_n]/I(X)$. Clearly there is an injection of $S \to \mathscr{O}_X(X)$ given by sending a polynomial to the corresponding regular function. Suppose that $F \in \mathscr{O}_X(X)$. Let $J = \{g \in S \mid gF \in S\}$. This is an ideal, and it suffices to show that $1 \in J$. By the Nullstellensatz, it is enough to check that $Z(J') = \emptyset$, where $J' \subset k[x_0,\dots,x_n]$ is the preimage of J. By assumption, for any $a \in X$ there exist polynomials f, g such that $g(a) \neq 0$ and $F(x) = f(x)/g(x)$ for all x in a neighborhood of a. We have $\bar{g} \in J$, where $\bar{g}$ is the image of g in S. Therefore $a \notin Z(J')$. □

Thus an affine variety gives rise to a k-space $(X, \mathscr{O}_X)$. The ring of global regular functions $\mathscr{O}(X) = \mathscr{O}_X(X)$ is an integral domain called the *coordinate ring* of X. Its field of fractions $k(X)$ is called the *function field* of X. An element f/g of this field that determines a regular function on the open subset $D(g)$ is called a *rational function* on X.

As we will explain later, $\mathscr{O}(X)$ is a complete invariant for an affine variety, that is, it is possible to reconstruct X from its coordinate ring. For now, we will be content to recover the underlying topological space. Given a ring R, we define the maximal ideal spectrum $\mathrm{Spec}_m R$ as the set of maximal ideals of R. For any ideal $I \subset R$, let

$$V(I) = \{p \in \mathrm{Spec}_m R \,|\, I \subseteq p\}.$$

The verification of the following standard properties will be left as an exercise.

Lemma 2.3.5.

(a) $V(IJ) = V(I) \cup V(J)$.
(b) $V(\sum I_i) = \bigcap_i V(I_i)$.

As a corollary, it follows that the collection of sets of the form $V(I)$ constitutes the closed sets of a topology on $\mathrm{Spec}_m R$ called the Zariski topology once again. A basis of the Zariski topology on $\mathrm{Spec}_m R$ is given by $D(f) = X - V(f)$.

Lemma 2.3.6. *Suppose that X is an affine variety. Given $a \in X$, let*

$$m_a = \{f \in \mathscr{O}(X) \mid f(a) = 0\}.$$

Then m_a is a maximal ideal, and the map $a \mapsto m_a$ induces a homeomorphism

$$X \cong \mathrm{Spec}_m \mathscr{O}(X).$$

Proof. The set m_a is clearly an ideal. It is maximal because evaluation at a induces an isomorphism $\mathscr{O}(X)/m_a \cong k$. The bijectivity of $a \mapsto m_a$ follows from the Nullstellensatz. The pullback of $V(I)$ is precisely $Z(I)$. □

In the sequel, we will use the symbols Z and V interchangeably.

Exercises

2.3.7. Identify $\mathbb{A}_k^{n^2}$ with the space of square matrices. Determine the closures in the Zariski topology of

(a) The set of matrices of rank r.
(b) The set of diagonalizable matrices.
(c) The set of matrices A of finite order in the sense that $A^N = I$ for some N.

2.3.8. Prove Lemma 2.3.5.

2.3.9. Given affine varieties $X \subset \mathbb{A}_k^n$ and $Y \subset \mathbb{A}_k^m$. Define a map $F : X \to Y$ to be a morphism if

$$F(a_1,\dots,a_n) = (f_1(a_1,\dots,a_n),\dots,f_m(a_1,\dots,a_n))$$

for polynomials $f_i \in k[x_1,\dots,x_n]$. Show that a morphism F is continuous and F^*f is regular whenever f is regular function defined on $U \subset Y$. Conversely, show that any map with this property is a morphism. Finally, show that morphisms are closed under composition, so that they form a category.

2.3.10. Show that the map $X \to \mathscr{O}(X)$ determines a contravariant functor from the category of affine varieties to the category of affine domains, i.e., finitely generated k-algebras that are domains, and algebra homomorphisms. Show that this determines an antiequivalence of categories, which means

1. $\mathrm{Hom}(X,Y) \cong \mathrm{Hom}(\mathscr{O}(Y),\mathscr{O}(X))$.
2. Every affine domain is isomorphic to some $\mathscr{O}(X)$.

2.3.11. Given closed (irreducible) subsets $X \subset \mathbb{A}_k^n$ and $Y \subset \mathbb{A}_k^m$, show that $X \times Y \subset \mathbb{A}_k^{n+m}$ is closed (and irreducible). This makes $X \times Y$ into an affine variety called the product. Prove that $\mathscr{O}(X \times Y) \cong \mathscr{O}(X) \otimes_k \mathscr{O}(Y)$ as algebras, and deduce that $X \times Y \cong X' \times Y'$ if $X \cong X'$ and $Y \cong Y'$. So it does not depend on the embedding.

2.3.12. An (affine) algebraic group is an algebraic geometer's version of a Lie group. It is an affine variety G that is also a group such that the group multiplication $G \times G \to G$ and inversion $G \to G$ are morphisms. Show that the set $G = \mathrm{GL}_n(k)$ of $n \times n$ invertible matrices is an algebraic group (embed this into $\mathbb{A}_k^{n^2+1}$ by $A \mapsto (A, \det A)$).

2.4 Algebraic Varieties

Fix an algebraically closed field k once again. In analogy with manifolds, we can define an (abstract) algebraic variety as a k-space that is locally isomorphic to an affine variety and that satisfies some version of the Hausdorff condition. It will be convenient to ignore this last condition for the moment. The resulting objects are dubbed prevarieties.

Definition 2.4.1. A prevariety over k is a k-space $(X, \mathcal{O}_X)$ such that X is connected and there exists a finite open cover $\{U_i\}$, called an affine open cover, such that each $(U_i, \mathcal{O}_X|_{U_i})$ is isomorphic, as a k-space, to an affine variety. A morphism of prevarieties is a morphism of the underlying k-spaces.

Before going further, let us consider the most important nonaffine example.

Example 2.4.2. Let $\mathbb{P}^n_k$ be the set of one-dimensional subspaces of k^{n+1}. Using the natural projection $\pi : \mathbb{A}^{n+1} - \{0\} \to \mathbb{P}^n_k$, give $\mathbb{P}^n_k$ the quotient topology ($U \subset \mathbb{P}^n_k$ is open if and only if $\pi^{-1}U$ is open). Equivalently, the closed sets of $\mathbb{P}^n_k$ are zeros of sets of homogeneous polynomials in $k[x_0, \ldots, x_n]$. Define a function $f : U \to k$ to be regular exactly when $f \circ \pi$ is regular. Such a function can be represented as the ratio

$$f \circ \pi(x_0, \ldots, x_n) = \frac{p(x_0, \ldots, x_n)}{q(x_0, \ldots, x_n)}$$

of two homogeneous polynomials of the same degree such that q has no zeros on $\pi^{-1}U$. Then the presheaf of regular functions $\mathcal{O}_{\mathbb{P}^n}$ is a sheaf, and the pair $(\mathbb{P}^n_k, \mathcal{O}_{\mathbb{P}^n})$ is easily seen to be a prevariety with affine open cover $\{U_i\}$ as in Example 2.2.13.

The Zariski topology is never actually Hausdorff except in the most trivial cases. However, there is a good substitute called the separation axiom. To motivate it, we make the following observation:

Lemma 2.4.3. *Let X be a topological space. Then the following statements are equivalent:*

(a) X is Hausdorff.
(b) If $f, g : Y \to X$ is a pair of continuous functions, the set $\{y \in Y \mid f(y) = g(y)\}$ is closed.
(c) The diagonal $\Delta = \{(x,x) \mid x \in X\}$ is closed in $X \times X$ with its product topology.

Proof. Suppose that X is Hausdorff. If $f, g : Y \to X$ are continuous functions with $f(y_0) \neq g(y_0)$, then $f(y) \neq g(y)$ for y in a neighborhood of y_0, so (b) holds. Assuming that (b), (c) is obtained by applying (b) to the projections $p_1, p_2 : X \times X \to X$. The final implication from (c) to (a) is clear. □

We use (b) of the previous lemma as the model for our separation axiom.

Definition 2.4.4. A prevariety $(X, \mathcal{O}_X)$ is said to be a variety, or is called separated, if for any prevariety Y and pair of morphisms $f, g : Y \to X$, the set $\{y \in Y \mid f(y) = g(y)\}$ is closed. A morphism of varieties is simply a morphism of prevarieties.

Example 2.4.5. Let $X = \mathbb{A}^1 \cup \mathbb{A}^1$ glued along $U = \mathbb{A}^1 - \{0\}$ via the identity, but with the origins not identified. Then X is a prevariety, but it is not a variety, because the identity $U \to X$ extends to $\mathbb{A}^1$ in two different ways by sending 0 to the first or second copy of $\mathbb{A}^1$.

Item (c) of Lemma 2.4.3 gives a more usable criterion for separation. Before we can formulate the analogous condition for varieties, we need products. These were constructed for affine varieties in Exercise 2.3.11, and the general case can be reduced to this. Full details can be found in [92, I§6].

Proposition 2.4.6. *Let $(X, \mathcal{O}_X)$ and $(Y, \mathcal{O}_Y)$ be prevarieties. Then the Cartesian product $X \times Y$ carries the structure of a prevariety such that the projections $p_1 : X \times Y \to X$ and $p_2 : X \times Y \to Y$ are morphisms, and if $(Z, \mathcal{O}_Z)$ is any prevariety that maps via morphisms f and g to X and Y, then the map $f \times g : Z \to X \times Y$ is a morphism of prevarieties.*

Lemma 2.4.7. *A prevariety is a variety if and only if the diagonal Δ is closed in $X \times X$.*

Proof. If X is a variety, then Δ is closed because it is the locus where p_1 and p_2 coincide. Conversely, if Δ is closed, then so is $\{y \mid f(y) = g(y)\} = (f \times g)^{-1}\Delta$. $\square$

Here are some basic examples.

Example 2.4.8. Affine spaces are varieties, since the diagonal $\Delta \subset \mathbb{A}_k^{2n} = \mathbb{A}_k^n \times \mathbb{A}_k^n$ is the closed set defined by $x_i = x_{i+n}$.

Example 2.4.9. Projective spaces are also varieties. The product can be realized as the image of the Segre map $\mathbb{P}_k^n \times \mathbb{P}_k^n \subset \mathbb{P}_k^{(n+1)(n+1)-1}$ given by

$$([x_0, \dots, x_n], [y_0, \dots, y_n]) \mapsto [x_0 y_0, x_0 y_1, \dots, x_n y_n].$$

The diagonal is given by explicit equations. See Exercise 2.4.14.

Further examples can be produced by taking suitable subsets. Let $(X, \mathcal{O}_X)$ be an algebraic variety over k. A closed irreducible subset $Y \subset X$ is called a *closed subvariety*. Given an open set $U \subset Y$, define $\mathcal{O}_Y(U)$ to be the set functions that are locally extendible to regular functions on X.

Proposition 2.4.10. *Suppose that $Y \subset X$ is a closed subvariety of an algebraic variety. Then $(Y, \mathcal{O}_Y)$ is an algebraic variety.*

Proof. Let $\{U_i\}$ be an open cover of X by affine varieties. Choose an embedding $U_i \subset \mathbb{A}_k^N$ as a closed subset. Then $Y \cap U_i \subset \mathbb{A}_k^N$ is also embedded as a closed set, and the restriction $\mathcal{O}_Y|_{Y \cap U_i}$ is the sheaf of functions on $Y \cap U_i$ that are locally extendible to regular functions on $\mathbb{A}_k^N$. Thus $(Y \cap U_i, \mathcal{O}_Y|_{Y \cap U_i})$ is an affine variety. This implies that Y is a prevariety. Denoting the diagonal of X and Y by Δ_X and Δ_Y respectively, we see that $\Delta_Y = \Delta_X \cap Y \times Y$ is closed in $X \times X$, and therefore in $Y \times Y$. $\square$

It is worth making the description of projective varieties, or closed subvarieties of projective space, more explicit. A nonempty subset of $\mathbb{A}_k^{n+1}$ is conical if it contains 0 and is stable under the action of $\lambda \in k^*$ given by $v \mapsto \lambda v$. Given $X \subset \mathbb{P}_k^n$, $CX = \pi^{-1}X \cup \{0\} \subset \mathbb{A}_k^{n+1}$ is conical, and all conical sets arise in this way. If $I \subseteq S = k[x_0,\dots,x_n]$ is a homogeneous ideal, then $Z(I)$ is conical and so corresponds to a closed subset of $\mathbb{P}_k^n$. From the Nullstellensatz, we obtain a dictionary similar to the earlier one.

Theorem 2.4.11. *Let $S_+ = (x_0,\dots,x_n)$ be the maximal ideal of the origin. Then there is a one-to-one correspondence as shown in the table below:*

Algebra	*Geometry*
homogeneous radical ideals in S other than S_+	*algebraic subsets of $\mathbb{P}^n$*
homogeneous prime ideals in S other than S_+	*algebraic subvarieties of $\mathbb{P}^n$*

Given a subvariety $X \subseteq \mathbb{P}_k^n$, the elements of $\mathscr{O}_X(U)$ are functions $f : U \to k$ such that $f \circ \pi$ is regular. Such a function can be represented locally as the ratio of two homogeneous polynomials of the same degree.

When $k = \mathbb{C}$, we can use the stronger classical topology on $\mathbb{P}_{\mathbb{C}}^n$ introduced in Example 2.2.13. This is inherited by subvarieties, and is also called the classical topology. When there is danger of confusion, we write X^{an} to indicate a variety X with its classical topology. (The superscript an stands for "analytic.")

Exercises

2.4.12. Given an open subset U of an algebraic variety X, let $\mathscr{O}_U = \mathscr{O}_X|_U$. Prove that $(U, \mathscr{O}_U)$ is a variety. An open subvariety of a projective (respectively affine) variety is called quasiprojective (respectively quasiaffine).

2.4.13. Let $X = \mathbb{A}_k^n - \{0\}$ with $n > 2$. Show that $\mathscr{O}(X) \cong k[x_1,\dots,x_n]$. Deduce that X is not isomorphic to an affine variety with the help of Exercise 2.3.10.

2.4.14. Verify that the image of Segre's embedding $\mathbb{P}^n \times \mathbb{P}^m \subset \mathbb{P}^{(n+1)(m+1)-1}$ is Zariski closed, and the diagonal Δ is closed in the product when $m = n$.

2.4.15. Prove that $\mathscr{O}(\mathbb{P}_k^n) = k$. Deduce that $\mathbb{P}_k^n$ is not affine unless $n = 0$.

2.4.16. Fix an integer $d > 0$ and let $N = \binom{n+d}{d} - 1$. The dth Veronese map $v_d : \mathbb{P}_k^n \to \mathbb{P}_k^N$ is given by sending $[x_0,\dots,x_n]$ to $[v]$, where v is the vector of degree-d monomials listed in some order. Show that this map is a morphism and that the image is Zariski closed.

2.4.17. Given a nonconstant homogeneous polynomial $f \in k[x_0,\dots,x_n]$, define $D(f)$ to be the complement of the hypersurface in $\mathbb{P}_k^n$ defined by $f = 0$. Prove that $(D(f), \mathscr{O}_{\mathbb{P}^n}|_{D(f)})$ is an affine variety. (Use the Veronese map to reduce to the case of a linear polynomial.)

2.4.18. Suppose that X is a prevariety such that any pair of points is contained in an affine open set. Prove that X is a variety.

2.4.19. Make the Grassmannian $\mathbb{G}_k(r,n)$, which is the set of r-dimensional subspaces of k^n, into a prevariety by imitating the constructions of Exercise 2.2.22. Check that $\mathbb{G}_k(r,n)$ is in fact a variety.

2.4.20. After identifying $k^6 \cong \wedge^2 k^4$, $\mathbb{G}_k(2,4)$ can be embedded in $\mathbb{P}^5_k$ by sending the span of $v, w \in k^4$ to the line spanned by $\omega = v \wedge w$. Check that this is a morphism and that the image is a subvariety given by the Plücker equation $\omega \wedge \omega = 0$. Write this out as a homogeneous quadratic polynomial equation in the coordinates of ω.

2.4.21. Given an algebraic group G (Exercise 2.3.12), an action on a variety X is a morphism $G \times X \to X$ denoted by "$\cdot$" such that $(gh)\cdot x = g\cdot(h\cdot x)$. A variety is called homogeneous if an algebraic group acts transitively on it. Check that affine spaces, projective spaces, and Grassmannians are homogeneous.

2.4.22. The blowup of the origin of $\mathbb{A}^n_k$ is the set

$$B = \mathrm{Bl}_0\mathbb{A}^n = \{(v,\ell) \in \mathbb{A}^n_k \times \mathbb{P}^{n-1}_k \mid v \in \ell\}.$$

Show that this is Zariski closed and irreducible. When $k = \mathbb{C}$, show that B is a complex submanifold of $\mathbb{C}^n \times \mathbb{P}^{n-1}_{\mathbb{C}}$. Show that the morphism $\pi : B \to \mathbb{A}^n_k$ given by projection is an isomorphism over $\mathbb{A}^n_k - \{0\}$.

2.4.23. Given an affine variety $X \subset \mathbb{A}^n_k$ containing 0, its blowup is given by

$$\mathrm{Bl}_0 X = \overline{\pi^{-1}(X - \{0\})} \subset \mathrm{Bl}_0\mathbb{A}^n \subset \mathbb{A}^n_k \times \mathbb{P}^{n-1}_k.$$

Given a variety $x \in X$ with affine open cover $\{U_i\}$, show that there exists a variety $\mathrm{Bl}_x X$ locally isomorphic to $\mathrm{Bl}_x(U_i)$.

2.5 Stalks and Tangent Spaces

Given two functions defined in possibly different neighborhoods of a point $x \in X$, we say they have the same *germ* at x if their restrictions to some common neighborhood agree. This is is an equivalence relation. The germ at x of a function f defined near X is the equivalence class containing f. We denote this by f_x.

Definition 2.5.1. Given a presheaf of functions $\mathscr{P}$, its stalk $\mathscr{P}_x$ at x is the set of germs of functions contained in some $\mathscr{P}(U)$ with $x \in U$.

It will be useful to give a more abstract characterization of the stalk using *direct limits* (which are also called inductive limits, or filtered colimits). We explain direct limits in the present context, and refer to [33, Appendix 6] or [76] for a more complete discussion. Suppose that a set L is equipped with a family of maps $\mathscr{P}(U) \to L$, where U ranges over open neighborhoods of x. We will say that the family is a compatible family if $\mathscr{P}(U) \to L$ factors through $\mathscr{P}(V)$ whenever $V \subset U$. For instance, the maps $\mathscr{P}(U) \to \mathscr{P}_x$ given by $f \mapsto f_x$ form a compatible family. A set L equipped with a compatible family of maps is called a direct limit of $\mathscr{P}(U)$ if for any M with a compatible family $\mathscr{P}(U) \to M$, there is a unique map $L \to M$ making the obvious diagrams commute. This property characterizes L up to isomorphism, so we may speak of *the* direct limit

$$\varinjlim_{x \in U} \mathscr{P}(U).$$

Lemma 2.5.2. $\mathscr{P}_x = \varinjlim_{x \in U} \mathscr{P}(U)$.

Proof. Suppose that $\phi : \mathscr{P}(U) \to M$ is a compatible family. Then $\phi(f) = \phi(f|_V)$ whenever $f \in \mathscr{P}(U)$ and $x \in V \subset U$. Therefore $\phi(f)$ depends only on the germ f_x. Thus ϕ induces a map $\mathscr{P}_x \to M$ as required. □

All the examples of k-spaces encountered so far (C^∞-manifolds, complex manifolds, and algebraic varieties) satisfy the following additional property.

Definition 2.5.3. We will say that a concrete k-space $(X, \mathscr{R})$ is *locally ringed* if $1/f \in \mathscr{R}(U)$ when $f \in \mathscr{R}(U)$ is nowhere zero.

Recall that a ring R is *local* if it has a unique maximal ideal, say m. The quotient R/m is called the *residue field.* We will often convey all this by referring to the triple $(R, m, R/m)$ as a local ring.

Lemma 2.5.4. *If $(X, \mathscr{R})$ is locally ringed, then for any $x \in X$, $\mathscr{R}_x$ is a local ring with residue field isomorphic to k.*

Proof. Let m_x be the set of germs of functions vanishing at x. For $\mathscr{R}_x$ to be local with maximal ideal m_x, it is necessary and sufficient that each $f \in \mathscr{R}_x - m_x$ be invertible. This is clear, since $1/f|_U \in R(U)$ for some $x \in U$.

To see that $\mathscr{R}_x / m_x = k$, it is enough to observe that the ideal m_x is the kernel of the evaluation map $\mathrm{ev} : \mathscr{R}_x \to k$ given by $\mathrm{ev}(f) = f(x)$, and the map is surjective, because $\mathrm{ev}(a) = a$ when $a \in k$. □

When $(X, \mathscr{O}_X)$ is an n-dimensional complex manifold, the local ring $\mathscr{O}_{X,x}$ can be identified with ring of convergent power series in n variables. When X is a variety, the local ring $\mathscr{O}_{X,x}$ is also well understood. We may replace X by an affine variety with coordinate ring $R = \mathscr{O}_X(X)$. Consider the maximal ideal

$$m_x = \{f \in R \mid f(x) = 0\}.$$

Lemma 2.5.5. *$\mathscr{O}_{X,x}$ is isomorphic to the localization*

$$R_{m_x} = \left\{ \frac{g}{f} \mid f, g \in R, f \notin m_x \right\}.$$

Proof. Let K be the field of fractions of R. A germ in $\mathcal{O}_{X,x}$ is represented by a regular function defined in a neighborhood of x, but this is the fraction $f/g \in K$ with $g \notin m_x$. □

By standard commutative algebra [8, Corollary 7.4], the local rings of algebraic varieties are Noetherian, since they are localizations of Noetherian rings. This is also true for complex manifolds, although the argument is a bit more delicate [46, p. 12]. By contrast, when X is a C^∞-manifold, the stalks are non-Noetherian local rings. This is easy to check by a theorem of Krull [8, pp. 110–111] that implies that a Noetherian local ring R with maximal ideal m satisfies $\cap_n m^n = 0$. When R is the ring of germs of C^∞ functions on $\mathbb{R}$, then the intersection $\cap_n m^n$ contains nonzero functions such as

$$\begin{cases} e^{-1/x^2} & \text{if } x > 0, \\ 0 & \text{otherwise.} \end{cases}$$

Nevertheless, the maximal ideals are finitely generated.

Proposition 2.5.6. *If R is the ring of germs at 0 of C^∞ functions on $\mathbb{R}^n$, then its maximal ideal m is generated by the coordinate functions $x_1, \dots, x_n$.*

Proof. See the exercises. □

In order to talk about tangent spaces in this generallity, it will be convenient to introduce the following axioms:

Definition 2.5.7. We will say that a local ring R with maximal ideal m and residue field k satisfies the tangent space conditions if

1. There is an inclusion $k \subset R$ that gives a splitting of the natural map $R \to k$.
2. The ideal m is finitely generated.

For stalks of C^∞ and complex manifolds and algebraic varieties over k, the residue fields are respectively $\mathbb{R}, \mathbb{C}$, and k. The inclusion of germs of constant functions gives the first condition in these examples, and the second was discussed above.

Definition 2.5.8. When (R, m, k) is a local ring satisfying the tangent space conditions, we define its cotangent space as $T_R^* = m/m^2 = m \otimes_R k$, and its tangent space as $T_R = \mathrm{Hom}(T_R^*, k)$. When X is a manifold or variety, we write $T_x = T_{X,x}$ (respectively $T_x^* = T_{X,x}^*$) for $T_{\mathcal{O}_{X,x}}$ (respectively $T_{\mathcal{O}_{X,x}}^*$).

When (R, m, k) satisfies the tangent space conditions, R/m^2 splits canonically as $k \oplus T_x^*$.

Definition 2.5.9. Let (R, m, k) satisfy the tangent space conditions. Given $f \in R$, define its differential df as the projection of $(f \mod m^2)$ to T_x^* under the above decomposition.

To see why this terminology is justified, we compute the differential when R is the ring of germs of C^∞ functions on $\mathbb{R}^n$ at 0. Then $f \in R$ can be expanded using Taylor's formula,

$$f(x_1,\ldots,x_n) = f(0) + \sum \left.\frac{\partial f}{\partial x_i}\right|_0 x_i + r(x_1,\ldots,x_n),$$

where the remainder r lies in m^2. Therefore df coincides with the image of the second term on the right, which is the usual expression

$$df = \sum \left.\frac{\partial f}{\partial x_i}\right|_0 dx_i.$$

Lemma 2.5.10. *$d : R \to T_R^*$ is a k-linear derivation, i.e., it satisfies the Leibniz rule $d(fg) = f(x)dg + g(x)df$.*

Proof. See the exercises. □

As a corollary, it follows that a tangent vector $v \in T_R = T_R^{**}$ gives rise to a derivation $\delta_v = v \circ d : R \to k$.

Lemma 2.5.11. *The map $v \mapsto \delta_v$ yields an isomorphism between T_R and the vector space $\mathrm{Der}_k(R,k)$ of k-linear derivations from R to k.*

Proof. Given $\delta \in \mathrm{Der}_k(R,k)$, we can see that $\delta(m^2) \subseteq m$. Therefore it induces a map $v : m/m^2 \to R/m = k$, and we can check that $\delta = \delta_v$. □

Lemma 2.5.12. *When (R,m,k) is the ring of germs at 0 of C^∞ functions on $\mathbb{R}^n$ (or holomorphic functions on $\mathbb{C}^n$, or regular functions on $\mathbb{A}_k^n$). Then a basis for $\mathrm{Der}_k(R,k)$ is given by*

$$D_i = \left.\frac{\partial}{\partial x_i}\right|_0, \quad i = 1,\ldots,n.$$

A homomorphism $F : S \to R$ of local rings is called *local* if it takes the maximal ideal of S to the maximal ideal of R. Under these conditions, we get map of cotangent spaces $T_S^* \to T_R^*$ called the codifferential of F. When residue fields coincide, we can dualize this to get a map $dF : T_R \to T_S$. We study this further in the exercises.

A big difference between algebraic varieties and manifolds is that the former can be very complicated, even locally. We want to say that a variety over an algebraically closed field k is nonsingular or smooth if it looks like affine space (very) locally, and in particular if it is a manifold when $k = \mathbb{C}$. The implicit function suggests a way to make this condition more precise. Suppose that $X \subseteq \mathbb{A}_k^N$ is a closed subvariety defined by the ideal $(f_1,\ldots,f_r)$ and let $x \in X$. Then $x \in X$ should be nonsingular if the Jacobian matrix $(\frac{\partial f_i}{\partial x_j}|_x)$ has the expected rank $N - \dim X$, where $\dim X$ can be defined as the transcendence degree of the function field $k(X)$ over k. We can reformulate this in a more intrinsic fashion thanks to the following:

Lemma 2.5.13. *The vector space $T_{X,x}$ is isomorphic to the kernel of $(\frac{\partial f_i}{\partial x_j}|_x)$.*

Proof. Let $R = \mathscr{O}_{\mathbb{A}^N,x}$, $S = \mathscr{O}_{X,x} \cong R/(f_1,\dots,f_r)$ and $\pi : R \to S$ be the natural map. We also set $J = (\frac{\partial f_i}{\partial x_j}|_x)$. Then any element $\delta' \in \mathrm{Der}_k(S,k)$ gives a derivation $\delta' \circ \pi \in \mathrm{Der}_k(R,k)$, which vanishes only if δ' vanishes. A derivation in $\delta \in \mathrm{Der}_k(R,k)$ comes from S if and only if $\delta(f_i) = 0$ for all i. We can use the basis $\partial/\partial x_j|_x$ to identify $\mathrm{Der}_k(R,k)$ with k^N. Putting all of this together gives a commutative diagram

$$\begin{array}{ccccccc} 0 & \longrightarrow & \mathrm{Der}(S,k) & \longrightarrow & \mathrm{Der}(R,k) & \xrightarrow{\delta\mapsto\delta(f_i)} & k^r \\ & & & & \downarrow\cong & & \downarrow = \\ & & & & k^N & \xrightarrow{J} & k^r \end{array}$$

from which the lemma follows. □

Definition 2.5.14. A point x on a (not necessarily affine) variety X is called a nonsingular or smooth point if $\dim T_{X,x} = \dim X$; otherwise, x is called singular; X is nonsingular or smooth if every point is nonsingular.

The condition for nonsingularity of x is usually formulated as saying that the local ring $\mathscr{O}_{X,x}$ is a regular local ring, which means that $\dim \mathscr{O}_{X,x} = \dim T_x$ [8, 33]. But this is equivalent to what was given above, since $\dim X$ coincides with the Krull dimension [8, 33] of the ring $\mathscr{O}_{X,x}$. Affine and projective spaces are examples of nonsingular varieties.

Over $\mathbb{C}$, we have the following characterization.

Proposition 2.5.15. *Given a subvariety $X \subset \mathbb{C}^N$ and a point $x \in X$, the point x is nonsingular if and only if there exists a neighborhood U of x in $\mathbb{C}^N$ for the classical topology such that $X \cap U$ is a closed complex submanifold of $\mathbb{C}^N$, with dimension equal to* $\dim X$.

Proof. This follows from the holomorphic implicit function theorem [66, Theorem 2.1.2]. □

Corollary 2.5.16. *Given a nonsingular algebraic subvariety X of $\mathbb{A}^n_{\mathbb{C}}$ or $\mathbb{P}^n_{\mathbb{C}}$, the space X^{an} is a complex submanifold of $\mathbb{C}^n$ or $\mathbb{P}^n_C$.*

Finally, we note the following result:

Proposition 2.5.17. *The set of nonsingular points of an algebraic variety forms an open dense subset.*

Proof. See [60, II Corollary 81.6]. □

Exercises

2.5.18. Prove Proposition 2.5.6. (Hint: given $f \in m$, let

$$f_i = \int_0^1 \frac{\partial f}{\partial x_i}(tx_1, \ldots, tx_n)\, dt.$$

Show that $f = \sum f_i x_i$.)

2.5.19. Prove Lemma 2.5.10.

2.5.20. Let $F : (X, \mathscr{R}) \to (Y, \mathscr{S})$ be a morphism of locally ringed k-spaces. If $x \in X$ and $y = F(x)$, check that the homomorphism $F^* : \mathscr{S}_y \to \mathscr{R}_x$ taking a germ of f to the germ of $f \circ F$ is well defined and is local. Conclude that there is an induced linear map $dF : T_x \to T_y$, called the differential or derivative.

2.5.21. Let $F : \mathbb{R}^n \to \mathbb{R}^m$ be a C^∞ map taking 0 to 0. Calculate $dF : T_0 \to T_0$, constructed above, and show that this is given by a matrix of partial derivatives.

2.5.22. Check that with the appropriate identification given a C^∞ function on X viewed as a C^∞ map from $f : X \to \mathbb{R}$, df in the sense of Definition 2.5.9 and in the sense of the previous exercise coincide.

2.5.23. Check that the operation $(X, x) \mapsto T_x$ determines a functor on the category of C^∞-manifolds and base-point-preserving maps. (The definition of functor can be found in §3.1.) Interpret this as the chain rule.

2.5.24. Given a Lie group G with identity e, an element $g \in G$ acts on G by $h \mapsto ghg^{-1}$. Let $\mathrm{Ad}(g) : T_e \to T_e$ be the differential of this map. Show that Ad defines a homomorphism from G to $\mathrm{GL}(T_e)$ called the adjoint representation.

2.5.25. The ring of dual numbers D is defined as $k[\varepsilon]/(\varepsilon^2)$. Let (R, m, k) be a local ring satisfying the tangent space conditions. Show that T_R is isomorphic to the space of k-algebra homomorphisms $\mathrm{Hom}_{k\text{-alg}}(R, D)$.

2.5.26. Prove the identity $\det(I + \varepsilon A) = 1 + \mathrm{trace}(A)\varepsilon$ for square matrices over D. Use this to prove that the tangent space T_I to $\mathrm{SL}_n(k)$ is isomorphic to the space of trace-zero $n \times n$ matrices, where $\mathrm{SL} - n(k)$ is the group of matrices with determinant 1.

2.5.27. If $f(x_0, \ldots, x_n)$ is a homogeneous polynomial of degree d, prove Euler's formula $\sum x_i \frac{\partial f}{\partial x_i} = d \cdot f(x_0, \ldots, x_n)$. Use this to show that the point p on the projective hypersurface defined by f is singular if and only if all the partials of f vanish at p. Determine the set of singular points defined by $x_0^5 + \cdots + x_4^5 - 5x_0 \cdots x_4$ in $\mathbb{P}^4_{\mathbb{C}}$.

2.5.28. Prove that if $X \subset \mathbb{A}^N_k$ is a variety, then $\dim T_x \leq N$. Give an example of a curve in $\mathbb{A}^N_k$ for which equality is attained, for $N = 2, 3, \ldots$.

2.5.29. Give a direct proof of Proposition 2.5.17 for hypersurfaces in $\mathbb{A}^n_k$.

2.5.30. Show that a homogeneous variety is nonsingular.

2.5.31. Let $f(x_0,\dots,x_n)=0$ define a nonsingular hypersurface $X\subset\mathbb{P}^n_{\mathbb{C}}$. Show that there exists a hyperplane H such that $X\cap H$ is nonsingular. This is a special case of Bertini's theorem.

2.6 1-Forms, Vector Fields, and Bundles

A C^∞ vector field on a manifold X is essentially a choice $v_x\in T_x$, for each $x\in X$, that varies in a C^∞ fashion. The dual notion, called a covector field, a differential form of degree 1, or simply a 1-form, is easier to make precise. So we start with this. Given a C^∞ function f on X, we can define df as the collection of local derivatives $df_x\in T_x^*$. This is the basic example of a 1-form.

Definition 2.6.1. A C^∞ 1-form on X is a finite linear combination $\sum g_i df_i$ with $f_i, g_i\in C^\infty(X)$. Let $\mathscr{E}^1(X)$ denote the space of these.

Clearly, $\mathscr{E}^1(X)$ is a module over the ring $C^\infty(X)$. Also, given an open set $U\subset X$, a 1-form can be restricted to U as a $\bigcup T_x^*$-valued function. In this way, $U\mapsto\mathscr{E}^1(U)$ becomes a presheaf and in fact a sheaf. If U is a coordinate neighborhood with coordinates $x_1,\dots,x_n$, then any 1-form on U can be expanded uniquely as $\sum f_i(x_1,\dots,x_n)dx_i$ with C^∞ coefficients. In other words, $\mathscr{E}^1(U)$ is a free module with basis dx_i. The module $\mathscr{E}^1(X)$ is generally not free.

Now we can define vector fields as the dual in the appropriate sense. Let $\langle,\rangle$ denote the pairing between T_x and T_x^*.

Definition 2.6.2. A C^∞ vector field on X is a collection of vectors $v_x\in T_x$, $x\in X$, such that the map $x\mapsto\langle v_x, df_x\rangle$ lies in $C^\infty(U)$ for each open $U\subseteq X$ and $f\in C^\infty(U)$. Let $\mathscr{T}(X)$ denote the set of vector fields.

The definition is rigged to ensure that any $D\in\mathscr{T}(X)$ defines a derivation $C^\infty(U)\to C^\infty(U)$ by $f\mapsto\langle D, df\rangle$. It can be seen that $\mathscr{T}$ is a sheaf of $\bigcup T_x$-valued functions. If U is a coordinate neighborhood with coordinates $x_1,\dots,x_n$, then any vector fields on U are given by $\sum f_i\partial/\partial x_i$.

There is another standard approach to defining vector fields on a manifold X. The disjoint union of the tangent spaces $T_X=\bigcup_x T_x$ can be assembled into a manifold called the tangent bundle T_X, which comes with a projection $\pi: T_X\to X$ such that $T_x=\pi^{-1}(x)$. We define the manifold structure on T_X in such a way that the vector fields correspond to C^∞ cross sections. The tangent bundle is an example of a structure called a vector bundle. In order to give the general definition simultaneously in several different categories, we will fix a choice of:

(a) a C^∞-manifold X and the standard C^∞-manifold structure on $k=\mathbb{R}$,
(b) a C^∞-manifold X and the standard C^∞-manifold structure on $k=\mathbb{C}$,
(c) a complex manifold X and the standard complex manifold structure on $k=\mathbb{C}$,
(d) an algebraic variety X with an identification $k\cong\mathbb{A}^1_k$.

A rank-n vector bundle is a map $\pi: V\to X$ that is locally a product $X\times k^n\to X$. Here is the precise definition.

Definition 2.6.3. A rank-n vector bundle on X is a morphism $\pi : V \to X$ such that there exist an open cover $\{U_i\}$ of X and commutative diagrams

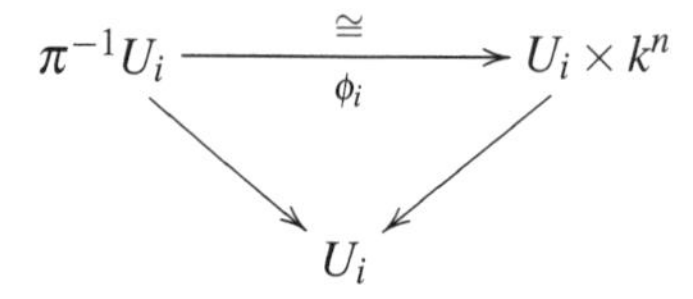

such that the isomorphisms

$$\phi_i \circ \phi_j^{-1} : U_i \cap U_j \times k^n \cong U_i \cap U_j \times k^n$$

are k-linear on each fiber. A bundle is called C^∞ real in case (a), C^∞ complex in case (b), holomorphic in case (c), and algebraic in case (d). A rank-1 vector bundle will also be called a line bundle.

The product $X \times k^n$ is an example of a vector bundle, called the *trivial bundle* of rank n. A simple nontrivial example to keep in mind is the Möbius strip, which is a real line bundle over the circle. The datum $\{(U_i, \phi_i)\}$ is called a local trivialization. Given a C^∞ real vector bundle $\pi : V \to X$, define the presheaf of sections

$$\mathscr{V}(U) = \{s : U \to \pi^{-1}U \mid s \text{ is } C^\infty, \pi \circ s = \mathrm{id}_U\}.$$

This is easily seen to be a sheaf. When $V = X \times \mathbb{R}^n$ is the trivial vector bundle, a section is given by $(x, f(x))$, where $f : X \to \mathbb{R}^n$, so that $\mathscr{V}(X)$ is isomorphic to the space of vector-valued C^∞ functions on X. In general, a section $s \in \mathscr{V}(U)$ is determined by the collection of vector-valued functions on $U_i \cap U$ given by projecting $\phi_i \circ s$ to $\mathbb{R}^n$. Thus $\mathscr{V}(U)$ has a natural $\mathbb{R}$-vector space structure. Later on, we will characterize the sheaves, called locally free sheaves, that arise from vector bundles in this way.

Theorem 2.6.4. *Given an n-dimensional manifold X, there exists a C^∞ real vector bundle T_X of rank n, called the tangent bundle, whose sheaf of sections is exactly $\mathscr{T}_X$.*

Proof. Complete details for the construction of T_X can be found in [110, 111, 117]. We outline a construction when $X \subset \mathbb{R}^N$ is a submanifold. (According to Whitney's embedding theorem [111], every manifold embeds into a Euclidean space. So in fact, this is no restriction at all.) Fix standard coordinates $y_1, \dots, y_N$ on $\mathbb{R}^N$. We define $T_X \subset X \times \mathbb{R}^N$ such that $(p; v_1, \dots, v_N) \in T_X$ if and only if

$$\sum v_i \left. \frac{\partial f}{\partial y_j} \right|_p = 0$$

whenever f is a C^∞ function defined in a neighborhood of p in $\mathbb{R}^N$ such that $X \cap U \subseteq f^{-1}(0)$. We have the obvious projection $\pi : T_X \to X$. A sum $v = \sum_j g_j \partial/\partial y_j$, with

$g_j \in C^\infty(U)$, defines a vector field on $U \subseteq X$ precisely when $(p; g_1(p), \ldots, g_N(p)) \in T_X$ for a $p \in U$. In other words, vector fields are sections of T_X.

It remains to find a local trivialization. We can find an open cover $\{U_i\}$ of X by coordinate neighborhoods. Choose local coordinates $x_1^{(i)}, \ldots, x_n^{(i)}$ in each U_i. Then the map

$$(p; w) \mapsto \left(p; \left(\frac{\partial y_j}{\partial x_k^{(i)}} \bigg|_p \right) w \right)$$

identifies $U_i \times \mathbb{R}^n$ with $\pi^{-1} U_i$. □

Tangent bundles also exist for complex manifolds and nonsingular algebraic varieties. However, we will postpone the construction. An example of an algebraic vector bundle of fundamental importance is given below.

Example 2.6.5. Projective space $\mathbb{P}_k^n$ is the set of lines $\{\ell\}$ in k^{n+1} through 0, and we can choose each line as a fiber of L. That is,

$$L = \{(x, \ell) \in k^{n+1} \times \mathbb{P}_k^n \,|\, x \in \ell\}.$$

Let $\pi : L \to \mathbb{P}_k^n$ be given by projection onto the second factor. Then L is a rank-one algebraic vector bundle, or line bundle, over $\mathbb{P}_k^n$. It is called the tautological line bundle. When $k = \mathbb{C}$, this can also be regarded as a holomorphic line bundle or a C^∞ complex line bundle.

L is often called the universal line bundle for the following reason:

Theorem 2.6.6. *If X is a compact C^∞ manifold with a C^∞ complex line bundle $p : M \to X$, then for $n \gg 0$, there exists a C^∞ map $f : X \to \mathbb{P}_{\mathbb{C}}^n$, called a classifying map, such that M is isomorphic (as a bundle) to the pullback*

$$f^* L = \{(v, x) \in L \times X \,|\, \pi(v) = f(x)\} \to X.$$

Proof. We just sketch the proof. Here we consider the dual line bundle M^* (see Exercise 2.6.14). Sections of this correspond to $\mathbb{C}$-valued functions on M that are linear on the fibers. Choose a local trivialization $\phi_i : M|_{U_i} \cong U_i \times \mathbb{C}$. A section of $M^*(U_i)$ can be identified with a function by $M^*(U_i) = C^\infty(U_i)\phi_i^{-1}(1)$. For each point $x \in U_i$, we can choose a C^∞ function f with compact support in U_i such that $f(x) \neq 0$ (which exists by Exercise 2.6.9). This can be extended by 0 to a global section. Thus by compactness, we can find finitely many sections $f_0, \ldots, f_n \in M^*(X)$ that do not simultaneously vanish at any point $x \in X$. Therefore we get an injective bundle map $M \hookrightarrow X \times \mathbb{C}^n$ given by $v \mapsto (f_0(v), \ldots, f_n(v))$. Or in more explicit terms, if we view $f_j|_{U_i}$ as functions, $M|_{U_i}$ can be identified with the span of $(f_0(x), \ldots, f_n(x))$ in $U_i \times \mathbb{C}^n$.

The maps

$$x \mapsto [f_0(x), \ldots, f_n(x)] \in \mathbb{P}^n, \quad x \in U_i,$$

are independent of the choice of trivialization. So this gives a map $f : X \to \mathbb{P}^n$. The pullback f^*L can also be described as the sub-line bundle of $X \times \mathbb{C}^n$ spanned by $(f_0(x), \ldots, f_n(x))$. So this coincides with M. □

Remark 2.6.7. When $M \to X$ is a holomorphic bundle on a complex manifold, the map $f : X \to \mathbb{P}^n_{\mathbb{C}}$ need not be holomorphic. This will follow from Exercise 2.7.13.

Exercises

2.6.8. Show that $v = \sum f_i(x) \frac{\partial}{\partial x_i}$ is a C^∞ vector field on $\mathbb{R}^n$ in the sense of Definition 2.6.2 if and only if the coefficients f_i are C^∞.

2.6.9. Show that the function

$$f(x) = \begin{cases} \exp\left(\frac{1}{\|x\|^2 - 1}\right) & \text{if } \|x\| < 1, \\ 0 & \text{otherwise,} \end{cases}$$

defines a nonzero C^∞-function on $\mathbb{R}^n$ with support in the unit ball. Conclude that any C^∞-manifold possesses nonconstant C^∞-functions.

2.6.10. Check that L is an algebraic line bundle.

2.6.11. Given a vector bundle $\pi : V \to X$ over a manifold and a C^∞ map $f : Y \to X$, show that the set

$$f^*V = \{(y, v) \in Y \times V \mid \pi(v) = f(y)\}$$

with its first projection to Y determines a vector bundle.

2.6.12. Given a vector bundle $V \to X$ with local trivialization $\phi_i : V|_{U_i} \xrightarrow{\sim} U_i \times k^n$, check that the matrix-valued functions $g_{ij} = \phi_i^{-1} \circ \phi_j$ on $U_i \cap U_j$ satisfy the cocycle identity $g_{ik} = g_{ij} g_{jk}$ on $U_i \cap U_j \cap U_k$. Conversely, check that any collection g_{ij} satisfying this identity arises from a vector bundle.

2.6.13. Given a vector bundle V with cocycle g_{ij} (as in the previous exercise), show that a section can be identified with a collection of vector-valued functions f_i on U_i satisfying $f_i = g_{ij} f_j$.

2.6.14. Suppose we are given a vector bundle $V = \bigcup V_x \to X$ with cocycle g_{ij}. Show that the union of dual spaces $V^* = \bigcup V_x^*$ can be made into a vector bundle with cocycle $(g_{ij}^{-1})^T$. Show that the sections of the dual of the tangent bundle T_X^*, called the cotangent bundle, are exactly the 1-forms.

2.6.15. Let $G = \mathbb{G}(r, n)$ be the Grassmannian of r-dimensional subspaces of k^n. This is an algebraic variety by Exercise 2.4.19. Let $S = \{(x, V) \in k^n \times G \mid x \in V\}$. Show that the projection $S \to G$ is an algebraic vector bundle of rank 2. This is called the universal bundle of rank r on G.

2.6.16. Prove the analogue of Theorem 2.6.6 for rank-r vector bundles: Any rank-two C^∞ complex vector bundle on a compact C^∞ manifold X is isomorphic to the pullback of the universal bundle for some C^∞ map $X \to \mathbb{G}(r,n)$ with $n \gg 0$. You may assume without proof that for any vector bundle $\pi : V \to X$, there exist finitely many sections f_i that span the fibers $V_x = \pi^{-1}(x)$.

2.7 Compact Complex Manifolds and Varieties

Up to this point, we have been treating C^∞ and complex manifolds in parallel. However, there are big differences, owing to the fact that holomorphic functions are much more rigid than C^∞ functions. We illustrate this in a couple of ways. In particular, we will see that the (most obvious) holomorphic analogue to Theorem 2.6.6 would fail. We start by proving some basic facts about holomorphic functions in many variables.

Theorem 2.7.1. *Let Δ^n be an open polydisk, that is, a product of disks, in $\mathbb{C}^n$.*

(1) If two holomorphic functions on Δ^n agree on a nonempty open set, then they agree on all of Δ^n.
(2) The maximum principle: If the absolute value of a holomorphic function f on Δ^n attains a maximum in Δ^n, then f is constant on Δ^n.

Proof. This can be reduced to the corresponding statements in one variable [1]. We leave the first statement as an exercise and we do the second. Suppose that $|f|$ attains a maximum at $(a_1,\ldots,a_n) \in \Delta$. The maximum principle in one variable implies that $f(z,a_2,\ldots,a_n)$ is constant. Fixing $z_1 \in \Delta$, we see that $f(z_1,z,a_3,\ldots)$ is constant, and so on. □

We saw in the exercises that all C^∞-manifolds carry nonconstant global C^∞-functions. By contrast we have the following:

Proposition 2.7.2. *If X is a compact connected complex manifold, then all holomorphic functions on X are constant.*

Proof. Let $f : X \to \mathbb{C}$ be holomorphic. Since X is compact, $|f|$ attains a maximum somewhere, say at $x_0 \in X$. The set $S = f^{-1}(f(x_0))$ is closed by continuity. It is also open by the maximum principle. So $S = X$. □

Corollary 2.7.3. *A holomorphic function is constant on a nonsingular complex projective variety.*

Proof. $\mathbb{P}^n_{\mathbb{C}}$ with its classical topology is compact, since the unit sphere in $\mathbb{C}^{n+1}$ maps onto it. Therefore any submanifold of it is also compact. □

We want to prove a version of Corollary 2.7.3 for algebraic varieties over arbitrary fields. We first need a good substitute for compactness. To motivate it, we make the following observation:

Lemma 2.7.4. *If X is a compact metric space, then for any metric space Y, the projection $p : X \times Y \to Y$ is closed.*

Proof. Given a closed set $Z \subset X \times Y$ and a sequence $y_i \in p(Z)$ converging to $y \in Y$, we have to show that y lies in $p(Z)$. By assumption, we have a sequence $x_i \in X$ such that $(x_i, y_i) \in Z$. Since X is compact, we can assume that x_i converges to say $x \in X$ after passing to a subsequence. Then we see that (x, y) is the limit of (x_i, y_i), so it must lie in Z because it is closed. Therefore $y \in p(Z)$. □

Definition 2.7.5. An algebraic variety X is called complete or proper if for any variety Y, the projection $p : X \times Y \to Y$ is closed, i.e., p takes closed sets to closed sets.

Theorem 2.7.6. *Projective varieties are complete.*

Proof. Complete proofs can be found in [60, 92, 104]. We give an outline, leaving the details for the exercises. First, we can reduce the theorem to showing that $\pi : \mathbb{P}^n_k \times \mathbb{A}^m_k \to \mathbb{A}^m_k$ is closed for each m, n. This case can be handled by classical elimination theory [22, §8.5]. A closed set of the product is defined by a collection of polynomials $f_i(x_0, \ldots, x_n, y_1, \ldots, y_m)$ homogeneous in the x's. Let I be the ideal in $k[x_0, \ldots, x_n, y_1, \ldots, y_m]$ generated by the f_i. The elimination ideal is defined by

$$J = \{g \in k[y_1, \ldots, y_m] \mid \forall i \exists e_i, x_i^{e_i} g \in I\}.$$

Then the image of the projection $\pi(V(I))$ is $V(J)$, hence closed. □

There is an analogue of Proposition 2.7.2.

Proposition 2.7.7. *If X is a complete algebraic variety, then all regular functions on X are constant.*

Proof. Let f be a regular function on X. We can view it as a morphism $f : X \to \mathbb{A}^1_k$. Since $\mathbb{A}^1_k \subset \mathbb{P}^1_k$, we also have a morphism $F : X \to \mathbb{P}^1_k$ given by composition. The image $f(X)$ is closed, since it coincides with the image of the graph $\{(x, f(x)) \mid x \in X\}$ under projection. Similarly, $F(X)$ is closed. Since X is irreducible, the same is true of $f(X)$. The only irreducible closed subsets of $\mathbb{A}^1_k$ are points and $\mathbb{A}^1_k$ itself. If $f(X) = \mathbb{A}^1_k$, then we would be forced to conclude that $F(X)$ was not closed. Therefore $f(X)$ must be a point. □

The converse is false (Exercise 2.7.15).

Corollary 2.7.8. *An affine variety is not complete, unless it is a point.*

Proof. If $p, q \in X$ are distinct, there exists a regular function such that $f(p) = 1$ and $f(q) = 0$. □

We can form the sheaf of regular sections of the tautological line bundle L (Example 2.6.5) on $\mathbb{P}^n_k$, which we denote by $\mathscr{O}_{\mathbb{P}^n_k}(-1)$ or often simply by $\mathscr{O}(-1)$. We write $\mathscr{O}_{\mathbb{P}^n_{an}}(-1)$, or simply $\mathscr{O}_{\mathbb{P}^n}(-1)$ when there is no danger of confusion, for the sheaf of holomorphic sections over the complex manifold $\mathbb{P}^n_{\mathbb{C}}$. These examples are central to algebraic geometry. For any open set $U \subset \mathbb{P}^n_k$ viewed as a set of lines, we have

$$\mathscr{O}(-1)(U) = \{f : U \to k^{n+1} \mid f \text{ regular and } f(\ell) \in \ell\}.$$

We saw in the course of proving Theorem 2.6.6 that any line has many C^∞-sections. The corresponding statements in the analytic and algebraic worlds are false:

Lemma 2.7.9. *The spaces of global sections* $\mathscr{O}(-1)(\mathbb{P}^n_k)$ *and* $\mathscr{O}_{\mathbb{P}^n_{an}}(-1)(\mathbb{P}^n_{\mathbb{C}})$ *are both equal to 0.*

Proof. We prove the first statement. The second is similar. A global section is given by a regular function $f : \mathbb{P}^n \to k^{n+1}$ satisfying $f(\ell) \in \ell$. However, f is constant with value, say, v. Thus $v \in \bigcap \ell = \{0\}$. □

Exercises

2.7.10. Finish the proof of Theorem 2.7.1 (1).

2.7.11. Check that Theorem 2.7.6 can be reduced to showing that $\pi : \mathbb{P}^n_k \times \mathbb{A}^m_k \to \mathbb{A}^m_k$ is closed.

2.7.12. In the notation of the proof of Theorem 2.7.6, show that the elimination ideal J is an ideal and that $V(J) = \pi(V(I))$. For the "hard" direction use the projective Nullstellensatz that $V(K) \subset \mathbb{P}^n_k$ is empty if and only if K contains a power of $(x_0, \dots, x_n)$.

2.7.13. If $f : X \to \mathbb{P}^n_{\mathbb{C}}$ is a nonconstant holomorphic function between compact manifolds, prove that $f^* \mathscr{O}_{\mathbb{P}^n}(-1)$ has no nonzero holomorphic sections.

2.7.14. Let $\mathscr{O}(1)$ denote the sheaf of holomorphic sections of the dual L^* (Exercise 2.6.14) of the tautological bundle on $\mathbb{P}^n_{\mathbb{C}}$. Show that $\mathscr{O}(1)$ has at least $n+1$ independent nonzero sections. Using the previous exercise, deduce that Theorem 2.6.6 will fail for holomorphic maps and line bundles.

2.7.15. Let $X = \mathbb{P}^n_k - \{p\}$ with $n > 1$. Show that X has no nonconstant regular functions, and that it is not complete.

2.7.16. Given a variety X, the collection of *constructible* subsets of X is the Boolean algebra generated by Zariski open sets. In other words, it is the smallest collection containing open sets and closed under finite unions, intersections, and complements. Prove Chevalley's theorem that a projection $p : \mathbb{A}^n_k \to \mathbb{A}^m_k$ takes constructible sets to constructible sets.

Chapter 3
More Sheaf Theory

We introduced sheaves of functions in the previous chapter as a convenient language for defining manifolds and varieties. However, as we will see, there is much more to this story. In this chapter, we develop sheaf theory in a more systematic fashion. Presheaves and sheaves are somewhat more general notions than what we described earlier. We give the full definitions here, and then explore their formal properties. We define the notion of an exact sequence in the category of sheaves. Exact sequences and the associated cohomology sequences, given in the next chapter, form one of the basic tools used throughout the rest of the book. We also give a brief introduction to Grothendieck's theory of schemes. A scheme is a massive generalization of an algebraic variety, and quite a bit of sheaf theory is required just to give the definition.

3.1 The Category of Sheaves

It will be convenient to define presheaves of things other than functions. For instance, one might consider sheaves of equivalence classes of functions, distributions, and so on. For this more general notion of presheaf, the restriction maps have to be included as part of the data:

Definition 3.1.1. A presheaf $\mathscr{P}$ of sets (respectively groups or rings) on a topological space X consists of a set (respectively group or ring) $\mathscr{P}(U)$ for each open set U, and maps (respectively homomorphisms) $\rho_{UV} : \mathscr{P}(U) \to \mathscr{P}(V)$ for each inclusion $V \subseteq U$ such that:

1. $\rho_{UU} = \mathrm{id}_{\mathscr{P}(U)}$;
2. $\rho_{VW} \circ \rho_{UV} = \rho_{UW}$ if $W \subseteq V \subseteq U$.

We will usually write $f|_V = \rho_{UV}(f)$. Here is a simple example of a presheaf given abstractly.

Example 3.1.2. Let X be topological space. Then

$$\mathscr{P}(U) = \begin{cases} \mathbb{Z} & \text{if } U = X, \\ 0 & \text{otherwise,} \end{cases}$$

with all $\rho_{UV} = 0$, is a presheaf.

A more natural class of examples, which arises frequently, is given by the quotient construction.

Example 3.1.3. Let $\mathscr{P}$ be a presheaf of abelian groups. Then a subpresheaf $\mathscr{P}' \subseteq \mathscr{P}$ is a collection of subgroups $\mathscr{P}'(U) \subseteq \mathscr{P}(U)$ stable under the restriction maps ρ_{UV}. The *presheaf quotient* is given by

$$(\mathscr{P}/\mathscr{P}')^P(U) = \mathscr{P}(U)/\mathscr{P}(U)'$$

with the induced restrictions. (This somewhat clumsy notation is used to distinguish this from the quotient sheaf to be defined later on.)

The definition of a sheaf carries over verbatim.

Definition 3.1.4. A sheaf $\mathscr{P}$ is a presheaf such that for any open cover $\{U_i\}$ of U and $f_i \in \mathscr{P}(U_i)$ satisfying $f_i|_{U_i \cap U_j} = f_j|_{U_i \cap U_j}$, there exists a unique $f \in \mathscr{P}(U)$ with $f|_{U_i} = f_i$. We also require that $\mathscr{P}(\emptyset)$ consist of a single element (which is necessarily 0 for sheaves of abelian groups).

In English, this says that a collection of local sections can be patched together uniquely provided they agree on the intersections. We have already seen plenty of examples of presheaves that are not sheaves. With this more general definition comes more pathologies. When X has a nontrivial open cover $\{U_i \neq X\}$, Example 3.1.2 is not a sheaf, because the local sections $0 \in \mathscr{P}(U_i)$ can be patched in many ways. More examples of nonsheaves arise as presheaf quotients (see the exercises).

Definition 3.1.5. Given presheaves of sets (respectively groups) $\mathscr{P}, \mathscr{P}'$ on the same topological space X, a morphism $f : \mathscr{P} \to \mathscr{P}'$ is a collection of maps (respectively homomorphisms) $f_U : \mathscr{P}(U) \to \mathscr{P}'(U)$ that commute with the restrictions. Given morphisms $f : \mathscr{P} \to \mathscr{P}'$ and $g : \mathscr{P}' \to \mathscr{P}''$, the compositions $g_U \circ f_U$ determine a morphism $\mathscr{P} \to \mathscr{P}''$. The collection of presheaves of abelian groups and morphisms with this notion of composition constitutes a category $\text{PAb}(X)$.

Definition 3.1.6. The category $\text{Ab}(X)$ is the full subcategory of $\text{PAb}(X)$ generated by sheaves of abelian groups on X. In other words, objects of $\text{Ab}(X)$ are sheaves, and morphisms are defined in the same way as for presheaves.

A special case of a morphism is the notion of a *subpresheaf* of a presheaf $\mathscr{P}' \subseteq \mathscr{P}$ defined above. We call this a subsheaf if both objects are sheaves.

Example 3.1.7. Given a sheaf of rings of functions $\mathscr{R}$ over X, and $f \in \mathscr{R}(X)$, the collection of maps $\mathscr{R}(U) \to \mathscr{R}(U)$ given by multiplication by $f|_U$ is a morphism.

Example 3.1.8. Let X be a C^∞ manifold. Then the differential $d : C^\infty_X \to \mathscr{E}^1_X$ given in Section 2.6 is a morphism of sheaves.

Example 3.1.9. Let Y be a closed subset of a k-space $(X, \mathscr{O}_X)$. Then the ideal sheaf associated to Y,

$$\mathscr{I}_Y(U) = \{f \in \mathscr{O}_X(U) \mid f|_{Y \cap U} = 0\},$$

is a subsheaf of $\mathscr{O}_X$.

Example 3.1.10. Let $(X, \mathscr{O}_X)$ be a C^∞ or complex manifold or algebraic variety and $Y \subset X$ a submanifold or subvariety. There is a morphism from the presheaf quotient $(\mathscr{O}_X / \mathscr{I}_Y)^P \to \mathscr{O}_Y$ induced by $f \mapsto f|_Y$.

We now introduce the notion of a *covariant functor* (or simply functor) $F : \mathscr{C}_1 \to \mathscr{C}_2$ between categories. This consists of a map $F : \mathrm{Obj}\mathscr{C}_1 \to \mathrm{Obj}\mathscr{C}_2$ and maps

$$F : \mathrm{Hom}_{\mathscr{C}_1}(A, B) \to \mathrm{Hom}_{\mathscr{C}_2}(F(A), F(B))$$

for each pair $A, B \in \mathrm{Obj}\mathscr{C}_1$ such that

(F1) $F(f \circ g) = F(f) \circ F(g)$;
(F2) $F(\mathrm{id}_A) = \mathrm{id}_{F(A)}$.

In most cases, we will be content to just describe the map on objects, and leave the map on morphisms implicit. There are many examples of functors in mathematics: the map that takes a ring to its group of units, the map that takes a group to its group ring, the map that takes a topological space with a distinguished point to its fundamental group, etc.

Contravariant functors are defined similarly but with

$$F : \mathrm{Hom}_{\mathscr{C}_1}(A, B) \to \mathrm{Hom}_{\mathscr{C}_2}(F(B), F(A))$$

and the rule (F1) for composition adjusted accordingly. For example, the dual $V \mapsto V^*$ gives a contravariant functor from the category of vector spaces to itself.

Presheaves are themselves functors (Exercise 3.1.14). We describe a couple of additional examples involving presheaves.

Example 3.1.11. Let Ab denote the category of abelian groups. Given a space X, the global section functor $\Gamma : \mathrm{PAb}(X) \to \mathrm{Ab}$ is defined by $\Gamma(\mathscr{P}) = \Gamma(X, \mathscr{P}) = \mathscr{P}(X)$ for objects, $\Gamma(f) = f_X$ for morphisms $f : \mathscr{P} \to \mathscr{P}'$.

Fix a space X and a point $x \in X$. We define the *stalk* $\mathscr{P}_x$ of $\mathscr{P}$ at x as the direct limit $\varinjlim \mathscr{P}(U)$ over neighborhoods of x. The elements can be represented by germs of sections of $\mathscr{P}$, as in Section 2.5. Given a morphism $\phi : \mathscr{P} \to \mathscr{P}'$, we get an induced map $\mathscr{P}_x \to \mathscr{P}'_x$ taking the germ of f to the germ of $\phi(f)$.

Example 3.1.12. The map $\mathscr{P} \mapsto \mathscr{P}_x$ gives a functor from $\mathrm{PAb}(X) \to \mathrm{Ab}$.

We have a functor generalizing the construction given in Example 2.1.10 that is of fundamental importance.

Theorem 3.1.13. *There is a functor* $\mathscr{P} \mapsto \mathscr{P}^+$ *from* $\mathrm{PAb}(X)$ *to* $\mathrm{Ab}(X)$ *called the sheafification or the associated sheaf, with the following properties:*

(a) There is a canonical morphism $\mathscr{P} \to \mathscr{P}^+$.
(b) The map $\mathscr{P} \to \mathscr{P}^+$ *induces an isomorphism on stalks.*
(c) If $\mathscr{P}$ *is a sheaf, then the morphism* $\mathscr{P} \to \mathscr{P}^+$ *is an isomorphism (see the exercises for the precise meaning).*
(d) Any morphism from $\mathscr{P}$ *to a sheaf factors uniquely through* $\mathscr{P} \to \mathscr{P}^+$.

Proof. We sketch the construction of $\mathscr{P}^+$. We do this in two steps. First, we construct a presheaf of functions $\mathscr{P}^\#$. Set $Y = \prod_{x \in X} \mathscr{P}_x$. We define a sheaf $\mathscr{P}^\#$ of Y-valued functions and a morphism $\mathscr{P} \to \mathscr{P}^\#$ as follows. There is a map $\sigma_x : \mathscr{P}(U) \to \mathscr{P}_x$ given by

$$\sigma_x(f) = \begin{cases} \text{the germ } f_x & \text{if } x \in U, \\ 0 & \text{otherwise.} \end{cases}$$

Then $f \in \mathscr{P}(U)$ determines a function $f^\# : U \to Y$ given by $f^\#(x) = \sigma_x(f)$. Let

$$\mathscr{P}^\#(U) = \{f^\# \mid f \in \mathscr{P}(U)\}.$$

This yields a presheaf. We have a morphism $\mathscr{P}(U) \to \mathscr{P}^\#(U)$ given by $f \mapsto f^\#$. Now apply the construction given earlier in Example 2.1.10 to produce a sheaf $\mathscr{P}^+ = (\mathscr{P}^\#)^s$. The composition $\mathscr{P}(U) \to \mathscr{P}^\#(U) \subset \mathscr{P}^+(U)$ yields the desired morphism $\mathscr{P} \to \mathscr{P}^+$. This construction is clearly functorial. Parts (b) and (c) are given in the exercises. For (d), suppose we are given a morphism from a presheaf to a sheaf $\mathscr{P} \to \mathscr{F}$. Then we have a morphism $\mathscr{P}^+ \to \mathscr{F}^+ \cong \mathscr{F}$. □

Sheafification is a very useful operation. Many constructions incorporate it as a finishing step. For example, given a sheaf $\mathscr{S}$ and a subsheaf $\mathscr{S}' \subseteq \mathscr{S}$, we define the *quotient sheaf* by $\mathscr{S}/\mathscr{S}' = ((\mathscr{S}/\mathscr{S}')^P)^+$. Note that $(\mathscr{S}/\mathscr{S}')^P$ by itself need not be a sheaf (see the exercises).

In the exercises, we give an alternative construction of $\mathscr{P}^+$. This is based on the so called étalé space (not to be confused with étale map), which was popular in older treatments of sheaf theory such as [45].

Exercises

3.1.14. Let X be a topological space. Fix a symbol $*$. Construct a category $\mathrm{Open}(X)$, whose objects are open subsets of X, and let

$$\mathrm{Hom}_{\mathrm{Open}(X)}(U,V) = \begin{cases} * & \text{if } U \subseteq V, \\ \emptyset & \text{otherwise.} \end{cases}$$

Show that a presheaf of sets (or groups...) on $\mathrm{Open}(X)$ is the same thing as a contravariant functor to the category of sets (or groups...).

3.1.15. Given a presheaf $\mathscr{F} \in \mathrm{PAb}(X)$ and $f \in \mathscr{F}(X)$, prove that the support $\mathrm{supp}(f) = \{x \in X \mid f_x \neq 0\}$ is closed. Give an example where the support of $\mathscr{F}$, $\mathrm{supp}(\mathscr{F}) = \{x \in X \mid \mathscr{F}_x \neq 0\}$, is not closed.

3.1.16. Given a presheaf $\mathscr{P}$, show that the morphisms $\mathscr{P} \to \mathscr{P}^{\#}$ and $\mathscr{P}^{\#} \to \mathscr{P}^{+}$ given in the proof of Theorem 3.1.13 are isomorphisms on stalks.

3.1.17. A morphism $f : \mathscr{P} \to \mathscr{Q}$ in the category of presheaves or sheaves is called an *isomorphism* if there exists a morphism $f^{-1} : \mathscr{Q} \to \mathscr{P}$ such that $f \circ f^{-1}$ and $f^{-1} \circ f$ are both the identity. Show that f is an isomorphism if and only if each f_U is a bijection. If $\mathscr{P}, \mathscr{Q}$ are sheaves, show that f is an isomorphism if and only if each $f_x : \mathscr{P}_x \to \mathscr{Q}_x$ is a bijection. This implies Theorem 3.1.13 (c).

3.1.18. A presheaf is called *separated* if for any $f_i \in \mathscr{P}(U_i)$ satisfying $f_i|_{U_i \cap U_j} = f_j|_{U_i \cap U_j}$, there exists at most one $f \in \mathscr{P}(U)$ with $f|_{U_i} = f_i$. Example 3.1.2 is not separated. Show that $\mathscr{P}$ is separated if and only if $\mathscr{P} \cong \mathscr{P}^{\#}$.

3.1.19. Let B be a basis for the topology of X. Fix a presheaf $\mathscr{F}$ on B, by which we mean a contravariant functor on B to sets. Suppose that $\mathscr{F}$ satisfies the following version of the sheaf axiom: for any open covering $\{U_i\} \subset B$ of $U \in B$, if $f_i \in F(U_i)$ are compatible in the sense that $f_i|_V = f_j|_V$ for all $U_i \cap U_j \supseteq V \in B$, then there exists a unique $f \in \mathscr{F}(U)$ with $f_i = f|_{U_i}$. Prove that there is at most one extension of $\mathscr{F}$ to a sheaf on X.

3.1.20. Continuing the notion from the previous exercise, prove that the rule

$$\mathscr{F}^e(U) = \{\text{families } f_V \in \mathscr{F}(V), U \supseteq V \in B, \text{ such that } f_W = f_V|_W \text{ when } W \subseteq V\}$$

determines the (unique) sheaf on X extending $\mathscr{F}$. Show that this construction is functorial in the obvious sense.

3.1.21. Given a submanifold or subvariety $Y \subset X$, show that $\mathscr{O}_Y$ is isomorphic to the quotient sheaf $\mathscr{O}_X / \mathscr{I}_Y$. Give an example to show that $(\mathscr{O}_X / I_X)^P$ is not a sheaf in general.

3.1.22. The étalé space of a presheaf $\mathscr{P}$ is the disjoint union of stalks $\mathrm{Et}(\mathscr{P}) = \bigcup \mathscr{P}_x$. This carries a projection to $\pi : \mathrm{Et}(\mathscr{P}) \to X$ by sending $\mathscr{P}_x$ to x. A section $f \in \mathscr{P}(U)$ defines a cross section $\mathrm{Et}(f) = (f_x)$ to π over U. We give $E(\mathscr{P})$ the weakest topology that makes all the $\mathrm{Et}(f)$'s continuous. Prove that the sheaf of continuous cross sections of π is isomorphic to $\mathscr{P}^+$.

3.2 Exact Sequences

The categories $\mathrm{PAb}(X)$ and $\mathrm{Ab}(X)$ are *additive*, which means among other things that $\mathrm{Hom}(A, B)$ has an abelian group structure such that composition is bilinear.

Actually, more is true. These categories are *abelian* [44, 82, 118], which implies that they possess many of the basic constructions and properties of the category of abelian groups. In particular, given a morphism, we can form kernels, cokernels, and images, characterized by the appropriate universal properties. This is spelled out more fully in the exercises. Here we just define these operations. Given a morphism of presheaves $f : \mathscr{A} \to \mathscr{B}$, we define the presheaf kernel, image, and cokernel by

$$(\mathrm{pker}\, f)(U) = \ker f_U : [\mathscr{A}(U) \to \mathscr{B}(U)],$$
$$(\mathrm{pim}\, f)(U) = \mathrm{im}\, f_U : [\mathscr{A}(U) \to \mathscr{B}(U)],$$
$$(\mathrm{pcoker}\, f)(U) = \mathrm{coker}\, f_U : [\mathscr{A}(U) \to \mathscr{B}(U)].$$

This is an *isomorphism* if f_U is an isomorphism for every U, or equivalently if $\mathrm{pker}\, f = \mathrm{pcoker}\, f = 0$.

For a morphism of sheaves $f : \mathscr{A} \to \mathscr{B}$, the sheaf kernel, etc. is given by

$$\ker f = (\mathrm{pker}\, f)^+, \quad \mathrm{im}\, f = (\mathrm{pim}\, f)^+, \quad \mathrm{coker}\, f = (\mathrm{pcoker}\, f)^+.$$

We may get a better sense of these by looking at the stalks:

$$(\ker f)_x = (\mathrm{pker}\, f)_x = \ker f_x : [\mathscr{A}_x \to \mathscr{B}_x],$$
$$(\mathrm{im}\, f)_x = (\mathrm{pim}\, f)_x = \mathrm{im}\, f_x : [\mathscr{A}_x \to \mathscr{B}_x],$$
$$(\mathrm{coker}\, f)_x = (\mathrm{pcoker}\, f)_x = \mathrm{coker}\, f_x : [\mathscr{A}_x \to \mathscr{B}_x].$$

We say that f is an *isomorphism of sheaves* if it is an isomorphism of presheaves. It is more natural to require that $\ker f = \mathrm{coker}\, f = 0$, but we will see later that this is equivalent (Lemma 3.2.6).

We can now introduce the notion of exactness in $\mathrm{Ab}(X)$ by copying the usual definition. But first a few words of warning. The notion of exactness is sensitive to the category in which we work. Below we give the notion of exactness in the category $\mathrm{Ab}(X)$, since this is the most important case. Exactness in $\mathrm{PAb}(X)$, even when the objects are sheaves, is different. This will be clarified in the exercises.

A sequence

$$\mathscr{A} \xrightarrow{f} \mathscr{B} \xrightarrow{g} \mathscr{C}$$

of sheaves on X is called a complex if $g \circ f = 0$. There is a morphism $\mathrm{pim}(f) \to \mathrm{pker}(g)$ given by the canonical morphism $\mathrm{im}(f)_U \to \ker(g)_U$. This induces a morphism $\mathrm{im}(f) \to \ker(g)$.

Definition 3.2.1. A sequence

$$\mathscr{A} \xrightarrow{f} \mathscr{B} \xrightarrow{g} \mathscr{C}$$

of sheaves on X is called exact if it is a complex and the canonical morphism $\mathrm{im}(f) \to \ker(g)$ is an isomorphism of sheaves.

As special cases, we have natural analogues of injective and surjective maps. Although, it is better to use neutral terminology, we say that a morphism $f : \mathscr{A} \to \mathscr{B}$

of sheaves is a *monomorphism* (respectively *epimorphism*) if $\ker f = 0$ (respectively $\operatorname{im} f = \mathscr{B}$).

Lemma 3.2.2. *Given a sequence*

$$\mathscr{A} \xrightarrow{f} \mathscr{B} \xrightarrow{g} \mathscr{C}$$

of sheaves on X, the following are equivalent:

(1) The sequence is exact.
(2) For every U:

  (a) $g_U \circ f_U = 0$.
  (b) Given $b \in \mathscr{B}(U)$ with $g(b) = 0$, there exist an open cover $\{U_i\}$ of U and $a_i \in \mathscr{A}(U_i)$ such that $f_{U_i}(a_i) = b|_{U_i}$.

(3) The sequence of stalks

$$\mathscr{A}_x \xrightarrow{f_x} \mathscr{B}_x \xrightarrow{g_x} \mathscr{C}_x$$

is exact for every $x \in X$.

Proof. The equivalence of (1) and (2) is clear after unraveling the definition of exactness. Condition (a) is equivalent to $g \circ f = 0$ and (b) to $\operatorname{im}(f) = \ker(g)$.

We will prove that (3) implies (2), leaving the remaining direction as an exercise. Suppose that $\mathscr{A} \to \mathscr{B} \to \mathscr{C}$ is exact. To simply notation, we suppress the subscript U. Given $a \in \mathscr{A}(U)$, we have $g(f(a)) = 0$, since $g(f(a))_x = g(f(a_x)) = 0$ for all $x \in U$. This proves (a).

Given $b \in \mathscr{B}(U)$ with $g(b) = 0$, then for each $x \in U$, b_x is the image of a germ in $\mathscr{A}_x$. Choose a representative a_i for this germ in some $\mathscr{A}(U_i)$, where U_i is a neighborhood of x. After shrinking U_i if necessary, we have $f(a_i) = b|_{U_i}$. As x varies, we get an open cover $\{U_i\}$ and sections $a_i \in \mathscr{A}(U_i)$, as required. □

Corollary 3.2.3. *If $\mathscr{A}(U) \to \mathscr{B}(U) \to \mathscr{C}(U)$ is exact for every open set U, then $\mathscr{A} \to \mathscr{B} \to \mathscr{C}$ is exact.*

Exactness can be extended to longer sequences by requiring that the conditions of Definition 3.2.1 hold for any two adjacent arrows.

Corollary 3.2.4. *A sequence of sheaves on X*

$$\cdots \to \mathscr{A} \to \mathscr{B} \to \mathscr{C} \to \cdots$$

is exact if and only if

$$\cdots \to \mathscr{A}_x \to \mathscr{B}_x \to \mathscr{C}_x \to \cdots$$

is exact for every $x \in X$. In particular, $f : \mathscr{A} \to \mathscr{B}$ is a monomorphism (respectively epimorphism) if and only if $\mathscr{A}_x \to \mathscr{B}_x$ is injective (respectively surjective) for all $x \in X$.

The key point of this result is that no matter how complicated X and $\mathscr{A},\mathscr{B},\mathscr{C},\ldots$ appear in the large, exactness is a local issue, and this is what gives the notion its power. We will let the symbols $\mathscr{A},\mathscr{B},\mathscr{C}$ stand for sheaves for the remainder of this section unless stated otherwise.

The converse to Corollary 3.2.3 is false, but we do have the following:

Lemma 3.2.5. *If*

$$0 \to \mathscr{A} \to \mathscr{B} \to \mathscr{C} \to 0$$

is an exact sequence of sheaves, then

$$0 \to \mathscr{A}(U) \to \mathscr{B}(U) \to \mathscr{C}(U)$$

is exact for every open set U.

Proof. Let $f : \mathscr{A} \to \mathscr{B}$ and $g : \mathscr{B} \to \mathscr{C}$ denote the maps. By Lemma 3.2.2, $g_u \circ f_U = 0$. Suppose $a \in \mathscr{A}(U)$ maps to 0 under f. Then $f(a_x) = f(a)_x = 0$ for each $x \in U$ (we are suppressing the subscript U once again). Therefore $a_x = 0$ for each $x \in U$, and this implies that $a = 0$.

Suppose $b \in \mathscr{B}(U)$ satisfies $g(b) = 0$. Then by Lemma 3.2.2, there exist an open cover $\{U_i\}$ of U and $a_i \in \mathscr{A}(U_i)$ such that $f(a_i) = b|_{U_i}$. Then $f(a_i|_{U_i\cap U_j} - a_j|_{U_i\cap U_j}) = 0$, which implies $a_i|_{U_i\cap U_j} - a_j|_{U_i\cap U_j} = 0$ by the first paragraph. Therefore $\{a_i\}$ patch together to yield an element of $a \in \mathscr{A}(U)$ such that $f(a) = b$. □

Lemma 3.2.6. *Given a morphism $f : \mathscr{A} \to \mathscr{B}$ of sheaves, the following statements are equivalent:*

(a) f is an isomorphism of sheaves, i.e., $f_U : \mathscr{A}(U) \cong \mathscr{B}(U)$ for each U.
(b) $\ker f = \operatorname{coker} f = 0$.
(c) f induces an isomorphism of abelian groups $\mathscr{A}_x \to \mathscr{B}_x$ for each $x \in X$.

Proof. (a) implies that $\operatorname{pker} f = \operatorname{pcoker} f = 0$, and therefore $\ker f = \operatorname{coker} f = 0$. Suppose that (b) holds. Then

$$\ker f_x = (\ker f)_x = 0$$

and

$$\operatorname{coker} f_x = (\operatorname{coker} f)_x = 0.$$

Therefore $f_x : \mathscr{A}_x \cong \mathscr{B}_x$. Finally, (c) implies $\mathscr{A}(U) \cong \mathscr{B}(U)$ by Lemma 3.2.5. □

We give some *natural* examples to show that $\mathscr{B}(X) \to \mathscr{C}(X)$ is not usually surjective when $\mathscr{B} \to \mathscr{C}$ is an epimorphism.

Example 3.2.7. Consider the circle $S^1 = \mathbb{R}/\mathbb{Z}$. Then

$$0 \to \mathbb{R}_{S^1} \to C^\infty_{S^1} \xrightarrow{d} \mathscr{E}^1_{S^1} \to 0$$

is exact. However, $C^\infty(S^1) \to \mathscr{E}^1(S^1)$ is not surjective.

To see the first statement, let $U \subset S^1$ be an open set diffeomorphic to an open interval. Then the sequence

$$0 \to \mathbb{R} \to C^\infty(U) \xrightarrow{f \to f'} C^\infty(U)dx \to 0$$

is exact by calculus. Thus we get exactness on stalks. The 1-form dx gives a global section of $\mathscr{E}^1(S^1)$, since it is translation-invariant. However, it is not the differential of a periodic function. Therefore $C^\infty(S^1) \to \mathscr{E}^1(S^1)$ is not surjective.

Example 3.2.8. Let $(X, \mathscr{O}_X)$ be a C^∞ or complex manifold or algebraic variety and let $Y \subset X$ be a submanifold or subvariety, with $i : Y \to X$ denoting inclusion. Define the sheaf $i_*\mathscr{O}_Y$ on X by $i_*\mathscr{O}_Y(U) = \mathscr{O}_Y(U \cap Y)$. Then

$$0 \to \mathscr{I}_Y \to \mathscr{O}_X \to i_*\mathscr{O}_Y \to 0$$

is exact, but the map $\mathscr{O}_X(X) \to \mathscr{O}_Y(Y)$ need not be surjective. For example, let $X = \mathbb{P}^1_{\mathbb{C}}$ with $\mathscr{O}_X$ the sheaf of holomorphic functions. Let $Y = \{p_1, p_2\} \subset \mathbb{P}^1$ be a set of distinct points. Then the function $f \in \mathscr{O}_Y(X)$ that takes the value 1 on p_1 and 0 on p_2 cannot be extended to a global holomorphic function on $\mathbb{P}^1$, since all such functions are constant by Liouville's theorem.

Exercises

3.2.9. Finish the proof of Lemma 3.2.2.

3.2.10. Let f be a morphism of sheaves. Check that the sheafification step is unnecessary for $\ker f$, i.e., that $\mathrm{pker} f$ is a sheaf. Is this true for $\mathrm{coker} f$ and $\mathrm{im} f$?

3.2.11. Let $f : \mathscr{A} \to \mathscr{B}$ be a morphism of sheaves. Show that if $g : \mathscr{C} \to \mathscr{A}$ is a morphism such that $f \circ g = 0$, then it factors uniquely through $\ker f$. Dually show that if $g : \mathscr{B} \to \mathscr{C}$ is a morphism such that $g \circ f = 0$, then it factors uniquely through $\mathrm{coker} f$. Show that $\mathrm{im} f$ is isomorphic to $\mathrm{coker}[\ker f \to \mathscr{A}]$ and to $\ker[\mathscr{B} \to \mathrm{coker} f]$.

3.2.12. Define a sequence of presheaves $\mathscr{A} \xrightarrow{f} \mathscr{B} \xrightarrow{g} \mathscr{C}$ to be exact in $\mathrm{PAb}(X)$ if $g \circ f = 0$ and $\mathrm{pim}(f) \cong \mathrm{pker}(g)$. Show that this is equivalent to the exactness of $\mathscr{A}(U) \to \mathscr{B}(U) \to \mathscr{C}(U)$ for every U. Show that if a sequence of sheaves is exact in $\mathrm{PAb}(X)$, then it is exact in the sense of Definition 3.2.1, but that the converse can fail.

3.2.13. Let $\Gamma_c(\mathscr{F}) \subseteq \mathscr{F}(X)$ consist of sections f whose support $\mathrm{supp}(f)$ is compact. Prove that a short exact sequence $\mathscr{A} \to \mathscr{B} \to \mathscr{C}$ of sheaves gives an exact sequence $0 \to \Gamma_c(\mathscr{A}) \to \Gamma_c(\mathscr{B}) \to \Gamma_c(\mathscr{C})$.

3.2.14. Let $i : Y \to X$ be as in Example 3.2.8. Show that the stalk $(i_*\mathscr{O}_Y)_x$ is equal to $\mathscr{O}_{Y,x}$ if $x \in Y$, and $(i_*\mathscr{O}_Y)_x = 0$ if $x \notin Y$. Verify that the sequence in the example is exact. Show that the restriction to Y given by $i_*\mathcal{O}_Y|_Y(U\cap Y) = i_*\mathcal{O}_Y(U)$ is well defined and coincides with $\mathcal{O}_Y$.

3.2.15. Let $U \subseteq \mathbb{C}$ be open. Show that there is an exact sequence of sheaves

$$0 \to \mathbb{Z}_U \to \mathcal{O}_U \to \mathcal{O}_U^* \to 1,$$

where $\mathcal{O}_U$ (respectively $\mathcal{O}_U^*$) is the sheaf of (nowhere zero) holomorphic functions, and the second map sends f to $\exp(2\pi i f)$. Give an example where the last map on global sections is not surjective.

3.2.16. Let $U \subseteq \mathbb{R}^2$ be open. Characterize $\mathrm{im}(\nabla)$, where $\nabla : C_U^\infty \to (C_U^\infty)^2$ is the gradient $\nabla(f) = (\frac{\partial f}{\partial x}, \frac{\partial f}{\partial y})$. Show that the map on global sections $\Gamma(U, C_U^\infty) \to \Gamma(U, \mathrm{im}\,\nabla)$ need not be surjective.

3.3 Affine Schemes

A scheme is a massive generalization of the notion of an algebraic variety. The canonical reference is Grothendieck and Dieudonné's "Eléments de géométrie algébrique" or EGA [55]. In spite of the title, it is certainly not casual reading. Hartshorne's book [60] has become the standard introduction to these ideas for most people. Somewhat less austere (and less comprehensive) treatments can be found in [34, 92, 104]. We will give a brief introduction to scheme theory in order to gain a deeper understanding of algebraic varieties. Since we do not want to give a full-blown treatment of the subject, we will use a much more limited notion of scheme in this book. To be clear, schemes in our sense are equivalent to Grothendieck's schemes that are locally of finite type over an algebraically closed field k.

Perhaps a few words of motivation are in order. In classical algebraic geometry, we want to consider the line $y = 0$ as somehow different from the degenerate conic $y^2 = 0$, even though they are the same as sets; the latter is the limiting case of an honest conic. We can do this by keeping track of multiplicities. However, in many situations it is useful to keep track of finer information. It will be instructive to look at another example. Imagine a collection of n distinct points $p_1(t), \dots, p_n(t) \in X = \mathbb{A}_k^d$ converging to a common value q as $t \to 0$. Then as a first approximation, the limit of the set can be considered as nq. But we can get a finer portrait by considering the limit of the defining ideal of the set $\{p_i(t)\}$, and then forming the corresponding geometric object, which would be *subscheme* of X. To make the comparison of these two approaches clearer, let us form a new affine variety

$$S^n X = \underbrace{X \times \cdots \times X}_{n \text{ times}} / S_n, \tag{3.3.1}$$

called the nth symmetric product. (The precise construction will be given in the exercises.) The points of this variety can be thought of as formal sums, called cycles,

$m_1p_1 + \cdots + m_rp_r$ with $p_i \in X$, $m_i \geq 0$, and $m_1 + \cdots + m_r = n$. When $d = 1$, these cycles are in bijection with the set of ideals $I \subset k[x]$ with vector space codimension n. The correspondence is given by associating $I = (\prod(x - p_i)^{m_i})$ with the cycle above. For any d, let $\mathrm{Hilb}^n X$ denote the set of ideals $I \subset k[x_1,\ldots,x_d]$ with $\mathrm{codim}\, I = n$. The ideals I define sets of points $Z(I) = \{p_1,\ldots,p_r\}$, and more. We have multplicities $m_i = \dim \mathscr{O}_{p_i}/I\mathscr{O}_{p_i}$, where $\mathscr{O}_{p_i}$ are the localizations of $k[x_1,\ldots,x_d]$ at the maximal ideals at the p_i. Thus we can map $\mathrm{Hilb}^n X \to S^n X$ by sending I to the set of points $Z(I) = \{p_1,\ldots,p_r\}$ with multplicities m_i. As we saw, this map is a bijection when $d = 1$, but in general $\mathrm{Hilb}^n X$ has a much richer structure; its points correspond to subschemes of $\mathbb{A}^d_k$ supported at the points $Z(I)$. What is remarkable is that $\mathrm{Hilb}^n X$ itself carries the structure of a scheme. This nonclassical object is very important in algebraic geometry; we will encounter it again much later on.

We start with an ad hoc construction. In Examples 3.1.9 and 3.2.8 and Exercise 3.2.14, we saw how to define the ideal sheaf $\mathscr{I}_Y$ of a subvariety $Y \subset \mathbb{A}^n_k$, and use it to recover $\mathscr{O}_Y$ as the restriction of $(\mathscr{O}_{\mathbb{A}^n}/\mathscr{I}_Y)|_Y$. The space of global sections $\mathscr{I}_Y(\mathbb{A}^n_k)$ is the radical ideal defining Y. We can refine this picture by starting with an arbitrary ideal $I \subset k[x_1,\ldots,x_n]$. Define a sheaf

$$\tilde{I}(U) = I\mathscr{O}_{\mathbb{A}^n}(U) = \left\{ \frac{f}{g} \mid g \in k[x_1,\ldots,x_n] \text{ is nowhere 0 on } U, f \in I \right\}$$

on $\mathbb{A}^n_k$. This is an ideal sheaf in the sense that $\tilde{I}(U) \subset \mathscr{O}_{\mathbb{A}^n}(U)$ is an ideal for each U. The quotient $\mathscr{O}_{\mathbb{A}^n}/\tilde{I}$ is an abstract sheaf of rings on $\mathbb{A}^n_k$ supported on $Z(I)$, so we identify this with its restriction to $Z(I)$. The pair $(Z(I), \mathscr{O}_{\mathbb{A}^n}/\tilde{I})$, which is an example of an abstract ringed space, is the *subscheme* of $\mathbb{A}^n_k$ defined by I.

Although it is not obvious from the way we obtained it, the isomorphism class of the pair $(Z(I), \mathscr{O}_{\mathbb{A}^n}/\tilde{I})$ depends only on the affine algebra, by which we mean a finitely generated algebra, $k[x_1,\ldots,x_n]/I$. This is called the *affine scheme* associated with this algebra. Since this is quite an important point, we give an alternative construction starting from the algebra. We use the case of varieties as a guide. As we saw in Section 2.3, the underlying topological space associated with an affine variety X can be identified with $\mathrm{Spec}_m R$, where $R = \mathscr{O}(X)$ is the coordinate ring. It remains to describe the sheaf. We have some flexibility in doing this, since we will no longer insist that it be a sheaf of k-valued functions. We will realize it as a subsheaf of the constant sheaf associated with the function field $k(X)$.

Lemma 3.3.1. *Suppose that X is an affine variety, which we identify with* $\mathrm{Spec}_m \mathscr{O}(X)$. *Then $\mathscr{O}_X(U)$ is simply the intersection of the localizations*

$$\bigcap_{m \in U} \mathscr{O}(X)_m = \{f \in k(X) \mid \forall m \in U, f \notin m\}.$$

Proof. Since regular functions are rational, we have an embedding $\mathscr{O}_X(U) \subset k(X)$. A rational function $F \in k(X)$ lies in $\mathscr{O}_X(U)$ exactly when for every $x \in X$, we can express $F = f/g$ such that $f, g \in \mathscr{O}(X)$ and $g(x) \neq 0$. This is equivalent to requiring that $F \in \mathscr{O}(X)_{m_x}$ for each x. □

The above description of the structure sheaf does not generalize well to rings that are not integral domains. So we need an alternative construction. First note that localizations $R[1/f] = f^{-1}R$ can be defined for any ring by formally inverting f [8]. The elements of this ring are equivalence classes of fractions r/f^n, with $r \in R, n \in \mathbb{N}$, where $r/f^n \sim rf^m/f^{n+m}$. The canonical map $R \to R[1/f]$ given by $r \mapsto r/1$ need no longer be injective; its kernel is the set of elements annihilated by powers of f. More generally, one can form localizations with respect to any set of elements closed under multiplication [8]. In particular, if $p \subset R$ is a prime ideal, we define $R_p = (R - p)^{-1}R$.

Proposition 3.3.2. *Let R be an affine k-algebra. There exists a sheaf of rings $\mathcal{O}_X$ on $X = \mathrm{Spec}_m R$ such that*

(a) $\mathcal{O}_X(D(f)) \cong R\left[\frac{1}{f}\right]$ and the restrictions $\mathcal{O}_X(D(f)) \to \mathcal{O}_X(D(gf))$ can be identified with the natural maps $R\left[\frac{1}{f}\right] \to R\left[\frac{1}{gf}\right]$.
(b) If R is the coordinate ring of an affine algebraic variety X, then $\mathcal{O}_X(U)$ is isomorphic to the ring of regular functions on U.
(c) The stalk $\mathcal{O}_{X,p}$ at $p \in X$ is isomorphic to the localization R_p.

Proof. The rule $\mathcal{O}_X(D(f)) = R[1/f]$ given in (a) determines a presheaf on the basis $\{D(f)\}$ in the sense of Exercise 3.1.19. We want to show that it is a sheaf. Suppose that $\{D(f_i)\}_{i\in I}$ covers $D(f)$, and that $g_i \in R[1/f_i]$ is a collection of elements such that the images of g_i and g_j in $R[1/f_if_j]$ coincide. Our goal is to find a unique element of $R[1/f]$ whose images are the g_i. To simplify notation, we assume that $f = 1$. Also there is no loss in assuming that the index set $I = \{1,\dots,n\}$ is finite, because X is quasicompact. Then we can write $g_i = h_i/f_i^N$ for some fixed N. Thus we have

$$f_j^N h_i = f_i^N h_j.$$

By assumption, $Z(f_1^N,\dots,f_n^N) = X - \cup D(f_i) = \emptyset$. Therefore by the Nullstellensatz, $(f_1^N,\dots,f_n^N)$ is the unit ideal, or equivalently

$$1 = \sum p_i f_i^N \tag{3.3.2}$$

for some $p_i \in R$. Set $g = \sum p_i h_i \in R$. Then by (3.3.2),

$$f_j^N g = \sum_i f_j^N p_i h_i = \sum_i f_i^N p_i h_j = h_j = f_j^N g_j.$$

Therefore g maps to the g_j for each j. To see that g is unique, suppose that $g' \in R$ is another such element. Then $g - g'$ is annihilated by the powers f_i^N (after increasing N if necessary). Therefore $g - g' = 0$ by (3.3.2).

By the previous paragraph and Exercise 3.1.20, $\mathcal{O}_X(D(f)) = R[1/f]$ extends uniquely to a sheaf on X. This proves (a). When X is an affine variety, the sheaf of regular functions satisfies the conditions of (a). Therefore by uniqueness, it must coincide with the extension just given.

For the last statement, note that when $f \notin p$, there are natural maps $R[1/f] \to R_p$ that induce a homomorphism

$$\varinjlim_{f \notin p} R[1/f] \to R_p,$$

which is easily checked to be an isomorphism. □

Definition 3.3.3. Let R be an affine k-algebra. The k-ringed space $(\mathrm{Spec}_m R, \mathscr{O}_{\mathrm{Spec}_m R})$ is the affine scheme associated to R.

Examples include affine varieties. A nonclassical example is the "fat point" $\mathrm{Spec}_m k[\varepsilon]/(\varepsilon^2)$. As a topological space, it is not particularly interesting, since it consists of one point. However, as a scheme, it is very interesting. As we shall see, morphisms from it to other schemes are the same thing as tangent vectors.

Exercises

3.3.4. For a general commutative ring R, $\mathrm{Spec}_m R$ is not well behaved. For instance, it is not functorial in general. It is better to consider the set of all prime ideals $\mathrm{Spec}\, R$. Show that the sets of the form $V(I) = \{p \in \mathrm{Spec}\, R \mid I \subseteq p\}$, where $I \subset R$ is an ideal, form the closed sets of a topology, again called the Zariski topology.

3.3.5. Show that the inclusion $\mathrm{Spec}_m R \subset \mathrm{Spec}\, R$ is continuous, and that a prime ideal p is maximal if and only if $\{p\}$ is closed in $\mathrm{Spec}\, R$.

3.3.6. Define a sheaf $\mathscr{O}_{\mathrm{Spec}\, R}$ such that $\mathscr{O}_{\mathrm{Spec}\, R}((D(f)) \cong R[1/f]$, where $D(f) = \mathrm{Spec}\, R - V(f)$. The pair $(\mathrm{Spec}\, R, \mathscr{O}_{\mathrm{Spec}\, R})$ is Grothendieck's affine scheme associated to R.

3.3.7. Let $X = \mathbb{A}^d_k = \mathrm{Spec}_m k[x_1, \dots, x_d]$, and let $R = k[x_{ij}]$ with $1 \le i \le d$ $1 \le j \le n$. The symmetric group S_n acts by $\sigma(x_{ij}) = x_{i,\sigma(j)}$. The nth symmetric product $S^n X$ is the affine scheme $\mathrm{Spec}_m k[x_{ij}]^{S_n}$ associated with the ring of invariant polynomials. Show that the points of $S^n X$ are exactly as described in (3.3.1).

3.3.8. Show that $S^n \mathbb{A}^1 \cong \mathbb{A}^n$ and that $S^n \mathbb{A}^d$ is singular when $d > 1$.

3.3.9. Let $\langle , \rangle$ denote the standard inner product on $\mathbb{R}^2$. A strongly convex rational cone in $\mathbb{R}^2$ is a subset of the form

$$\sigma = \{t_1 \mathbf{v}_1 + t_2 \mathbf{v}_2 \mid t_i \in \mathbb{R}, t_i \ge 0\}$$

with nonproportional vectors $\mathbf{v}_i \in \mathbb{Z}^2$ called generators. For each such cone σ, define R_σ to be the subspace of $k[x, x^{-1}, y, y^{-1}]$ spanned by $x^m y^n$ for all $(m, n) \in \sigma \cap \mathbb{Z}^2$. Show that this an affine domain. The variety $\mathrm{Spec}_m S_\sigma$ is called a two-dimensional affine *toric* variety. Show that $\mathbb{A}^2, \mathbb{A}^1 \times k^*$ and $k^* \times k^*$ are of this form. We will continue this in the next set of exercises. A general reference for toric varieties is [41].

3.4 Schemes and Gluing

Once again, we fix an algebraically closed field k. Our goal is to define more general schemes and methods for constructing them, with an eye toward gaining a better understanding of algebraic varieties. Just as the building blocks of general varieties are affine varieties, schemes are built from affine schemes.

Definition 3.4.1. Let $\mathscr{R}$ be a sheaf of (commutative) k-algebras over a space X. The pair $(X, \mathscr{R})$ is called a k-ringed (or usually just ringed) space. It is called locally ringed if in addition all the stalks are local rings.

For example, any concrete locally ringed k-space (§2.5) or an affine scheme is a locally ringed k-space. We will formulate the notion of morphisms between such spaces later on. For the moment, we will be content to explain isomorphisms in a somewhat ad hoc manner.

Definition 3.4.2. An isomorphism of ringed spaces $f:(X,\mathscr{R}) \to (Y,\mathscr{S})$ is a homeomorphism $f: X \to Y$ together with a collection of algebra isomorphisms $f' : \mathscr{R}(U) \to \mathscr{S}(f(U))$ compatible with restriction.

Definition 3.4.3. A k-scheme (in our sense) is a k-ringed space that is locally isomorphic to an affine scheme $(\mathrm{Spec}_m R, \mathscr{O}_{\mathrm{Spec}_m R})$. (Note that it is necessarily locally ringed.)

Example 3.4.4. Prevarieties are schemes.

Example 3.4.5. Any affine scheme is a scheme.

By definition schemes, and hence varieties, can be specified by gluing a collection of affine schemes together. This is similar to giving a manifold by specifying an atlas for it. Let us describe the process explicitly for a pair of affine schemes. Let $X_1 = \mathrm{Spec}_m R_1$ and $X_2 = \mathrm{Spec}_m R_2$, and suppose we have an isomorphism

$$\phi : R_1\left[\frac{1}{r_1}\right] \cong R_2\left[\frac{1}{r_2}\right]$$

for some $r_i \in R_i$. Then we have a corresponding diagram of schemes

$$X_2 \supset D(r_2) \overset{\Phi}{\cong} D(r_1) \subset X_1.$$

We can define the set $X = X_1 \cup_\Phi X_2$ as the disjoint union modulo the equivalence relation $x \sim \Phi(x)$ for $x \in D(r_2)$. We can equip X with the quotient topology, and then define

$$\mathscr{O}_X(U) = \{(s_1, s_2) \in \mathscr{O}(U \cap U_1) \times \mathscr{O}(U \cap U_2) \mid \phi(s_1) = s_2\}.$$

Then X becomes a scheme with $\{X_i\}$ as an open cover.

Example 3.4.6. Setting

$$R_1 = k[x], \quad r_1 = x, R_2 = k[y], \quad r_2 = y, \quad \phi(x) = y^{-1},$$

or more suggestively

$$R_1 = k[x], \quad R_2 = k[x^{-1}],$$

yields $X = \mathbb{P}^1_k$.

Example 3.4.7. Take two copies of the plane

$$X_1 = \mathrm{Spec}_m\, k[x,t],$$
$$X_2 = \mathrm{Spec}_m\, k[xt,t^{-1}],$$

glued along $\mathrm{Spec}_m\, k[x,t,t^{-1}]$ via the obvious identification. We obtain the *blowup* of $\mathbb{A}^2_k$ at $(0,0)$, which was described more geometrically in Exercise 2.4.22.

More than two schemes can be handled in a similar way, although the datum is a bit more cumbersome. It consists of rings R_i and isomorphisms

$$\phi_{ji} : R_i\left[\frac{1}{r_{ij}}\right] \cong R_j\left[\frac{1}{r_{ji}}\right]. \tag{3.4.1}$$

These are subject to a compatibility, or *cocycle*, condition that the isomorphisms

$$R_i\left[\frac{1}{r_{ij}r_{ik}}\right] \cong R_k\left[\frac{1}{r_{ki}r_{kj}}\right]$$

induced by ϕ_{ik} and $\phi_{ij}\phi_{jk}$ coincide, and $\phi_{ii} = id$, $\phi_{ji} = \phi_{ij}^{-1}$. Then the spectra $\mathrm{Spec}_m\, R_i$ can then be glued as above. This is described in the exercises. For an abstractly given scheme X to arise from such a construction it suffices to have a finite affine open cover $\{U_i\}$ such that $U_i \cap U_j$ is also affine. For this, it would be sufficient to assume that X is quasicompact and separated. The last condition means that the diagonal in $X \times X$ is closed. In particular, this applies to varieties, which is the case we are really interested in.

Example 3.4.8. Let

$$R_i = k[x_{0i}, \ldots, \hat{x}_{ii}, \ldots, x_{ni}]$$

with gluing isomorphisms determined by

$$\phi_{ji}(x_{ai}) = x_{ij}^{-1} x_{aj}.$$

Then the glued scheme is precisely $\mathbb{P}^n_k$. This becomes much clearer if we identify $x_{ai} = x_a/x_i$ for homogeneous coordinates x_i. Then $\{\mathrm{Spec}_m\, R_i\}$ is the standard covering by the affine spaces defined by $x_i \neq 0$.

The double cover of $\mathbb{A}^n_k$ branched along $f(x_1, \ldots, x_n) = 0$ is simply given by $y^2 - f(x_1, \ldots, x_n) = 0$ in $\mathbb{A}^{n+1}_k$. The projective version is a bit more complicated. It can be described conveniently by a gluing construction.

Example 3.4.9. Let $f(x_0,\dots,x_n)\in k[x_0,\dots,x_n]$ be a homogeneous polynomial of degree $2d$. In outline, the double cover of $\mathbb{P}^n$ branched along $f=0$ is given by

$$R_i = k\left[\frac{x_0}{x_i},\dots,\frac{x_n}{x_i},\frac{y}{x_i^d}\right] / \left(\left(\frac{y}{x_i^d}\right)^2 - f\left(\frac{x_0}{x_i},\dots,\frac{x_n}{x_i}\right)\right)$$

with the obvious gluing.

The gluing method can be used to construct normalizations mentioned in Theorem 1.5.1.

Proposition 3.4.10. *If X is a variety, then there exist a variety $\tilde{X}$ and a continuous map $\pi:\tilde{X}\to X$ such that for any affine open $U\subseteq X$, $\mathcal{O}_{\tilde{X}}(\pi^{-1}U)$ is the integral closure of X.*

Remark 3.4.11. π is in fact a morphism, although we defer this to the exercises at the end of §3.7.

Proof. As pointed out above, X can be obtained by a gluing construction applied to a finite number of affine domains R_i. Let $\tilde{R}_i$ be their integral closures. These are again affine by [33, Corollary 13.13]. The gluing isomorphisms

$$\phi_{ji}: R_i\left[\frac{1}{r_{ij}}\right] \cong R_j\left[\frac{1}{r_{ji}}\right]$$

extend to isomorphisms

$$\tilde{\phi}_{ij}: \tilde{R}_i\left[\frac{1}{r_{ij}}\right] \cong \tilde{R}_j\left[\frac{1}{r_{ji}}\right]$$

satisfying the compatability conditions. Thus $\tilde{X}$ can be constructed by gluing $\operatorname{Spec}\tilde{R}_i$. The maps $\operatorname{Spec}_m\tilde{R}\to\operatorname{Spec}_m R_i$ glue to yield π. □

Further important examples of gluing are discussed in the exercises.

Exercises

3.4.12. Suppose we are given gluing data $\{R_i,\phi_{ij}\}$ as described in (3.4.1). Let $\sim$ be the equivalence relation on the disjoint union $\bigcup\operatorname{Spec}_m R_i$ generated by $x\sim\phi_{ij}^*(x)$. Show that $X=\bigcup\operatorname{Spec}_m R_i/\sim$ can be made into a scheme with $\operatorname{Spec}_m R_i$ as an open affine cover.

3.4.13. An r-fold cyclic cover of affine space is described by the equation $y^r = f(x_1,\dots,x_n)$.

3.4.14. An n-jet on a scheme X is a morphism from $\operatorname{Spec} k[\varepsilon]/(\varepsilon^{n+1})$ to X. Give an interpretation of these similar to Lemma 3.7.8.

Fill in the details for Example 3.4.9, and generalize this to describe cyclic covers of $\mathbb{P}^n_k$.

3.4.15. We refer to Exercise 3.3.9 for notation and terminology. Let

$$\sigma = \{t_1\mathbf{v}_1 + t_2\mathbf{v}_2 \mid t_i \in \mathbb{R}, t_i \geq 0\}$$

be a strongly convex rational cone. Show that the dual cone

$$\sigma^\vee = \{\mathbf{v} \mid \langle \mathbf{v}, \mathbf{w} \rangle \geq 0, \forall \mathbf{w} \in \sigma\}$$

is also strongly rational convex. Let τ denote either the degenerate cone spanned by $\mathbf{v}_i$ or $\{0\}$. Show that $R_{\tau^\vee}$ is a localization of $R_{\sigma^\vee}$ at a single element $r_{\sigma\tau}$. Given a second strongly convex rational cone σ' with $\sigma \cap \sigma' = \tau$, show that $R_{\sigma^\vee}[1/r_{\sigma\tau}]$ and $R_{\sigma'^\vee}[1/r_{\sigma'\tau}]$ can be identified. Therefore their spectra can be glued to yield the toric variety associated with the *fan* $\{\sigma, \sigma'\}$.

3.4.16. Show that the blowup of $\mathbb{A}^2_k$ at 0 can be constructed from the fan

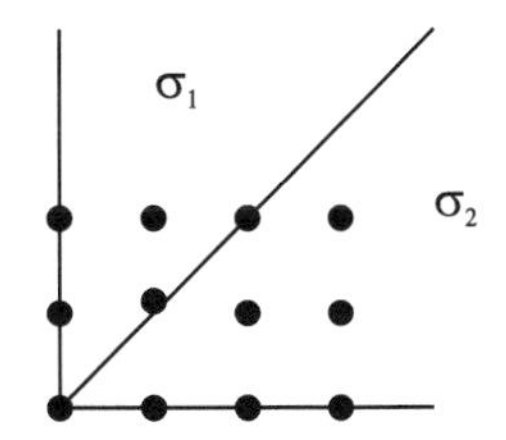

where σ_1 and σ_2 are generated by $\{(0,1),(1,1)\}$ and $\{(1,0),(1,1)\}$ respectively.

3.4.17. A fan Δ in $\mathbb{R}^2$ is a finite collection of nonoverlapping rational strongly convex cones. Generalize Exercise 3.4.15 to glue together all the affine toric varieties $\operatorname{Spec}_m R_{\sigma^\vee}$, with $\sigma \in \Delta$.

3.4.18. Show that the toric variety corresponding to the fan

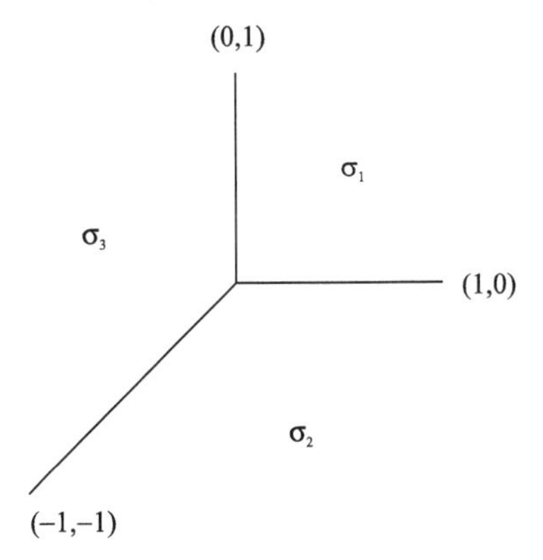

is $\mathbb{P}^2$.

3.5 Sheaves of Modules

Subvarieties or subschemes of affine spaces are given by ideals in polynomial rings. It is convenient to extend this idea.

Definition 3.5.1. Let $(X,\mathscr{R})$ be a ringed space. A subsheaf $\mathscr{I}\subset\mathscr{R}$ is called an ideal sheaf if $\mathscr{I}(U)\subset\mathscr{R}(U)$ is an ideal for every U.

An ideal sheaf determines a sheaf of rings $\mathscr{R}/\mathscr{I}$ and a subspace called the support,

$$\mathrm{supp}(\mathscr{R}/\mathscr{I})=\{x\in X\mid(\mathscr{R}/\mathscr{I})_x\neq 0\}.$$

Example 3.5.2. The sheaf $\mathscr{I}_Y\subset\mathscr{O}_X$ introduced in Example 3.2.8 is an ideal sheaf. The support of $\mathscr{O}_X/\mathscr{I}$ is precisely Y.

Example 3.5.3. Given an ideal $I\subset k[x_1,\ldots,x_n]$, we get an associated ideal sheaf

$$\tilde{I}(U)=I\mathscr{O}_{\mathbb{A}^n}(U)=\left\{\frac{f}{g}\mid g\in k[x_1,\ldots,x_n]\text{ is nowhere 0 on }U,\,f\in I\right\}$$

defined in §3.3. The support of the sheaf $\mathscr{O}_{\mathbb{A}^n}/\mathscr{I}$ is $Z(I)$, and this sheaf is the structure sheaf of the scheme $\mathrm{Spec}_m(k[x_1,\ldots,x_n]/I)$ extended to $\mathbb{A}^n_k$.

We can construct ideal sheaves on $\mathbb{P}^n_k$ fairly concretely as well.

Example 3.5.4. Let $I\subset S=k[x_0,\ldots,x_n]$ be a homogeneous ideal. Let $S_d\subset$ be the set of degree-d polynomials, and let $I_d=I\cap S_d$. The corresponding ideal sheaf on $\mathbb{P}^n$ is given by

$$\tilde{I}(U)=\left\{\frac{f}{g}\mid\exists d,f\in I_d,g\in S_d\,\&\,g\neq 0\text{ on }U\right\}.$$

Note that the support of $\mathscr{O}_{\mathbb{P}^n}/\tilde{I}$ is $Z(I)$.

Just as in ordinary algebra, it is useful to pass from ideals to more general modules:

Definition 3.5.5. A sheaf of $\mathscr{R}$-modules or simply an $\mathscr{R}$-module is a sheaf $\mathscr{M}$ such that each $\mathscr{M}(U)$ is an $\mathscr{R}(U)$-module and the restrictions $\mathscr{M}(U)\to\mathscr{M}(V)$ are $\mathscr{R}(U)$-linear.

Clearly ideal sheaves are examples, but there are plenty of others.

Example 3.5.6. Let (X,C^∞_X) be a C^∞ manifold. Then the space of vector fields on $U\subset X$ is a module over $C^\infty(U)$ in a way compatible with restriction. Therefore the tangent sheaf $\mathscr{T}_X$ is a C^∞_X-module, as is the sheaf of 1-forms $\mathscr{E}^1_X$.

Sheaves of modules form a category $\mathscr{R}$-Mod where the morphisms are morphisms of sheaves $\mathscr{M} \to \mathscr{N}$ such that each map $\mathscr{M}(U) \to \mathscr{N}(U)$ is $\mathscr{R}(U)$-linear. This is in fact an abelian category. The notion of exactness in this category coincides with the notion introduced in Section 3.2.

Before giving more examples, we recall the tensor product and related constructions [8, 33, 76].

Theorem 3.5.7. *Given modules M_1, M_2 over a commutative ring R, there exist an R-module $M_1 \otimes_R M_2$ and a map $(m_1, m_2) \mapsto m_1 \otimes m_2$ of $M_1 \times M_2 \to M_1 \otimes_R M_2$ that is bilinear and universal:*

(a) $(rm + r'm') \otimes m_2 = r(m \otimes m_2) + r'(m' \otimes m_2)$.
(b) $m_1 \otimes (rm + r'm') = r(m_1 \otimes m) + r'(m_1 \otimes m')$.
(c) Any map $M_1 \times M_2 \to N$ satisfying the first two properties is given by $\phi(m_1 \otimes m_2)$ for a unique homomorphism $\phi : M_1 \otimes M_2 \to N$.

Definition 3.5.8. Given a module M over a commutative ring R,

$$T^*(M) = R \oplus M \oplus (M \otimes_R M) \oplus (M \otimes_R M \otimes_R M) \cdots$$

becomes a noncommutative associative R-algebra called the tensor algebra with product induced by $\otimes$. The exterior algebra $\wedge^* M$ (respectively symmetric algebra S^*M) is the quotient of $T^*(M)$ by the two-sided ideal generated $m \otimes m$ (respectively $(m_1 \otimes m_2 - m_2 \otimes m_1)$).

The product in $\wedge^* M$ is denoted by $\wedge$. We denote by $\wedge^p M$ ($S^p M$) the submodule generated by products of p elements. If V is a finite-dimensional vector space, $\wedge^p V^*$ (respectively $S^p V^*$) can be identified with the set of alternating (respectively symmetric) multilinear forms on V in p-variables. After choosing a basis for V, one sees that $S^p V^*$ are degree-p polynomials in the coordinates.

These operations can be carried over to sheaves:

Definition 3.5.9. Let $\mathscr{M}$ and $\mathscr{N}$ be two $\mathscr{R}$-modules.

1. The direct sum $\mathscr{M} \oplus \mathscr{N}$ is the sheaf $U \mapsto \mathscr{M}(U) \oplus \mathscr{N}(U)$.
2. The tensor product $\mathscr{M} \otimes \mathscr{N}$ is the sheafification of the presheaf $U \mapsto \mathscr{M}(U) \otimes_{\mathscr{R}(U)} \mathscr{N}(U)$.
3. The dual $\mathscr{M}^*$ of $\mathscr{M}$ is the sheafification of the presheaf $U \mapsto \mathrm{Hom}_{\mathscr{R}|_U}(\mathscr{M}|_U, \mathscr{R}|_U)$.
4. The pth exterior power $\wedge^p \mathscr{M}$ is the sheafification of $U \mapsto \wedge^p \mathscr{M}(U)$.
5. The pth symmetric power $S^p \mathscr{M}$ is the sheafification of $U \mapsto S^p \mathscr{M}(U)$.

We will see in the exercises that the sheafification step in the definition of $\mathscr{M}^*$ is unnecessary. We have already encountered this construction implicitly: the sheaf of 1-forms $\mathscr{E}_X^1$ on a manifold is the dual of the tangent sheaf $\mathscr{T}_X$. The objects of classical tensor analysis are sections of the sheaf $T^*(\mathscr{T}_X \oplus \mathscr{E}_X^1)$. A special case of fundamental importance is the following:

Definition 3.5.10. The sheaf of p-forms on a C^∞-manifold is $\mathscr{E}_X^p = \wedge^p \mathscr{E}_X^1$.

We want to describe some examples in the algebraic setting. First, we describe some standard methods for transferring modules between rings. If $f : R \to S$ is a homomorphism of commutative rings, then any R-module M gives rise to an S-module $S \otimes_R M$, with S acting by $s(s_1 \otimes m) = ss_1 \otimes m$. This construction is called *extension of scalars*. In the opposite direction, an S-module can always be regarded as an R-module via f. This is *restriction of scalars*.

Definition 3.5.11. Let R be an affine k-algebra, and M an R-module. Let $X = \mathrm{Spec}_m R$. We define an $\mathscr{O}_X$-module $\tilde{M}$ by extending scalars $\tilde{M}(U) = \mathscr{O}_X(U) \otimes_R M$. Such a module is called quasicoherent.

When $M = I$ is an ideal in a polynomial ring, $\tilde{M}$ coincides with the ideal sheaf constructed in Example 3.5.3.

A final source of examples will come from vector bundles. These examples have a special property.

Definition 3.5.12. A module $\mathscr{M}$ over $(X, \mathscr{R})$ is locally free (of rank n) if every point has a neighborhood $i : U \to X$ such that $i^{-1}\mathscr{M}$ is isomorphic to a finite (n-fold) direct sum $i^{-1}\mathscr{R}^n = i^{-1}\mathscr{R} \oplus \cdots \oplus i^{-1}\mathscr{R}$. (We will usually write $\mathscr{M}|_U$ instead of $i^{-1}\mathscr{M}$ in the future.)

Let $\pi : V \to X$ be a rank-n C^∞ real (or holomorphic or algebraic) vector bundle on a manifold or variety $(X, \mathscr{R})$ with a local trivialization $\{(U_i, \phi_i)\}$, as discussed in §2.6. Recall that we have a sheaf

$$\mathscr{V}(U) = \{s : U \to \pi^{-1}U \text{ is a morphism} \mid \pi \circ s = \mathrm{id}_U\}.$$

Via the trivialization, $\mathscr{V}(U)$ can be identified with a submodule of $\prod_i \mathscr{R}(U_i \cap U)^n$. In this way, $\mathscr{V}$ becomes a sheaf of $\mathscr{R}$-modules. The map ϕ_i induces an isomorphism $\mathscr{V}|_{U_i} \cong \mathscr{R}|_{U_i}^n$. Thus $\mathscr{V}$ is locally free of rank n. Conversely, we will see in Section 7.3 that every such sheaf arises in this way. Further examples can be obtained by applying the linear algebra operations of Definition 3.5.9.

Locally free modules over affine varieties have a simple characterization:

Theorem 3.5.13. *Let M be a finitely generated module over an affine domain R. The following are equivalent:*

(a) $\tilde{M}$ is locally free.
(b) M_p is free for all $p \in \mathrm{Spec}_m R$.
(c) M is projective, i.e., A direct summand of a finitely generated free module.

Under these conditions, the function $p \mapsto \dim k(p) \otimes_R M$ on $\mathrm{Spec}_m R$ is constant with value equal to the rank, where $k(p)$ is the quotient field of R/p.

Proof. See [15, Chapter II §5.2]. □

The last statement in the theorem can be strengthened slightly to give a numerical characterization. Let R be an affine domain with field of fractions K. Any finitely generated R-module determines a finite-dimensional K-vector space $K \otimes_R M$.

Lemma 3.5.14. *A finitely generated R-module M yields a locally free sheaf $\tilde{M}$ if and only if*

$$\dim k(p) \otimes_R M = \dim K \otimes_R M$$

for all $p \in \operatorname{Spec}_m R$.

Proof. If M is locally free, the function $p \mapsto \dim k(p) \otimes_R M$ is constant. Furthermore, M_p is free, so that

$$\dim k(p) \otimes_R M = \dim k(p) \otimes_{R_p} M_p = \dim K \otimes_{R_p} M_p = \dim K \otimes_R M.$$

Conversely, suppose that $\dim k(p) \otimes_R M_p = \dim K \otimes_R M = n$. Choose a set of elements $m_1, \ldots, m_n \in M$ that maps onto a basis of $k(p) \otimes_R M_p$. Then these generate M_p by Nakayama's lemma [33, Corollary 4.8]. So we get a surjection $\phi : R_p^n \to M_p$. We will check that $\ker \phi = 0$, which will imply that M_p is free. Since R_p is a domain, it suffices to check this after tensoring by K. After doing so, we get a surjection of n-dimensional vector spaces, which is necessarily an isomorphism. □

Exercises

3.5.15. Show that $\tilde{M}$ given in Definition 3.5.11 really is a sheaf, and that its stalk at p is precisely the localization M_p.

3.5.16. Show that duals, direct sums, tensor products, and exterior and symmetric powers of locally free sheaves are locally free.

3.5.17. Given two $\mathscr{R}$-modules $\mathscr{M}, \mathscr{N}$, define the presheaf $\mathscr{H}om(\mathscr{M}, \mathscr{N})$ by $U \mapsto \operatorname{Hom}_{\mathscr{R}|_U}(\mathscr{M}|_U, \mathscr{N}|_U)$. Show that this is a sheaf. Show that $\mathscr{H}om(\mathscr{M}, \mathscr{N}) \cong \mathscr{M}^* \otimes \mathscr{N}$ when $\mathscr{M}, \mathscr{N}$ are locally free.

3.5.18. Given an $\mathscr{R}$-module $\mathscr{M}$ over a locally ringed space $(X, \mathscr{R})$, the stalk $\mathscr{M}_x$ is an $\mathscr{R}_x$-module for any $x \in X$. If $\mathscr{M}$ is a locally free $\mathscr{R}$-module, show that then each stalk $\mathscr{M}_x$ is a free $\mathscr{R}_x$-module of finite rank. Show that the converse can fail.

3.5.19. Suppose that X is an affine algebraic variety. Let M be a finitely generated $\mathscr{O}(X)$-module. Show that there exists a nonempty open set $U \subseteq X$ such that $\tilde{M}|_U$ is locally free of rank $\dim K \otimes M$, where K is the field of fractions of $\mathscr{O}(X)$.

3.5.20. Show that an ideal is free if and only if it is principal. Prove that the ideal $M = (x, y)$ in $R = \mathbb{C}[x,y]/(y^2 - x(x-1)(x-2))$ is locally free but not free.

3.5.21. If $\mathscr{M}, \mathscr{N}$ are locally free of rank one on $(X, \mathscr{R})$, show that $\mathscr{M}^* \otimes \mathscr{N} \cong \mathscr{R}$. Use this to show that the set of isomorphism classes of such sheaves forms an abelian group denoted by $\operatorname{Pic}(X)$. In the notation of the exercise, show that M is a 2-torsion element of $X = \operatorname{Spec}_m R$.

3.6 Line Bundles on Projective Space

In this section we define and study certain key examples of sheaves that are of fundamental importance in algebraic geometry. Fix a positive integer n. Let k be an algebraically closed field, $\mathbb{P} = \mathbb{P}^n_k$, and $V = k^{n+1}$. Let $R = k[x_0, \ldots, x_n] = \bigoplus R_d$, where R_d is the subspace of homogeneous polynomials of degree d. Given an integer d, define the subsheaf $\mathscr{O}_{\mathbb{P}}(d)$ of the constant sheaf of rational functions on $\mathbb{P}$ by

$$\mathscr{O}_{\mathbb{P}}(d)(U) \cong \left\{ \frac{f}{g} \mid \exists e, g \in R_e, f \in R_{e+d}, g \text{ nowhere zero on } U \right\}. \tag{3.6.1}$$

Of course, we have $\mathscr{O}_{\mathbb{P}}(0) = \mathscr{O}_{\mathbb{P}}$ immediately from the definition. We have products

$$\mathscr{O}_{\mathbb{P}}(a) \otimes \mathscr{O}_{\mathbb{P}}(b) \to \mathscr{O}_{\mathbb{P}}(a+b).$$

given by

$$\frac{f}{g} \otimes \frac{f'}{g'} \mapsto \frac{ff'}{gg'}.$$

In particular, this shows that $\mathscr{O}_{\mathbb{P}}(d)$ is an $\mathscr{O}_{\mathbb{P}}$-module.

Lemma 3.6.1.

(a) When $d \geq 0$, the space of global sections of $\mathscr{O}_{\mathbb{P}}(d)$ is R_d.
(b) When $d < 0$, $\mathscr{O}_{\mathbb{P}}(d)$ has no nonzero global sections.
(c) Each $\mathscr{O}_{\mathbb{P}}(a)$ is locally free of rank one.
(d) There are natural isomorphisms $\mathscr{O}_{\mathbb{P}}(a) \otimes \mathscr{O}_{\mathbb{P}}(b) \cong \mathscr{O}_{\mathbb{P}}(a+b)$ and $\mathscr{O}_{\mathbb{P}}(a)^* \cong \mathscr{O}_{\mathbb{P}}(-a)$.

Proof. Statements (a) and (b) are immediate consequences of the definition (3.6.1), since the denominator g is necessarily a nonzero constant. From the definition, we can also identify

$$\mathscr{O}_{\mathbb{P}}(d)(U_i) = x_i^d \mathscr{O}_{\mathbb{P}}(U_i), \tag{3.6.2}$$

where U_i denotes the complement of $x_i = 0$. The remaining statements follows easily from this. □

Part of the importance of these sheaves stems from the following fact:

Lemma 3.6.2. *The ideal sheaf of any degree-$d > 0$ hypersurface in $\mathbb{P}$ (see Example 3.5.4) is isomorphic to $\mathscr{O}_{\mathbb{P}}(-d)$.*

Proof. Given a nonzero degree-d homogeneous polynomial $h \in R_d$, we can see from (3.6.1) that we have an embedding $\mathscr{O}_{\mathbb{P}}(-d) \to \mathscr{O}_{\mathbb{P}}$ given by $f/g \mapsto fh/g$. The image is precisely the ideal sheaf defined by h. □

We claimed earlier that locally free sheaves arise from vector bundles. We can see this explicitly in these cases. We treat degree -1 here, and the leave the remaining cases as exercises. Recall that the tautological line bundle (Example 2.6.5) on projective space $\mathbb{P}$ is given by

$$L = \{(x,\ell) \in V \times \mathbb{P} \mid x \in \ell\}$$

with its natural projection to $\mathbb{P}$.

Lemma 3.6.3. *$\mathcal{O}_{\mathbb{P}}(-1)$ is isomorphic to the sheaf $\mathcal{L}$ of sections of L.*

Proof. By definition,

$$\mathcal{L}(U) = \{H : U \to k^{n+1} \mid H \text{ is regular and } H(\ell) \in \ell\}.$$

Thus the components h_i of H are regular functions satisfying

$$h_i(x_0,\dots,x_n) = \lambda(x_0,\dots,x_n)x_i$$

for some function λ. Rewriting $\lambda = h_i/x_i$, we see that it lies in $\mathcal{O}_{\mathbb{P}}(-1)(U)$. This process can be reversed. □

When $k = \mathbb{C}$, we can pass to the associated complex analytic line bundle associated with L. Instead of trying to mimic equation (3.6.1) in the analytic category, we simply define $\mathcal{O}_{\mathbb{P}_{an}}(-1)$ as the sheaf of holomorphic sections of L, and

$$\mathcal{O}_{\mathbb{P}_{an}}(d) = \begin{cases} \mathcal{O}_{\mathbb{P}_{an}}(-1) \otimes \cdots \otimes \mathcal{O}_{\mathbb{P}_{an}}(-1) & (|d|\text{-times}) \quad \text{if } d \le 0, \\ \mathcal{O}_{\mathbb{P}_{an}}(-1)^* \otimes \cdots \otimes \mathcal{O}_{\mathbb{P}_{an}}(-1)^* & (d\text{-times}) \quad \text{otherwise.} \end{cases}$$

Note that in the sections where we will be working exclusively in the analytic category, we usually suppress the subscript an.

Exercises

3.6.4. Show that $\mathcal{O}_{\mathbb{P}}(a) \cong \mathcal{O}_{\mathbb{P}}(b)$ implies that $a = b$. Using the notation of Exercise 3.5.21, conclude that $\mathbb{Z} \subset \mathrm{Pic}(\mathbb{P})$.

3.6.5. Let $U = \mathbb{P}^{n+1} - \{[0,\dots,0,1]\}$ and let $\pi : U \to \mathbb{P}$ denote projection. Show that this is a line bundle, and that $\mathcal{O}_{\mathbb{P}}(1)$ is isomorphic to the sheaf of its sections.

3.6.6. Let $v : \mathbb{P} \to \mathbb{P}^N = \mathbb{P}(S^dV)$ denote the degree-d Veronese map. Show that $\{(x,\ell) \in S^dV \times \mathbb{P} \mid x \in v(\ell)\}$ determines a line bundle over $\mathbb{P}$, and that $\mathcal{O}_{\mathbb{P}}(-d)$ is its sheaf of sections.

3.6.7. Choose a point $p \in \mathbb{P}^1$. Prove that there is an exact sequence

$$0 \to \mathcal{O}_{\mathbb{P}^1_{an}}(d-1) \to \mathcal{O}_{\mathbb{P}^1_{an}}(d) \to \mathbb{C}_p \to 0,$$

where $\mathbb{C}_p(U) = \mathbb{C}$ if $p \in U$ and 0 otherwise. Use this to prove that $\dim \Gamma(\mathcal{O}_{\mathbb{P}^1_{an}}(d)) = d+1$ when $d \ge 0$.

3.7 Direct and Inverse Images

It is useful to transfer a sheaf from one topological space to another via a continuous map $f : X \to Y$. In fact, we have already done this in special cases. We start by explaining how to push a sheaf on X down to Y.

Definition 3.7.1. Given a presheaf $\mathscr{F}$ on X, the direct image $f_*\mathscr{F}$ is a presheaf on Y given by $f_*\mathscr{F}(U) = \mathscr{F}(f^{-1}U)$ with restrictions given by

$$\rho_{f^{-1}Uf^{-1}V} : \mathscr{F}(f^{-1}U) \to \mathscr{F}(f^{-1}V).$$

Lemma 3.7.2. *Direct images of sheaves are sheaves.*

Proof. Suppose that $f : X \to Y$ is a continuous map and $\mathscr{F}$ is a sheaf on X. Let $\{U_i\}$ be an open cover of $U \subseteq Y$, and $s_i \in f_*\mathscr{F}(U_i)$ a collection of sections that agree on the intersections. Then $\{f^{-1}U_i\}$ is an open cover of $f^{-1}U$, and we can regard $s_i \in \mathscr{F}(f^{-1}U_i)$ as a compatible collection of sections for it. Thus we can patch s_i to get a uniquely defined $s \in f_*\mathscr{F}(U) = \mathscr{F}(f^{-1}U)$ such that $s|_{U_i} = s_i$. This proves that $f_*\mathscr{F}$ is a sheaf. □

Now we want to consider the opposite direction. Suppose that $\mathscr{G}$ is a sheaf on Y. We would like to pull it back to X. We will denote this by $f^{-1}\mathscr{G}$, since f^* is reserved for something related to be defined later on. Naively, we can simply try to define

$$f^{-1}\mathscr{G}(U) = \mathscr{G}(f(U)).$$

However, this does not yet make sense unless $f(U)$ is open. So as a first step, given any subset $S \subset Y$ of a topological space and a presheaf $\mathscr{G}$, define

$$\mathscr{G}(S) = \varinjlim \mathscr{G}(V) \tag{3.7.1}$$

as V ranges over all open neighborhoods of S. When S is a point, $\mathscr{G}(S)$ is just the stalk. An element of $\mathscr{G}(S)$ can be viewed as germ of a section defined in a neighborhood of S, where two sections define the same germ if their restrictions agree in a common neighborhood. If $S' \subset S$, there is a natural restriction map $\mathscr{G}(S) \to \mathscr{G}(S')$ given by restriction of germs. So our naive attempt can now be made precise.

Definition 3.7.3. If $\mathscr{G}$ is a presheaf on Y, the presheaf inverse image $f^P\mathscr{G}$ is a presheaf on X given by $f^P\mathscr{G}(U) = \mathscr{G}(f(U))$ with restrictions as above. The sheaf inverse image is the sheafification $f^{-1}\mathscr{G} = (f^P\mathscr{G})^+$.

Although the definition of f^{-1} seems much more complicated compared to f_*, it is in fact just as natural. The picture becomes clearer if we view the inverse image in terms of its stalks. We then have

$$(f^{-1}\mathscr{G})_x \cong \mathscr{G}_{f(x)}, \tag{3.7.2}$$

and there is more. As explained in Exercise 3.1.22, the sheaf $\mathscr{G}$ determines the étalé space $\pi : \mathrm{Et}(\mathscr{G}) = \bigcup \mathscr{G}_y \to Y$. The étalé space for $f^{-1}\mathscr{G}$ is the fiber product

$$\mathrm{Et}(f^{-1}\mathscr{G}) = \{(x,g) \in X \times \mathrm{Et}(\mathscr{G}) \mid f(x) = \pi(g)\},$$

so that $f^{-1}\mathscr{G}$ is the sheaf of continuous cross sections of the projection $\mathrm{Et}(f^{-1}\mathscr{G}) \to X$.

If $f : X \to Y$ is the inclusion of a closed set, we also call $f_*\mathscr{F}$ the extension of $\mathscr{F}$ by 0 and $f^{-1}\mathscr{G}$ the restriction of $\mathscr{G}$. We often identify $\mathscr{F}$ with its extension, and denote the restriction by $\mathscr{G}|_X$.

These operations extend to functors $f_* : \mathrm{Ab}(X) \to \mathrm{Ab}(Y)$ and $f^{-1} : \mathrm{Ab}(Y) \to \mathrm{Ab}(X)$ in an obvious way. While these operations are generally not inverses, there is a relationship. There are canonical morphisms, called adjunctions, $\alpha : \mathscr{F} \to f_*f^{-1}\mathscr{F}$ and $\beta : f^{-1}f_*\mathscr{G} \to \mathscr{G}$ induced by the restriction maps

$$\mathscr{F}(U) \to \mathscr{F}(f(f^{-1}(U))) = f_*f^P\mathscr{F}(U) \to f_*f^{-1}\mathscr{F}(U), \quad U \subseteq Y,$$
$$f^P f_*\mathscr{G}(V) = \mathscr{G}(f^{-1}(f(V))) \to \mathscr{G}(V), \quad V \subseteq X.$$

Lemma 3.7.4. *There is a natural isomorphism*

$$\mathrm{Hom}_{\mathrm{Ab}(X)}(f^{-1}\mathscr{F},\mathscr{G}) \cong \mathrm{Hom}_{\mathrm{Ab}(Y)}(\mathscr{F},f_*\mathscr{G})$$

given by $\eta \mapsto (f_*\eta)\circ\alpha$*. The inverse is given by* $\xi \mapsto \beta\circ(f^{-1}\xi)$*.*

In the language of category theory, this says that the functor f^{-1} is left adjoint to f_*. The proof will be left as an exercise.

The collection of ringed spaces will form a category. To motivate the definition of morphism, observe that from a morphism F of concrete k-spaces $(X,\mathscr{R}) \to (Y,\mathscr{S})$, we get a morphism of sheaves of rings $\mathscr{S} \to F_*\mathscr{R}$ given by $s \mapsto s\circ F$.

Definition 3.7.5. A morphism of (k-)ringed spaces $(X,\mathscr{R}) \to (Y,\mathscr{S})$ is a continuous map $F : X \to Y$ together with a morphism of sheaves of rings (or algebras) $F' : \mathscr{S} \to F_*\mathscr{R}$.

By Lemma 3.7.4, to give F' is equivalent to giving the adjoint map $F^{-1}\mathscr{S} \to \mathscr{R}$. Therefore we have the induced ring homomorphism $\mathscr{S}_{F(x)} \to \mathscr{R}_x$ for each $x \in X$.

Definition 3.7.6. If $(X,\mathscr{R})$ and $(Y,\mathscr{S})$ are locally ringed, a morphism $F : (X,\mathscr{R}) \to (Y,\mathscr{S})$ is called a morphism of locally ringed spaces if it satisfies the additional requirement that the induced maps $\mathscr{S}_{F(x)} \to \mathscr{R}_x$ be local homomorphisms. A morphism of schemes is simply a morphism of locally ringed spaces

For affine schemes, we have the following description.

Proposition 3.7.7. *Let R and S be affine algebras. There is a bijection*

$$\mathrm{Hom}_{\mathrm{Schemes}}((\mathrm{Spec}_m S, \mathscr{O}_{\mathrm{Spec}_m S}), (\mathrm{Spec}_m R, \mathscr{O}_{\mathrm{Spec}_m R})) \cong \mathrm{Hom}_{k\text{-alg}}(R,S).$$

Therefore there is an antiequivalence between the category of affine schemes and affine algebras.

Proof. Let $X = \mathrm{Spec}_m S$ and $Y = \mathrm{Spec}_m R$. Given a homomorphism $f : R \to S$, we just construct the corresponding morphism $F = f^* : X \to Y$. If $m \subset S$ is a maximal ideal, then $f^{-1}m$ is again a maximal ideal by Hilbert's Nullstellensatz. So we can define $F(m) = f^{-1}m$. It is easy to see that F is continuous. The induced homomorphisms $R[1/r] \to S[1/f(r)]$ yield morphisms $\mathscr{O}_Y \to F_*\mathscr{O}_X$ of sheaves of rings on basic open sets, and therefore on their extensions. The maps on stalks are the induced local homomorphisms $f_m : R_{F(m)} \to S_m$. Therefore F is a morphism of schemes.

In the opposite direction, a morphism $F : X \to Y$ induces a homomorphism $R = \mathscr{O}(Y) \to \mathscr{O}(X) = S$. This yields the inverse. The remaining details will be left as an exercise. □

The simplest example of a scheme that is not a variety is the fat point $\mathrm{Spec}_m k[\varepsilon]/(\varepsilon^2)$. This has the following interesting property.

Lemma 3.7.8. *If X is a scheme, the set of morphisms from $\mathrm{Spec}_m k[\varepsilon]/(\varepsilon^2)$ to X is in one-to-one correspondence with the set of vectors in the disjoint union of tangent spaces $\cup T_x$.*

Proof. The scheme $D = \mathrm{Spec}_m k[\varepsilon]/(\varepsilon^2)$ has a unique point $m = (\varepsilon)$. We show that the set of morphisms $i : D \to X$ with $i(m) = x$ is in one to one correspondence with T_x. To give such a morphism is equivalent to giving a morphism of D to a neighborhood of x. Therefore, we may replace X by an affine scheme $\mathrm{Spec}_m R$. By Proposition 3.7.7, $D \to \mathrm{Spec}_m R$ corresponds to an algebra homomorphism $i^* : R \to k[\varepsilon]/(\varepsilon^2)$ such that $(i^*)^{-1}m = m_x$. The map i^* factors through R/m_x^2. We can see that any homomorphism

$$h : R/m_x^2 = k \oplus m_x/m_x^2 \to k[\varepsilon]/(\varepsilon^2) = k \oplus k\varepsilon$$

is given by $h(a,b) = a + v(b)\varepsilon$ for $v \in T_x = (m_x/m_x^2)^*$. The argument can be reversed to show that conversely, any vector in $v \in T_x$ determines an algebra map $i^* : R \to k[\varepsilon]/(\varepsilon^2)$ for which $(i^*)^{-1}m = m_x$. □

Suppose that $f : (X, \mathscr{R}) \to (Y, \mathscr{S})$ is a morphism of ringed spaces. Given an $\mathscr{R}$-module $\mathscr{M}$, $f_*\mathscr{M}$ is naturally an $f_*\mathscr{R}$-module, and hence an $\mathscr{S}$-module by restriction of scalars via $\mathscr{S} \to f_*\mathscr{R}$. Similarly, given an $\mathscr{S}$-module $\mathscr{N}$, $f^{-1}\mathscr{N}$ is naturally an $f^{-1}\mathscr{S}$-module. We define the $\mathscr{R}$-module

$$f^*\mathscr{N} = \mathscr{R} \otimes_{f^{-1}\mathscr{S}} f^{-1}\mathscr{N},$$

where the $\mathscr{R}$ is regarded as an $f^{-1}\mathscr{S}$-module under the adjoint map $f^{-1}\mathscr{S} \to \mathscr{R}$. When f is injective, we often write $\mathscr{N}|_X$ instead of $f^*\mathscr{N}$.

The inverse image of a locally free sheaf is easily seen to be locally free. This has an interpretation in the context of vector bundles. If $\pi : V \to Y$ is a vector bundle, the pullback $f^*V \to X$ is the vector bundle given set theoretically as the projection

$$f^*V = \{(v,x) \mid \pi(v) = f(x)\} \to X.$$

Then

$$f^*(\mathscr{V}) \cong (\text{sheaf of sections of } f^*V). \tag{3.7.3}$$

A special case is that in which $X = \{y\} \to Y$ is the inclusion of a point. Then the fiber $\pi^{-1}y$ can be identified with $f^*(\mathscr{V}) = (\mathscr{R}_y/m_y) \otimes \mathscr{V}_y$.

Exercises

3.7.9. Verify (3.7.2).

3.7.10. Prove Lemma 3.7.4.

3.7.11. Give examples where $\mathscr{F} \to f_*f^{-1}\mathscr{F}$ and $f^{-1}f_*\mathscr{G} \to \mathscr{G}$ are not isomorphisms.

3.7.12. Complete the proof of Proposition 3.7.7.

3.7.13. Given an inclusion of an open set $j : U \to X$ and a sheaf $\mathscr{F}$ on U, define $j_!\mathscr{F}$ to be the sheafification of the presheaf

$$\mathscr{G}(V) = \begin{cases} \mathscr{F}(V) & \text{if } V \subseteq U, \\ 0 & \text{otherwise.} \end{cases}$$

If $\mathscr{S}$ is a sheaf on X, show that this fits into an exact sequence

$$0 \to j_!j^{-1}\mathscr{S} \to \mathscr{S} \to i_*i^{-1}\mathscr{S} \to 0,$$

where $i : X - U \to X$ is the inclusion of the complement.

3.7.14. Generalize Lemma 3.2.5 to show that an exact sequence $0 \to \mathscr{A} \to \mathscr{B} \to \mathscr{C} \to 0$ of sheaves gives rise to an exact sequence $0 \to f_*\mathscr{A} \to f_*\mathscr{B} \to f_*\mathscr{C}$. In other words, f_* is left exact.

3.7.15. Prove that f^{-1} is exact, i.e., that it takes exact sequences to exact sequences.

3.7.16. Verify the assertion given in (3.7.3).

3.7.17. Suppose we have gluing data (R_i, ϕ_{ij}) and (R'_i, ϕ'_{ij}) and homomorphisms $f_i : R_i \to R'_i$ such that $f_i\phi'_{ij} = \phi_{ij}f_j$. Show that there is an induced morphism of glued schemes extending $f_i^* : \mathrm{Spec}_m R'_i \to \mathrm{Spec}_m R_i$. Use this to justify Remark 3.4.11.

3.7.18. Let L_d be obtained by gluing $\mathrm{Spec}_m k[x,y]$ to $\mathrm{Spec}_m k[x^{-1}, x^d y]$. Show that L_d is a line bundle over $\mathbb{P}^1_k$ whose sheaf of sections is isomorphic to $\mathscr{O}(\pm d)$.

3.7.19. Let $f : \mathbb{P}^{n-1}_k \to \mathbb{P}^n_k$ be a linear embedding, given for example by setting $x_n = 0$. Show that $f^*\mathscr{O}_{\mathbb{P}^n}(d) = \mathscr{O}_{\mathbb{P}^{n-1}}(d)$.

3.7.20. Let $v : \mathbb{P}^n_k \to \mathbb{P}^N_k$ be the dth Veronese embedding. Show that $v^*\mathscr{O}_{\mathbb{P}^N}(1) = \mathscr{O}_{\mathbb{P}^n}(d)$.

3.7.21. An n-jet on a scheme X is a morphism from $\operatorname{Spec} k[\varepsilon]/(\varepsilon^{n+1})$ to X. Give an interpretation of these similar to Lemma 3.7.8.

3.8 Differentials

With basic sheaf theory in hand, we can now construct sheaves of differential forms on manifolds and varieties in a unified way. In order to motivate things, let us start with a calculation. Suppose that $X = \mathbb{R}^n$ with coordinates $x_1, \ldots, x_n$. Given a C^∞ function f on X, we can develop a Taylor expansion about $(y_1, \ldots, y_n)$:

$$f(x_1,\ldots,x_n) = f(y_1,\ldots,y_n) + \sum \frac{\partial f}{\partial x_i}(y_1,\ldots,y_n)(x_i - y_i) + O((x_i - y_i)^2).$$

Thus the differential is given by

$$df = f(x_1,\ldots,x_n) - f(y_1,\ldots,y_n) \mod (x_i - y_i)^2.$$

We can view $x_1,\ldots,x_n,y_1,\ldots,y_n$ as coordinates on $X \times X = \mathbb{R}^{2n}$, so that $x_i - y_i = 0$ defines the diagonal Δ. Then df lies in the ideal of Δ modulo its square.

Let X be a C^∞ or complex manifold or an algebraic variety over a field k. We take $k = \mathbb{R}$ or $\mathbb{C}$ in the first two cases. We have a diagonal map $\delta : X \to X \times X$ given by $x \mapsto (x,x)$, and projections $p_i : X \times X \to X$. Let $\mathscr{I}_\Delta$ be the ideal sheaf of the image of δ, and let $\mathscr{I}_\Delta^2 \subseteq \mathscr{I}_\Delta$ be the sub–ideal sheaf locally generated by products of pairs of sections of $\mathscr{I}_\Delta$. Then we define the sheaf of 1-forms by

$$\Omega^1_X = (\mathscr{I}_\Delta/\mathscr{I}_\Delta^2)|_\Delta.$$

This has two different $\mathscr{O}_X$-module structures; we pick the first one, where $\mathscr{O}_X$ acts on $\mathscr{I}_\Delta$ through $p_1^{-1}\mathscr{O}_X \to \mathscr{O}_{X\times X}$. We define the sheaf of p-forms by $\Omega^p_X = \wedge^p \Omega^1_X$. We define a morphism $d : \mathscr{O}_X \to \Omega^1_X$ of sheaves (but not $\mathscr{O}_X$-modules) by $df = p_1^*(f) - p_2^*(f)$. This has the right formal properties because of the following:

Lemma 3.8.1. *d is a k-linear derivation.*

Proof. By direct calculation,

$$d(fg) - fdg - gdf = [p_1^*(f) - p_2^*(f)][p_2^*(g) - p_2^*(g)] \in \mathscr{I}_\Delta^2. \qquad \square$$

Differentials behave contravariantly with respect to morphisms. Given a morphism of manifolds or varieties $f : Y \to X$, we get a morphism $Y \times Y \to X \times X$

preserving the diagonal. This induces a morphism of sheaves $g^*\Omega^1_X \to \Omega^1_Y$ that fits into a commutative square:

$$\begin{array}{ccc} g^*\mathcal{O}_X & \to & \mathcal{O}_Y \\ d\downarrow & & \downarrow d \\ g^*\Omega^1_X & \to & \Omega^1_Y \end{array}$$

The calculations at the beginning of this section basically show the following:

Lemma 3.8.2. *If X is C^∞ manifold, then $\Omega^1_X \cong \mathcal{E}^1_X$ and d coincides with the derivative constructed in Section 2.6.*

Since complex manifolds are also C^∞-manifolds, there is potential for confusion. So in the sequel, we will use the symbol Ω^1_X only in the holomorphic or algebraic case. Sections of $\mathcal{E}^1_X$ over a coordinate neighborhood are given by sums $\sum f_i dx_i$ with C^∞ coefficients. A similar statement holds for sections of Ω^1_X on complex manifolds, except that x_i should be chosen as analytic coordinates, and the coefficients should be holomorphic.

Now suppose that $X \subset \mathbb{A}^N_k$ is a closed subvariety defined by the ideal $(f_1,\dots,f_r)$. Let $S = k[x_1,\dots,x_N]$ and $R = S/(f_1,\dots,f_r)$ be the coordinate rings on $\mathbb{A}^N_k$ and X. Then Ω^1_X is quasicoherent. In fact, it is given by $\tilde{\Omega}_{R/k}$, where the module of Kähler differentials [33, Chapter 16] is given by

$$\Omega_{R/k} = \frac{\ker[R\otimes_k R \to R]}{(\ker[R\otimes_k R \to R])^2}.$$

The differential is induced by the map $d : R \to \Omega_{R/k}$ given by $dr = r\otimes 1 - 1\otimes r$. An argument similar to the proof of Lemma 3.8.1 shows that this is a derivation. The pair $(\Omega_{R/k}, d)$ is characterized as the universal k-linear derivation [33, §16.8]. This means that any k-linear derivation $\delta : R \to M$ factors as $R \to \Omega_{R/k} \to M$ in a unique way. From this, we can extract an alternative description by generators and relations: $\Omega_{R/k}$ is isomorphic to the free module on symbols $\{dr \mid r \in R\}$ modulo the submodule generated by the relations $d(ar+bs) = adr + bds$ and $d(rs) - rds - sdr$ for $r,s \in R, a,b \in k$. With the second description, we easily verify that $\Omega_{S/k}$ is the free S-module on $dx_1,\dots,dx_N$ with

$$df = \sum \frac{\partial f}{\partial x_i} dx_i.$$

In general, we obtain the following isomorphism:

Lemma 3.8.3.

$$\Omega_{R/k} \cong \operatorname{coker}\left(\frac{\partial f_i}{\partial x_j}\right).$$

Proof. Given $r \in R$, dr can be expanded as a sum of dx_j, since d is a derivation. Thus there is a surjection

$$\bigoplus_j R dx_j \to \Omega_{R/k} \to 0.$$

Let K denote the kernel of this map. The expressions df_i necessarily lie in K. Let Ω' be the quotient of $\oplus Rdx_j$ by the span K' of the df_i. This is exactly the cokernel of the Jacobian matrix. The composition $S \to \Omega_{S/k} \to \Omega'$ descends to a well-defined k-linear derivation $d' : R \to \Omega'$. Therefore we obtain a map $\Omega_{R/k} \to \Omega'$, which can be seen to be compatible with the projections from $\oplus Rdx_j$. Thus $K \subseteq K'$, so that $\Omega_{R/k} = \Omega'$ as claimed. □

Theorem 3.8.4. *X is nonsingular if and only if Ω^1_X is locally free of rank equal to* $\dim X$.

Proof. X is nonsingular if and only if the function $x \mapsto \dim k(m_x) \otimes \Omega^1_{R/k}$ takes the constant value $\dim X$. So the result follows from Lemma 3.5.14. □

Exercises

3.8.5. The tangent sheaf to a variety or manifold can be defined by $\mathcal{T}_X = (\Omega^1_X)^*$. Show that a section $D \in \mathcal{T}_X(U)$ determines an $\mathcal{O}_X(U)$-linear derivation from $\mathcal{O}_X(U)$ to $\mathcal{O}_X(U)$.

3.8.6. Prove that $\mathcal{M} \cong \mathcal{M}^{**}$ if $\mathcal{M}$ is locally free. Show that Ω^1_X is not isomorphic to $\mathcal{T}^*_X$ if X is the cusp defined by $y^2 - x^3 = 0$. Thus Ω^1_X contains more information in general.

3.8.7. Check that $\Omega_{S/k}$ is a free module on dx_i as claimed above, where $S = k[x_1, \dots, x_N]$.

3.8.8. Given a subvariety or submanifold $i : X \subset Y$, show that there is an epimorphism $i^*\Omega^1_Y = \mathcal{O}_Y \otimes \Omega^1_Y \to \Omega^1_X$.

3.8.9. When $X \subset Y$ is a subvariety with ideal sheaf $\mathcal{I}$, show that there is an exact sequence

$$\mathcal{I} / \mathcal{I}^2 \to \mathcal{O}_Y \otimes \Omega^1_Y \to \Omega^1_X \to 0.$$

The sheaf $\mathcal{I} / \mathcal{I}^2$ is called the conormal sheaf of X in Y.

3.8.10. Using the identity $dx^{-1} = -x^{-2}dx$, show that $\mathcal{O}_{\mathbb{P}^1} \cong \mathcal{O}_{\mathbb{P}^1}(-2)$ over any algebraically closed field.

Chapter 4
Sheaf Cohomology

As we saw in the previous chapter, a "surjection" or epimorphism $\mathscr{F} \to \mathscr{G}$ of sheaves need not induce a surjection of global sections. We would like to understand what further conditions are required to ensure this. Although this may seem like a fairly technical problem, it lies at the heart of many fundamental questions in geometry and function theory. Typically, we may want to know when some interesting class of functions extends from a subspace to the whole space, and this is a special case of the above problem.

In this chapter, we introduce sheaf cohomology, which gives a framework in which to answer such questions and more. Our basic approach is due to Godement [45], but we have set things up as an inductive definition. This will allow us to get to the main results rather quickly.

4.1 Flasque Sheaves

We start by isolating a class of sheaves for which the extension problem becomes trivial. These will form the building blocks for cohomology in general.

Definition 4.1.1. A sheaf $\mathscr{F}$ on X is called flasque (or flabby) if the restriction maps $\mathscr{F}(X) \to \mathscr{F}(U)$ are surjective for any nonempty open set U.

Sheaves are rarely flasque. Here are a few examples.

Example 4.1.2. Let X be a space with the property that any open set is connected (more specifically, suppose X is irreducible). Then any constant sheaf is flasque.

Example 4.1.3. For any abelian group, the so-called skyscraper sheaf

$$A_x(U) = \begin{cases} A & \text{if } x \in U, \\ 0 & \text{otherwise,} \end{cases}$$

is flasque.

The importance of flasque sheaves stems from the following:

Lemma 4.1.4. *If* $0 \to \mathscr{A} \to \mathscr{B} \to \mathscr{C} \to 0$ *is an exact sequence of sheaves with* $\mathscr{A}$ *flasque, then* $\mathscr{B}(X) \to \mathscr{C}(X)$ *is surjective.*

Proof. We will prove this by the unfashionable method of transfinite induction.[1] Let $\gamma \in \mathscr{C}(X)$. By assumption, there is an open cover $\{U_i\}_{i\in I}$ such that $\gamma|_{U_i}$ lifts to a section $\beta_i \in \mathscr{B}(U_i)$. By the well-ordering theorem, we can assume that the index set I is the set of ordinal numbers less than a given ordinal κ. We will define

$$\sigma_i \in \mathscr{B}\left(\bigcup_{j<i} U_j\right)$$

inductively, so that it maps to the restriction of γ. Set $\sigma_1 = \beta_0$. Now suppose that σ_i exists. Let $U = U_i \cap (\cup_{j<i} U_j)$. Then $\beta_i|_U - \sigma_i|_U$ is the image of a section $\alpha_i' \in \mathscr{A}(U)$. By hypothesis, α_i' extends to a global section $\alpha_i \in \mathscr{A}(X)$. Then set

$$\sigma_{i+1} = \begin{cases} \sigma_i \text{ on } \bigcup_{j<i} U_j, \\ \beta_i - \alpha_i|_{U_i} \text{ on } U_i. \end{cases}$$

If i is a limit (nonsuccessor) ordinal, then the previous σ_j's patch to define σ_i. Then σ_κ is a global section of B mapping to γ. □

Corollary 4.1.5. *The sequence* $0 \to \mathscr{A}(X) \to \mathscr{B}(X) \to \mathscr{C}(X) \to 0$ *is exact if* $\mathscr{A}$ *is flasque.*

Proof. Combine this with Lemma 3.2.5. □

Let $\mathscr{F}$ be a presheaf. Define the presheaf $\mathbf{G}(\mathscr{F})$ of discontinuous sections of $\mathscr{F}$ by

$$U \mapsto \prod_{x\in U} \mathscr{F}_x$$

with the obvious restrictions. This terminology is explained by thinking about this as the sheaf of discontinuous sections of the étalé space, although it is simpler to define it directly as we did. There is a canonical morphism $\gamma_{\mathscr{F}} : \mathscr{F} \to \mathbf{G}(\mathscr{F})$, sending $f \in \mathscr{F}(U)$ to the product of germs $(f_x)_{x\in U}$.

Lemma 4.1.6. $\mathbf{G}(\mathscr{F})$ *is a flasque sheaf, and the morphism* $\gamma_{\mathscr{F}} : \mathscr{F} \to \mathbf{G}(\mathscr{F})$ *is a monomorphism if* $\mathscr{F}$ *is a sheaf.*

The operation $\mathbf{G}$ clearly determines a functor from $\mathrm{Ab}(X)$ to itself.

Lemma 4.1.7. *The functor* $\mathbf{G}$ *is exact, which means that it takes exact sequences to exact sequences.*

[1] See [70, appendix] for background. However, for most cases of interest to us, X will have a countable basis, so ordinary induction will suffice.

Proof. Given an exact sequence $0 \to \mathscr{A} \to \mathscr{B} \to \mathscr{C} \to 0$, the sequence $0 \to \mathscr{A}_x \to \mathscr{B}_x \to \mathscr{C}_x \to 0$ is exact by definition. Therefore

$$0 \to \prod_{x \in U} \mathscr{A}_x \to \prod_{x \in U} \mathscr{B}_x \to \prod_{x \in U} \mathscr{C}_x \to 0$$

is exact. Therefore

$$0 \to \mathbf{G}(\mathscr{A}) \to \mathbf{G}(\mathscr{B}) \to \mathbf{G}(\mathscr{C}) \to 0$$

is an exact sequence of presheaves, and hence of sheaves by Exercise 3.2.12. □

Let $\Gamma : \mathrm{Ab}(X) \to \mathrm{Ab}$ denote the functor of global sections, $\Gamma(X, \mathscr{F}) = \Gamma(\mathscr{F}) = \mathscr{F}(X)$.

Lemma 4.1.8. *Then* $\Gamma \circ \mathbf{G} : \mathrm{Ab}(X) \to \mathrm{Ab}$ *is exact.*

Proof. Follows from Corollary 4.1.5 and Lemma 4.1.7. □

Exercises

4.1.9. Prove that any constant sheaf on an irreducible algebraic variety is flasque.

4.1.10. Find a proof of Lemma 4.1.4 that uses Zorn's lemma.

4.1.11. Given a possibly infinite family of flasque sheaves $\mathscr{G}_i$, prove that their product $(\prod \mathscr{G}_i)(U) = \prod \mathscr{G}_i(U)$ is flasque, and conclude that $\mathbf{G}(\mathscr{F})$ is flasque for any presheaf $\mathscr{F}$. Prove that the canonical map $\mathscr{F} \to \mathbf{G}(\mathscr{F})$ is a monomorphism when $\mathscr{F}$ is a sheaf.

4.1.12. Let X^{disc} denote the set X with its discrete topology, and let $\pi : X^{\mathrm{disc}} \to X$ be the identity map. Show that $\mathbf{G}(\mathscr{F}) = \pi_* \pi^* \mathscr{F}$, and that the above map $\mathscr{F} \to \mathbf{G}(\mathscr{F})$ is the adjunction map (§3.7).

4.1.13. Prove that the sheaf of bounded continuous real-valued functions on $\mathbb{R}$ is flasque.

4.1.14. Prove the same thing for the sheaf of bounded C^∞ functions on $\mathbb{R}$.

4.1.15. Prove that if $0 \to \mathscr{A} \to \mathscr{B} \to \mathscr{C}$ is exact and $\mathscr{A}$ is flasque, then $0 \to f_*\mathscr{A} \to f_*\mathscr{B} \to f_*\mathscr{C} \to 0$ is exact for any continuous map f.

4.2 Cohomology

Let $\mathscr{F}$ be a sheaf on a topological space X. Define the following operations by induction:

$$\mathbf{C}^i(\mathscr{F}) = \begin{cases} \mathscr{F} & \text{if } i = 0, \\ \operatorname{coker}[\gamma_{\mathscr{F}} : \mathscr{F} \to \mathbf{G}(\mathscr{F})] & \text{if } i = 1, \\ \mathbf{C}^1(\mathbf{C}^n(\mathscr{F})) & \text{if } i = n+1. \end{cases}$$

By definition, $\gamma_{\mathscr{F}}$ can be prolonged to an exact sequence

$$0 \to \mathscr{F} \to \mathbf{G}(\mathscr{F}) \to \mathbf{C}^1(\mathscr{F}) \to 0.$$

Thus we also have exact sequences

$$0 \to \mathbf{C}^n(\mathscr{F}) \to \mathbf{G}(\mathbf{C}^n(\mathscr{F})) \to \mathbf{C}^{n+1}(\mathscr{F}) \to 0.$$

Sheaf cohomology can now be defined inductively:

Definition 4.2.1.

$$H^i(X, \mathscr{F}) = \begin{cases} \Gamma(X, \mathscr{F}) & \text{if } i = 0, \\ \operatorname{coker}[\Gamma(X, \mathbf{G}(\mathscr{F})) \to \Gamma(X, \mathbf{C}^1(\mathscr{F}))] & \text{if } i = 1, \\ H^1(X, \mathbf{C}^n(\mathscr{F})) & \text{if } i = n+1. \end{cases}$$

The following can be deduced from the definition by induction.

Lemma 4.2.2. $\mathscr{F} \mapsto H^i(X, \mathscr{F})$ *is a functor from* $\mathrm{Ab}(X) \to \mathrm{Ab}$.

We now come to the key result.

Theorem 4.2.3. *Given an exact sequence of sheaves*

$$0 \to \mathscr{A} \to \mathscr{B} \to \mathscr{C} \to 0,$$

there is a long exact sequence

$$0 \to H^0(X, \mathscr{A}) \to H^0(X, \mathscr{B}) \to H^0(X, \mathscr{C}) \to H^1(X, \mathscr{A}) \to H^1(X, \mathscr{B}) \to \cdots.$$

The proof will be based on a couple of lemmas.

Lemma 4.2.4 (Snake lemma). *Given a commutative diagram of abelian groups*

$$\begin{array}{ccccccccc} 0 & \longrightarrow & A_1 & \longrightarrow & B_1 & \xrightarrow{f} & C_1 & \longrightarrow & 0 \\ & & \downarrow{\scriptstyle\alpha} & & \downarrow{\scriptstyle\beta} & & \downarrow{\scriptstyle\gamma} & & \\ 0 & \longrightarrow & A_2 & \longrightarrow & B_2 & \longrightarrow & C_2 & & \end{array}$$

with exact rows, there is a canonical exact sequence

$$0 \to \ker(\alpha) \to \ker(\beta) \to \ker(\gamma) \to \operatorname{coker}(\alpha) \to \operatorname{coker}(\beta) \to \operatorname{coker}(\gamma).$$

The last map is surjective if $B_2 \to C_2$ *is also surjective.*

The existence of the sequence is standard and can be proved by a diagram chase (or looked up in many places [8, 33, 76]). The only thing we want to explain is how the sequence can be made canonical, since this is not always clear from standard proofs. Canonicity is crucial for us.

Proof. Most of the maps are the obvious ones, and these are clearly natural. We explain how to construct a canonical connecting (or "snaking") homomorphism $\delta : \ker(\gamma) \to \mathrm{coker}(\alpha)$. Define $P = f^{-1}\ker(\gamma) \subseteq B_1$ and let K be the kernel of the surjective homomorphism $P \to \ker(\gamma)$. We note that f induces an isomorphism $F : P/K \cong \ker(\gamma)$. The image of P under β maps to 0 in C_2. Therefore, we have a canonical map $\delta' : P \to A_2$ through which $\beta|_P$ factors. A simple chase shows that $\delta'(K) \subseteq \mathrm{im}(\alpha)$. Therefore we get an induced map

$$\delta : \ker(\gamma) \xrightarrow{F^{-1}} P/K \xrightarrow{\delta'} \mathrm{coker}(\alpha). \tag{4.2.1}$$

□

The snake lemma and similar results hold in any abelian category [82]. For sheaves, we can see this directly.

Lemma 4.2.5 (Snake lemma II). *Given a commutative diagram of sheaves of abelian groups*

$$\begin{array}{ccccccccc} 0 & \longrightarrow & \mathscr{A}_1 & \longrightarrow & \mathscr{B}_1 & \longrightarrow & \mathscr{C}_1 & \longrightarrow & 0 \\ & & \downarrow{\scriptstyle\alpha} & & \downarrow{\scriptstyle\beta} & & \downarrow{\scriptstyle\gamma} & & \\ 0 & \longrightarrow & \mathscr{A}_2 & \longrightarrow & \mathscr{B}_2 & \longrightarrow & \mathscr{C}_2 & & \end{array}$$

with exact rows, there is a canonical exact sequence

$$0 \to \ker(\alpha) \to \ker(\beta) \to \ker(\gamma) \to \mathrm{coker}(\alpha) \to \mathrm{coker}(\beta) \to \mathrm{coker}(\gamma).$$

The last map is an epimorphism if $\mathscr{B}_2 \to \mathscr{C}_2$ is an epimorphism.

Proof. Since we have canonical maps in the previous lemma, these extend to sheaves. The exactness can be checked on stalks, and so this reduces to the previous statement. □

Lemma 4.2.6. *The exact sequence of sheaves given in the top row can be extended to a commutative diagram with exact rows and columns:*

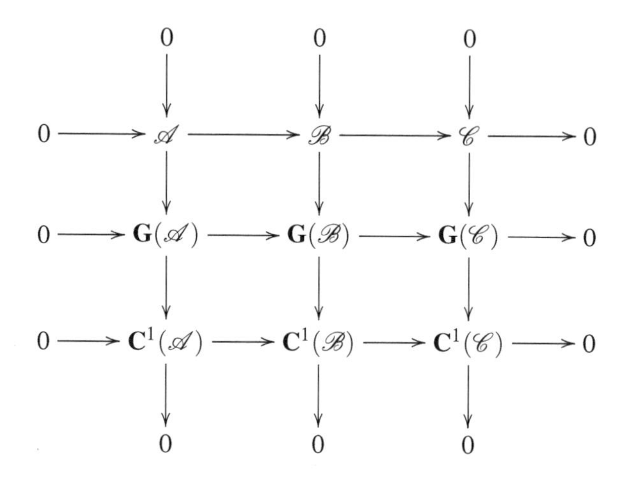

Proof. By Lemma 4.1.7, there is a commutative diagram with exact rows

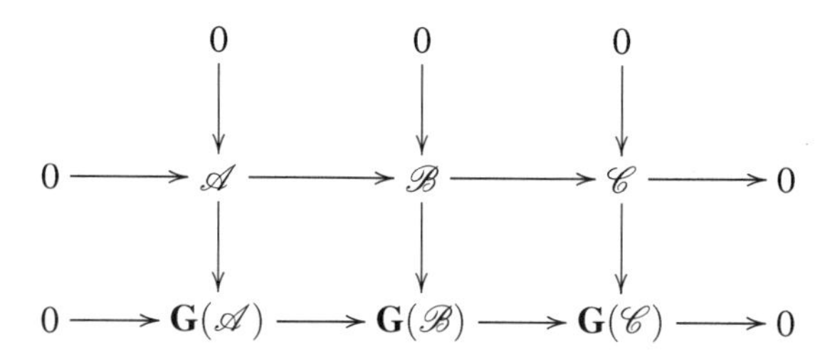

The snake lemma gives the rest. □

We are now ready to prove Theorem 4.2.3.

Proof. From Lemmas 3.2.5, 4.1.8, 4.2.6 we get a commutative diagram with exact rows:

$$\begin{array}{ccccccccc}
0 & \longrightarrow & \Gamma(\mathbf{G}(\mathscr{A})) & \longrightarrow & \Gamma(\mathbf{G}(\mathscr{B})) & \longrightarrow & \Gamma(\mathbf{G}(\mathscr{C})) & \longrightarrow & 0 \\
 & & \downarrow & & \downarrow & & \downarrow & & \\
0 & \longrightarrow & \Gamma(\mathbf{C}^1(\mathscr{A})) & \longrightarrow & \Gamma(\mathbf{C}^1(\mathscr{B})) & \longrightarrow & \Gamma(\mathbf{C}^1(\mathscr{C})) & &
\end{array}$$

From the snake lemma, we obtain a six-term exact sequence

$$\begin{array}{l}
0 \longrightarrow H^0(X,\mathscr{A}) \longrightarrow H^0(X,\mathscr{B}) \longrightarrow H^0(X,\mathscr{C}) \longrightarrow \\
\longrightarrow H^1(X,\mathscr{A}) \longrightarrow H^1(X,\mathscr{B}) \longrightarrow H^1(X,\mathscr{C})
\end{array} \tag{4.2.2}$$

which we need to extend. Applying (4.2.2) to

$$0 \to \mathbf{C}^1(\mathscr{A}) \to \mathbf{C}^1(\mathscr{B}) \to \mathbf{C}^1(\mathscr{C}) \to 0$$

yields a six-term sequence

$$\cdots \to H^0(X, \mathbf{C}^1(\mathscr{B})) \to H^0(X, \mathbf{C}^1(\mathscr{C})) \xrightarrow{\delta} H^2(X, \mathscr{A}) \to \cdots,$$

which is not quite what we want. We will show that the connecting map δ factors through a map $\bar{\delta}$ of

$$H^1(X, \mathscr{C}) = \frac{H^0(X, \mathbf{C}^1(\mathscr{C}))}{H^0(X, \mathbf{G}(\mathscr{C}))}.$$

From Lemma 4.2.6 and (4.2.2), we obtain a commutative diagram with exact rows and columns

$$\begin{array}{ccccc} H^0(\mathbf{G}(\mathscr{B})) & \longrightarrow & H^0(\mathbf{G}(\mathscr{C})) & \longrightarrow & 0 \\ \downarrow & & \downarrow & & \downarrow \\ H^0(\mathbf{C}^1(\mathscr{B})) & \longrightarrow & H^0(\mathbf{C}^1(\mathscr{C})) & \xrightarrow{\delta} & H^2(\mathscr{A}) \\ \downarrow & & \downarrow & \nearrow_{\bar{\delta}} & \\ H^1(\mathscr{B}) & \longrightarrow & H^1(\mathscr{C}) & & \end{array} \tag{4.2.3}$$

Given an element $\bar{\gamma} \in H^1(X, \mathscr{C})$, choose a lift $\gamma \in H^0(X, \mathbf{C}^1(\mathscr{C}))$ and set $\bar{\delta}\bar{\gamma} = \delta\gamma$. To see that this is well defined, let $\gamma' \in H^0(X, \mathbf{C}^1(\mathscr{C}))$ be another lift. Then $\gamma - \gamma'$ lies in $H^0(X, \mathbf{G}(\mathscr{C}))$, which can be lifted to $H^0(X, \mathbf{G}(\mathscr{B}))$ by Lemma 4.1.8. An easy diagram chase using (4.2.3) shows that $\delta\gamma = \delta\gamma'$ as desired. We also see immediately that $\bar{\delta}\bar{\gamma} = 0$ if and only if $\bar{\gamma}$ lies in the image of $H^1(X, \mathscr{B})$. Thus we have a six-term cohomology sequence

$$\cdots \to H^1(X, \mathscr{B}) \to H^1(X, \mathscr{C}) \xrightarrow{\delta} H^2(X, \mathscr{A}) \to \cdots,$$

which extends the earlier one. Applying this to

$$0 \to \mathbf{C}^i(\mathscr{A}) \to \mathbf{C}^i(\mathscr{B}) \to \mathbf{C}^i(\mathscr{C}) \to 0$$

shows that we can continue this sequence indefinitely. □

Corollary 4.2.7. *$\mathscr{B}(X) \to \mathscr{C}(X)$ is surjective if $H^1(X, \mathscr{A}) = 0$.*

We make an addendum to the theorem that says that the cohomology sequence is natural, or in words that the $H^i(X, -)$ form a "δ-functor" [51, 60]. By a map or morphism of diagrams (and particularly exact sequences), we will simply mean a map between objects in the same position that commute with the maps of the diagram.

Theorem 4.2.8. *A morphism of short exact sequences of sheaves gives rise to a map of long exact sequences of cohomology.*

Exercises

4.2.9. If $\mathscr{F}$ is flasque, prove that $H^i(X,\mathscr{F}) = 0$ for $i > 0$. (Prove this for $i = 1$, and that $\mathscr{F}$ flasque implies that $\mathbf{C}^1(\mathscr{F})$ is flasque.)

4.2.10. Prove that $H^i(X,\mathbb{Z}_X) = 0$ when $i > 0$ for any irreducible algebraic variety.

4.2.11. Finish all the details in the proof of the snake lemmas.

4.2.12. Give a proof of Theorem 4.2.8.

4.2.13. Given a sheaf $\mathscr{A}$ on X, let $\mathrm{Ext}^1(\mathbb{Z},\mathscr{A})$ denote the set of exact sequences

$$0 \to \mathscr{A} \to \mathscr{B} \to \mathbb{Z} \to 0$$

modulo the equivalence relation that two sequences are exact if and only if they fit into a commutative diagram

$$\begin{array}{ccccccccc} 0 & \longrightarrow & \mathscr{A} & \longrightarrow & \mathscr{B} & \longrightarrow & \mathbb{Z} & \longrightarrow & 0 \\ & & \downarrow{=} & & \downarrow & & \downarrow{=} & & \\ 0 & \longrightarrow & \mathscr{A} & \longrightarrow & \mathscr{C} & \longrightarrow & \mathbb{Z} & \longrightarrow & 0 \end{array}$$

Check that this is in fact an equivalence relation. Show that the class $\delta(1) \in H^1(X,\mathscr{A})$ associated to an exact sequence determines a well-defined map $Ext^1(\mathbb{Z},\mathscr{A}) \to H^1(X,\mathscr{A})$. Show that this is a bijection.

4.3 Soft Sheaves

We introduce the class of soft sheaves, whose definition is similar to the class of flasque sheaves. The main advantage is that unlike flasque sheaves, the soft sheaves often occur "in nature." *We assume throughout this section that X is a metric space,* although the results hold under the weaker assumption that X is paracompact and Hausdorff.

Definition 4.3.1. A sheaf $\mathscr{F}$ is called soft if the map $\mathscr{F}(X) \to \mathscr{F}(S)$ is surjective for all closed sets, where $\mathscr{F}(S)$ was defined in (3.7.1).

Lemma 4.3.2. *If* $0 \to \mathscr{A} \to \mathscr{B} \to \mathscr{C} \to 0$ *is an exact sequence of sheaves with* $\mathscr{A}$ *soft, then* $\mathscr{B}(X) \to \mathscr{C}(X)$ *is surjective.*

Proof. The proof is very similar to the proof of Lemma 4.1.4. We just indicate the modifications. Let $\gamma \in \mathscr{C}(X)$. We can assume that the open cover $\{U_i \mid i < \kappa\}$, such that $\gamma|_{U_i}$ lifts to $\mathscr{B}$, consists of open balls. We can also assume that this cover is *locally finite*, which means that every point has a neighborhood that intersects finitely many members of the cover. This follows from Stone's theorem [112, 70] that metric spaces are paracompact. In fact, we are mostly interested in the case of locally compact spaces, where this fact is trivial. Let $\{V_i\}$ be a new open cover where we shrink the radii of each ball, so that $\bar{V}_i \subset U_i$. We can define

$$\sigma_i \in \mathscr{B}(\cup_{j<i}\bar{V}_j)$$

inductively as before, so that σ_κ maps to γ. □

Corollary 4.3.3. *If $\mathscr{A}$ and $\mathscr{B}$ are soft, then so is $\mathscr{C}$.*

Proof. For any closed set S, $\mathscr{B}(X) \to \mathscr{B}(S)$ is surjective. The lemma shows that $\mathscr{B}(S) \to \mathscr{C}(S)$ is surjective. Therefore $\mathscr{B}(X) \to \mathscr{C}(S)$ is surjective, and this implies the same for $\mathscr{C}(X) \to \mathscr{C}(S)$. □

One trivially has the following:

Lemma 4.3.4. *A flasque sheaf is soft.*

Lemma 4.3.5. *If $\mathscr{F}$ is soft, then $H^i(X,\mathscr{F}) = 0$ for $i > 0$.*

Proof. Lemma 4.3.2 applied to $0 \to \mathscr{F} \to \mathbf{G}(\mathscr{F}) \to \mathbf{C}^1(\mathscr{F}) \to 0$ shows that $H^1(X,\mathscr{F}) = 0$. Corollary 4.3.3 and the previous lemma imply that $\mathbf{C}^1(\mathscr{F})$ is soft. By induction it follows that all the $\mathbf{C}^i(\mathscr{F})$ are soft. Hence $H^i(X,\mathscr{F}) = 0$ for all i. □

Theorem 4.3.6. *The sheaf* $\mathrm{Cont}_{X,\mathbb{R}}$ *of continuous real-valued functions on a metric space X is soft.*

Proof. The basic strategy of this proof and many of the subsequent softness proofs is the construction of a continuous "cutoff" function ρ that is 0 outside a given neighborhood U of a closed set S, and 1 close to S. Then given any continuous function $f : U \to \mathbb{R}$, ρf can be extended by 0 to all of X. Since f and ρf have the same germ along S, this would prove the surjectivity of $\mathrm{Cont}_{X,\mathbb{R}}(X) \to \mathrm{Cont}_{X,\mathbb{R}}(S)$ as required

To construct ρ we proceed as follows. Let $S_1 \subset U$ be a closed set containing S in its interior. This can be constructed by expressing U as a union of open balls, and taking the union of closed balls of half the radii. Let $S_2 = X - U$. Then Urysohn's lemma [70] guarantees the existence of a continuous ρ taking a value of 1 on S_1 and 0 on S_2. □

We get many more examples of soft sheaves with the following.

Lemma 4.3.7. *Let $\mathscr{R}$ be a soft sheaf of rings. Then any $\mathscr{R}$-module is soft.*

Proof. Let m be a section of an $\mathscr{R}$-module defined in a neighborhood of a closed set S, and let S_2 be the complement of this neighborhood. Since $\mathscr{R}$ is soft, the section that is 1 on S and 0 on S_2 extends to a global section ρ. Then ρm extends to a global section of the module. □

Let $S^1 \subset \mathbb{C}$ denote the unit circle, and let $e : \mathbb{R} \to S^1$ denote the normalized exponential $e(x) = \exp(2\pi i x)$. Let us say that X is *locally simply connected* if every neighborhood of every point contains a simply connected neighborhood.

Lemma 4.3.8. *If X is locally simply connected, then the sequence*

$$0 \to \mathbb{Z}_X \to \mathrm{Cont}_{X,\mathbb{R}} \xrightarrow{e} \mathrm{Cont}_{X,S^1} \to 1$$

is exact.

Proof. Exactness of the sequence of sections can be checked in any simply connected neighborhood of a point. This implies exactness of the sequence stalks. □

Lemma 4.3.9. *If X is simply connected and locally simply connected, then $H^1(X, \mathbb{Z}_X) = 0$.*

Note that in the future, we will usually write $H^i(X, \mathbb{Z})$ instead of $H^i(X, \mathbb{Z}_X)$, and likewise for other constant sheaves.

Proof. Since X is simply connected, any continuous map from X to S^1 can be lifted to a continuous map to its universal cover $\mathbb{R}$. In other words, $C_{\mathbb{R}}(X)$ surjects onto $C_{S^1}(X)$. Since $C_{\mathbb{R}}$ is soft, Lemma 4.3.8 implies that $H^1(X, \mathbb{Z}) = 0$. □

Corollary 4.3.10. $H^1(\mathbb{R}^n, \mathbb{Z}) = 0$.

Exercises

4.3.11. Show that the sheaf of piecewise linear continuous real-valued functions on $\mathbb{R}$ is soft.

4.3.12. Show that $\mathrm{Cont}_{\mathbb{R},\mathbb{R}}$ is not flasque. Conclude that a module over a flasque sheaf of rings need not be flasque.

4.3.13. Suppose that X is a locally compact metric space. If $0 \to \mathscr{A} \to \mathscr{B} \to \mathscr{C} \to 0$ is an exact sequence of sheaves with $\mathscr{A}$ soft, show that $\Gamma_c(\mathscr{B}) \to \Gamma_c(\mathscr{C})$ is surjective, where Γ_c are the sections with compact support.

4.3.14. Assuming that X is locally compact, we can define cohomology with compact support by setting $H_c^0(X, \mathscr{F}) = \Gamma_c(\mathscr{F})$, $H_c^1(X, \mathscr{F}) = \mathrm{coker}[\Gamma_c(\mathscr{F}) \to \Gamma_c(\mathbf{C}^1(\mathscr{F}))]$ and so on. Construct a long exact sequence (or at least the first six terms) analagous to Theorem 4.2.3.

4.3.15. Again assuming that X is locally compact, prove that the higher cohomology with compact support of a soft sheaf is trivial.

4.4 C^∞-Modules Are Soft

We want to prove that the sheaf of C^∞ functions on a manifold is soft. We start with a few lemmas.

Lemma 4.4.1. *Given $\varepsilon > 0$, there exists an $\mathbb{R}$-linear "smoothing" operator Σ_ε : $\mathrm{Cont}(\mathbb{R}^n) \to C^\infty(\mathbb{R}^n)$ such that if $f(x)$ is constant for all points x in an ε ball about x_0, then $\Sigma_\varepsilon(f)(x_0) = f(x_0)$.*

Proof. Let ψ be a C^∞ function on $\mathbb{R}^n$ with support in $\|x\| < 1$ such that $\int \psi dx = 1$. Rescale this by setting $\phi(x) = \varepsilon^n \psi(x/\varepsilon)$. Then the convolution

$$\Sigma_\varepsilon(f)(x) = \int_{\mathbb{R}^n} f(y)\phi(x-y)dy$$

will have the desired properties. □

Lemma 4.4.2. *Let $S \subset \mathbb{R}^n$ be a compact subset, and let U be an open neighborhood of S. Then there exists a C^∞ function $\rho : \mathbb{R}^n \to \mathbb{R}$ that is 1 on S and 0 outside U.*

Proof. We constructed a continuous function $\rho_1 : \mathbb{R}^n \to \mathbb{R}$ with these properties in the previous section. Now set $\rho = \Sigma_\varepsilon(\rho_1)$ with ε sufficiently small. □

We want to extend this to a manifold X. For this, we need the following construction. Let $\{U_i\}$ be a locally finite open cover of X, which means that every point of X is contained in a finite number of U_i's. A *partition of unity* subordinate to $\{U_i\}$ is a collection of C^∞ functions $\phi_i : X \to [0,1]$ such that

1. The support of ϕ_i lies in U_i.
2. $\sum \phi_i = 1$ (the sum is meaningful by local finiteness).

Partitions of unity always exist for any locally finite cover; see [110, 117] or the exercises.

Lemma 4.4.3. *Let $S \subset X$ be a closed subset, and let U be an open neighborhood of S. Then there exists a C^∞ function $\rho : X \to \mathbb{R}$ that is 1 on S and 0 outside U.*

Proof. Let $\{U_i\}$ be a locally finite open cover of X such that each U_i is diffeomorphic to a ball $\mathbb{R}^n$ (see the exercises). Then we have functions $\rho_i \in C^\infty(U_i)$ that are 1 on $S \cap U_i$ and 0 outside $U \cap U_i$ by the previous lemma. Choose a partition of unity $\{\phi_i\}$. Then $\rho = \sum \phi_i \rho_i$ will give the desired function. □

Theorem 4.4.4. *Given a C^∞ manifold X, C^∞_X is soft.*

Proof. Given a function f defined in a neighborhood of a closed set $S \subset X$, let ρ be given as in Lemma 4.4.3. Then ρf gives a global C^∞ function extending f. □

Corollary 4.4.5. *Any C^∞_X-module is soft.*

These arguments can be extended by introducing the notion of a fine sheaf, described below.

Exercises

4.4.6. Let X be an n dimensional C^∞ manifold. Prove that it has a locally finite open cover $\{U_i\}$ such that each U_i is diffeomorphic to $\mathbb{R}^n$.

4.4.7. Let X and $\{U_i\}$ be as in the previous exercise. Construct a partition of unity in this case, by first constructing a family of continuous functions satisfying these conditions, and then applying Σ_ε.

4.4.8. A sheaf $\mathscr{F}$ is called *fine* if for each locally finite cover $\{U_i\}$ there exists a partition of unity that is a collection of endomorphisms $\phi_i : \mathscr{F} \to \mathscr{F}$ such that

1. The endomorphism of $\mathscr{F}_x$ induced by ϕ_i vanishes for $x \notin U_i$.
2. $\sum \phi_i = 1$.

Prove that a fine sheaf on a paracompact Hausdorff space is soft.

4.4.9. Let $X = \mathbb{R}$ or S^1. Compute $H^*(X, \mathbb{R})$ using the complex $0 \to \mathbb{R} \to C_X^\infty \to C_X^\infty \to 0$, where the last map is the derivative.

4.5 Mayer–Vietoris Sequence

Given a continuous map $f : Y \to X$ and a sheaf $\mathscr{F} \in \mathrm{Ab}(X)$, we defined the inverse image $f^{-1}\mathscr{F} \in \mathrm{Ab}(Y)$ as the sheaf associated with

$$\Gamma(Y, f^P\mathscr{F}) = \mathscr{F}(f(Y)) = \varinjlim_{U \supseteq f(Y)} \mathscr{F}(U).$$

Thus we have a natural map

$$p_0 : \Gamma(X, \mathscr{F}) \to \Gamma(Y, f^{-1}\mathscr{F})$$

induced from restriction.

Theorem 4.5.1. *There exist natural maps $p_{i,\mathscr{F}} : H^i(X, \mathscr{F}) \to H^i(Y, f^{-1}\mathscr{F})$, agreeing with above map when $i = 0$, such that a short exact sequence*

$$0 \to \mathscr{F}_1 \to \mathscr{F}_2 \to \mathscr{F}_3 \to 0$$

gives rise to a commutative diagram

$$\begin{array}{ccccccccc}
\cdots \longrightarrow & H^i(X,\mathscr{F}_1) & \longrightarrow & H^i(X,\mathscr{F}_2) & \longrightarrow & H^i(X,\mathscr{F}_3) & \longrightarrow \cdots \\
& \downarrow & & \downarrow & & \downarrow & \\
\longrightarrow & H^i(Y,f^{-1}\mathscr{F}_1) & \longrightarrow & H^i(Y,f^{-1}\mathscr{F}_2) & \longrightarrow & H^i(Y,f^{-1}\mathscr{F}_3) & \longrightarrow
\end{array}$$

Proof. We define the map by induction. For $i = 0$ the map p_0 has already been defined. We construct a morphism $f^{-1}\mathbf{G}(\mathscr{F}) \to \mathbf{G}(f^{-1}\mathscr{F})$ as follows. Let $U \subset Y$. Then let

$$\beta : f^{-1}\mathbf{G}(\mathscr{F})(U) = \prod_{x\in f(U)} \mathscr{F}_x \to \prod_{y\in U} \mathscr{F}_{f(y)} = \mathbf{G}(f^{-1}\mathscr{F})(U)$$

be defined as the identity $\mathscr{F}_x \to \mathscr{F}_{f(y)}$ if $x = f(y)$, and 0 otherwise. Then there is a unique morphism

$$\alpha_1 : f^{-1}\mathbf{C}^1(\mathscr{F}) \to \mathbf{C}^1(f^{-1}\mathscr{F})$$

fitting into a commutative diagram.

$$\begin{array}{ccccccccc}
0 & \longrightarrow & f^{-1}\mathscr{F} & \longrightarrow & f^{-1}\mathbf{G}(\mathscr{F}) & \longrightarrow & f^{-1}\mathbf{C}^1(\mathscr{F}) & \longrightarrow & 0 \\
 & & \downarrow{\scriptstyle id} & & \downarrow{\scriptstyle \beta} & & \downarrow{\scriptstyle \alpha_1} & & \\
0 & \longrightarrow & f^{-1}\mathscr{F} & \longrightarrow & \mathbf{G}(f^{-1}\mathscr{F}) & \longrightarrow & \mathbf{C}^1(f^{-1}\mathscr{F}) & \longrightarrow & 0
\end{array}$$

We can iterate this to obtain a morphism

$$\alpha_i : f^{-1}\mathbf{C}^i(\mathscr{F}) \to \mathbf{C}^i(f^{-1}\mathscr{F}).$$

Then from what has been defined so far, we get the diagram below, marked with solid arrows:

$$\begin{array}{ccccccc}
\Gamma(X,\mathbf{G}(\mathscr{F})) & \longrightarrow & \Gamma(X,\mathbf{C}^1(\mathscr{F})) & \longrightarrow & H^1(X,\mathscr{F}) & \longrightarrow & 0 \\
\downarrow & & \downarrow & & \vdots{\scriptstyle p_1} & & \\
\Gamma(Y,\mathbf{G}(f^{-1}\mathscr{F})) & \longrightarrow & \Gamma(Y,\mathbf{C}^1(f^{-1}\mathscr{F})) & \longrightarrow & H^1(Y,f^{-1}\mathscr{F}) & \longrightarrow & 0
\end{array}$$

This gives rise to the dotted arrow p_1. For $i > 1$, we define p_i as the composite

$$H^1(X,\mathbf{C}^{i-1}(\mathscr{F})) \xrightarrow{p_1} H^1(Y,f^{-1}\mathbf{C}^{i-1}(\mathscr{F})) \xrightarrow{\alpha_{i-1}} H^1(Y,\mathbf{C}^{i-1}(f^{-1}\mathscr{F})).$$

The remaining properties are left as an exercise. □

When $f : Y \to X$ is an inclusion of an open set, we will usually write $H^i(Y,\mathscr{F})$ for $H^i(Y,f^{-1}\mathscr{F})$ and refer to the induced map on cohomology as restriction. We will introduce a basic tool for computing cohomology groups that is a prelude to Čech cohomology.

Theorem 4.5.2. *Let X be a union of two open sets $U \cup V$ and let $\mathscr{F}$ be a sheaf on X. Then there is a long exact sequence, called the Mayer–Vietoris sequence,*

$$\begin{aligned}\cdots &\to H^i(X,\mathscr{F}) \to H^i(U,\mathscr{F}) \oplus H^i(V,\mathscr{F}) \\ &\to H^i(U\cap V,\mathscr{F}) \to H^{i+1}(X,\mathscr{F}) \to \cdots,\end{aligned}$$

where the first indicated arrow is the sum of the restrictions, and the second is the difference.

Proof. The proof is very similar to the proof of Theorem 4.2.3, so we will just sketch it, leaving some of the details as an exercise. We may construct a commutative diagram with exact rows

$$\begin{array}{ccccccc} 0 \to \Gamma(X,\mathbf{G}(\mathscr{F})) & \to & \Gamma(U,\mathbf{G}(\mathscr{F}))\oplus\Gamma(V,\mathbf{G}(\mathscr{F})) & \to & \Gamma(U\cap V,\mathbf{G}(\mathscr{F})) & \to 0 \\ \downarrow & & \downarrow & & \downarrow & \\ 0 \to \Gamma(X,\mathbf{C}^1(\mathscr{F})) & \to & \Gamma(U,\mathbf{C}^1(\mathscr{F}))\oplus\Gamma(V,\mathbf{C}^1(\mathscr{F})) & \to & \Gamma(U\cap V,\mathbf{C}^1(\mathscr{F})) & \end{array}$$

Then applying the snake lemma yields the sequence of the first six terms:

$$\cdots \to H^0(U\cap V,\mathscr{F}) \to H^1(X,\mathscr{F}) \to \cdots.$$

When applied to $\mathbf{C}^1(\mathscr{F})$, this yields

$$\cdots \to H^0(U\cap V,\mathbf{C}^1(\mathscr{F})) \xrightarrow{\delta} H^2(X,\mathscr{F}) \to \cdots.$$

One checks that δ factors through a map

$$H^1(U\cap V,\mathscr{F}) \xrightarrow{\bar{\delta}} H^2(X,\mathscr{F})$$

such that

$$\ker(\bar{\delta}) = \mathrm{im}[H^1(U,\mathscr{F})\oplus H^1(V,\mathscr{F})].$$

Then repeat with $\mathbf{C}^i(\mathscr{F})$ in place of $\mathscr{F}$. □

Exercises

4.5.3. Let A be an abelian group. Use Mayer–Vietoris to prove that $H^1(S^1,A)\cong A$.

4.5.4. Show that $H^1(S^n,A)=0$ if $n\geq 2$.

4.5.5. Let us say that X has finite-dimensional cohomology if $\sum \dim H^i(X,\mathbb{R}) < \infty$. These dimensions are called the *Betti numbers*. Then we define the topological *Euler characteristic* $e(X)=\sum(-1)^i \dim H^i(X,\mathbb{R})$. Show that if $X=U\cup V$ and U and V have finite-dimensional cohomology, then the same is true of X, and furthermore $e(X)=e(U)+e(V)-e(U\cap V)$.

4.5.6. Let S be the topological space associated with a finite simplicial complex (jump ahead to Chapter 7 for the definition if necessary). Prove that $e(S)$ is the alternating sum of the number of simplices.

4.5.7. Let $X=X_1\cup X_2$ be a union of algebraic varieties such that $X_1\cap X_2$ consists of exactly two points. Prove that $H^1(X,\mathbb{Z})=\mathbb{Z}$ and $H^i(X,\mathbb{Z})=0$ for $i>1$. (Use Exercise 4.2.10.)

4.5.8. Complete the proof of Theorem 4.5.1.

4.5.9. Complete the proof of Theorem 4.5.2.

4.6 Products*

Since this section is a bit technical, it may be better to skip it on first reading.

Suppose that we have sheaves $\mathscr{F}$ and $\mathscr{G}$ of k-vector spaces, where k is a field. The tensor product $\mathscr{F} \otimes_k G$ is the sheafification of the presheaf

$$U \mapsto \mathscr{F}(U) \otimes_k \mathscr{G}(U).$$

So in particular, we have a product map

$$\Gamma(X,\mathscr{F}) \otimes_k \Gamma(X,\mathscr{G}) \to \Gamma(X,\mathscr{F} \otimes_k \mathscr{G}). \tag{4.6.1}$$

Our goal is to extend this to higher cohomology. The restriction to sheaves of vector spaces is not essential, but it simplifies the discussion. Let $\otimes = \otimes_k$ below.

We want a canonical family of maps

$$\cup : H^i(X,\mathscr{F}) \otimes H^j(X,\mathscr{G}) \to H^{i+j}(X,\mathscr{F} \otimes \mathscr{G}),$$

referred to collectively as the cup product, coinciding with the product in (4.6.1) when $i = j = 0$. This is not saying much, since the zero map will satisfy this (for $i \neq 0 \neq j$). So we impose a stronger condition. Suppose

$$0 \to \mathscr{F}_1 \to \mathscr{F}_2 \to \mathscr{F}_3 \to 0$$

is exact. Then

$$0 \to \mathscr{F}_1 \otimes \mathscr{G} \to \mathscr{F}_2 \otimes \mathscr{G} \to \mathscr{F}_3 \otimes \mathscr{G} \to 0$$

remains exact, because we are tensoring sheaves of vector spaces. Let

$$\delta : H^i(X,\mathscr{F}_3) \to H^{i+1}(X,\mathscr{F}_1),$$

$$\delta : H^j(X,\mathscr{F}_3 \otimes \mathscr{G}) \to H^{j+1}(X,\mathscr{F}_1 \otimes \mathscr{G}),$$

denote the connecting maps. Then we require that

$$\delta(\alpha \cup \beta) = \delta(\alpha) \cup \beta \tag{4.6.2}$$

hold for all choices. We make a similar requirement that

$$\delta(\alpha \cup \beta) = (-1)^i \alpha \cup \delta(\beta) \tag{4.6.3}$$

when $\mathscr{F}$ is fixed, and $\mathscr{G}$ is replaced by an exact sequence.

Theorem 4.6.1. *There exists a unique family of products satisfying the conditions* (4.6.2) *and* (4.6.3).

Proof. Since the proof is rather long, we treat only the low-degree cases in detail. For $i = 1, j = 0$, we proceed as follows. There is a natural morphism $\mathbf{G}(\mathscr{F}) \otimes \mathscr{G} \to \mathbf{G}(\mathscr{F} \otimes \mathscr{G})$ given by $(f_x)_x \otimes g \mapsto (f_x \otimes g_x)_x$. Therefore, we get a map

$$\mu_{10} : \mathbf{C}^1(\mathscr{F}) \otimes \mathscr{G} \to \mathbf{C}^1(\mathscr{F} \otimes \mathscr{G})$$

making the diagram

$$\begin{array}{ccccccccc}
0 & \longrightarrow & \mathscr{F} \otimes \mathscr{G} & \longrightarrow & \mathbf{G}(\mathscr{F}) \otimes \mathscr{G} & \longrightarrow & \mathbf{C}^1(\mathscr{F}) \otimes \mathscr{G} & \longrightarrow & 0 \\
 & & \downarrow = & & \downarrow & & \downarrow & & \\
0 & \longrightarrow & \mathscr{F} \otimes \mathscr{G} & \longrightarrow & \mathbf{G}(\mathscr{F} \otimes \mathscr{G}) & \longrightarrow & \mathbf{C}^1(\mathscr{F} \otimes \mathscr{G}) & \longrightarrow & 0
\end{array}$$

commute. Recalling that

$$H^1(X, \mathscr{F}) = \frac{\Gamma(X, \mathbf{C}^1(\mathscr{F}))}{\operatorname{im} \Gamma(X, \mathbf{G}(\mathscr{F}))},$$

$$H^1(X, \mathscr{F} \otimes \mathscr{G}) = \frac{\Gamma(X, \mathbf{C}^1(\mathscr{F} \otimes \mathscr{G}))}{\operatorname{im} \Gamma(X, \mathbf{G}(\mathscr{F} \otimes \mathscr{G}))},$$

it follows that the product map

$$\Gamma(X, \mathbf{C}^1(\mathscr{F})) \otimes \Gamma(X, \mathscr{G}) \to \Gamma(X, \mathbf{C}^1(\mathscr{F} \otimes \mathscr{G}))$$

determines a map on first cohomology

$$\cup : H^1(X, \mathscr{F}) \otimes H^0(X, \mathscr{G}) \to H^1(X, \mathscr{F} \otimes \mathscr{G}).$$

The construction for $i = 1, j = 0$ proceeds in the same way by first constructing a map

$$\mu_{01} : \mathscr{F} \otimes \mathbf{C}^1(\mathscr{G}) \to \mathbf{C}^1(\mathscr{F} \otimes \mathscr{G}).$$

Turning to $i = j = 1$, we have a commutative diagram

$$\begin{array}{ccc}
\mathbf{C}^1(\mathscr{F}) \otimes \mathbf{C}^1(\mathscr{G}) & \longrightarrow & \mathbf{C}^1(\mathscr{F} \otimes \mathbf{C}^1(\mathscr{G})) \\
\downarrow & \searrow^{\mu_{11}} & \downarrow \\
\mathbf{C}^1(\mathbf{C}^1(\mathscr{F}) \otimes \mathscr{G}) & \longrightarrow & \mathbf{C}^2(\mathscr{F} \otimes \mathscr{G})
\end{array} \tag{4.6.4}$$

where the sides are induced by μ_{01} and μ_{10}. The diagonal map μ_{11} induces a product

$$\Gamma(X, \mathbf{C}^1(\mathscr{F})) \otimes \Gamma(X, \mathbf{C}^1(\mathscr{G})) \to \Gamma(X, \mathbf{C}^2(\mathscr{F} \otimes \mathscr{G})).$$

Using the commutativity of the diagram (4.6.4), we readily verify that the induced map

$$\cup : H^1(X,\mathscr{F}) \otimes H^1(X,\mathscr{G}) \to H^2(X,\mathscr{F}\otimes\mathscr{G})$$

is well defined.

This can now be iterated to obtain a series of maps

$$\mu_{ij} : \mathbf{C}^i(\mathscr{F}) \otimes \mathbf{C}^j(\mathscr{G}) \to \mathbf{C}^{i+j}(\mathscr{F}\otimes\mathscr{G}).$$

The cup product can now be constructed in general using this map and the previous cases. For example, for $i > 1, j > 0$, we use the composition

$$\begin{aligned} H^1(X,\mathbf{C}^{i-1}(\mathscr{F})) \otimes H^1(X,\mathbf{C}^{j-1}(\mathscr{G})) &\to H^2(X,\mathbf{C}^{i-1}(\mathscr{F})\otimes\mathbf{C}^{j-1}(\mathscr{G})) \\ &\to H^2(X,\mathbf{C}^{i+j-2}(\mathscr{F}\otimes\mathscr{G})) \\ &\cong H^{i+j}(X,\mathscr{F}\otimes\mathscr{G})). \end{aligned}$$

The boundary formula (4.6.2) comes down to a diagram chase. We treat just the first case, which contains all the main ideas. We recall that

$$\delta : H^0(X,\mathscr{F}_3) \to H^1(X,\mathscr{F}_1),$$

$$\delta : H^i(X,\mathscr{F}_3 \otimes_k \mathscr{G}) \to H^{i+1}(X,\mathscr{F}_1 \otimes_k \mathscr{G}),$$

are constructed from the boundary operators given by the snake lemma (4.2.1) applied to

$$\begin{array}{ccccccccc} 0 & \longrightarrow & \Gamma(\mathbf{G}(\mathscr{F}_1)) & \longrightarrow & \Gamma(\mathbf{G}(\mathscr{F}_2)) & \longrightarrow & \Gamma(\mathbf{G}(\mathscr{F}_3)) & \longrightarrow & 0 \\ & & \downarrow & & \downarrow & & \downarrow & & \\ 0 & \longrightarrow & \Gamma(\mathbf{C}^1(\mathscr{F}_1)) & \longrightarrow & \Gamma(\mathbf{C}^1(\mathscr{F}_2)) & \longrightarrow & \Gamma(\mathbf{C}^1(\mathscr{F}_3)) & & \end{array} \tag{4.6.5}$$

and

$$\begin{array}{ccccccccc} 0 & \longrightarrow & \Gamma(\mathbf{G}(\mathscr{F}_1\otimes\mathscr{G})) & \longrightarrow & \Gamma(\mathbf{G}(\mathscr{F}_2\otimes\mathscr{G})) & \longrightarrow & \Gamma(\mathbf{G}(\mathscr{F}_3\otimes\mathscr{G})) & \longrightarrow & 0 \\ & & \downarrow & & \downarrow & & \downarrow & & \\ 0 & \longrightarrow & \Gamma(\mathbf{C}^1(\mathscr{F}_1\otimes\mathscr{G})) & \longrightarrow & \Gamma(\mathbf{C}^1(\mathscr{F}_2\otimes\mathscr{G})) & \longrightarrow & \Gamma(\mathbf{C}^1(\mathscr{F}_3\otimes\mathscr{G})) & & \end{array} \tag{4.6.6}$$

respectively. Tensoring (4.6.5) with $\Gamma(\mathscr{G})$ yields a new diagram (4.6.5)$\otimes\Gamma(\mathscr{G})$ with exact rows. The associated boundary operator is seen to be

$$\delta \otimes id_{\Gamma(\mathscr{G})} : H^0(X,\mathscr{F}_3) \otimes \Gamma(\mathscr{G}) \to H^1(X,\mathscr{F}_1) \otimes \Gamma(\mathscr{G}).$$

The diagram (4.6.5)$\otimes\Gamma(\mathscr{G})$ maps to (4.6.6) by the cup product map. The resulting compatibility of boundary operators means that formula (4.6.2) holds.

Now we treat a special case of uniqueness. Suppose that $\cup$ is a family of products satisfying the conditions of the theorem. Given $\alpha \in H^1(X,\mathscr{F})$ and $\beta \in H^0(X,\mathscr{G})$, let us calculate $\alpha \cup \beta$. Since the boundary operator

$$\delta : H^0(X, \mathbf{C}^1(\mathscr{F})) \to H^1(X,\mathscr{F})$$

is surjective, we can write $\alpha = \delta(\gamma)$, so that

$$\alpha \cup \beta = \delta(\gamma \cup \beta) = \delta(\gamma \otimes \beta).$$

The commutivity of the diagram

$$\begin{array}{ccc} H^0(\mathbf{C}^1(\mathscr{F})\otimes\mathscr{G}) & \xrightarrow{\delta} & H^1(\mathscr{F}\otimes\mathscr{G}) \\ \downarrow{\scriptstyle \mu_{10}} & & \downarrow{\scriptstyle =} \\ H^0(\mathbf{C}^1(\mathscr{F}\otimes\mathscr{G})) & \xrightarrow{\delta} & H^1(\mathscr{F}\otimes\mathscr{G}) \end{array}$$

shows that

$$\alpha \cup \beta = \delta(\mu_{10}(\gamma \otimes \beta))$$

as in the original construction. □

The theorem gives an axiomatic characterization of products that allows comparison with the more flexible construction given in [45, Chapter II, §6]. As a corollary, we see that $H^*(X,k)$ carries a product, which is known to make this into a graded commutative associative algebra.

Exercises

4.6.2. Verify the commutivity of (4.6.4).

4.6.3. Check that the product with values in $H^*(X,\mathscr{F}\otimes\mathscr{G}\otimes\mathscr{L})$ is associative.

4.6.4. Finish the proof of uniqueness.

Chapter 5
De Rham Cohomology of Manifolds

In this chapter, we study the topology of C^∞-manifolds. We define the de Rham cohomology of a manifold, which is the vector space of closed differential forms modulo exact forms. After sheafifying the construction, we see that the de Rham complex forms a so-called acyclic resolution of the constant sheaf $\mathbb{R}$. We prove a general result that sheaf cohomology can be computed using such resolutions, and deduce a version of de Rham's theorem that de Rham cohomology is sheaf cohomology with coefficients in $\mathbb{R}$. It follows that de Rham cohomology depends only on the underlying topology. Using a different acyclic resolution that is dual to the de Rham complex, we prove Poincaré duality. This duality makes cohomology, which is normally contravariant, into a covariant theory. We devote a section to explaining these somewhat mysterious covariant maps, called Gysin maps. We end this chapter with the remarkable Lefschetz trace formula, which in principle, calculates the number of fixed points for a map of a manifold to itself.

A systematic development of topology from the de Rham point of view is given in Bott and Tu [14].

5.1 Acyclic Resolutions

We start by reviewing some standard notions from homological algebra.

Definition 5.1.1. A complex of (sheaves of) abelian groups is a possibly infinite sequence

$$\cdots \to F^i \xrightarrow{d_i} F^{i+1} \xrightarrow{d_{i+1}} \cdots$$

of (sheaves of) groups and homomorphisms satisfying $d_{i+1}d_i = 0$.

These conditions guarantee that $\mathrm{im}(d_i) \subseteq \ker(d_{i+1})$. We denote a complex by $F^\bullet$, and we often suppress the indices on d. The ith *cohomology* of $F^\bullet$ is defined by

$$\mathscr{H}^i(F^\bullet) = \frac{\ker(d_i)}{\mathrm{im}(d_{i-1})}.$$

(We reserve the regular font "H" for sheaf cohomology.) These groups are zero precisely when the complex is exact.

Definition 5.1.2. A sheaf $\mathscr{F}$ is called acyclic if $H^i(X,\mathscr{F}) = 0$ for all $i > 0$.

For example, flasque sheaves and soft sheaves on a metric space are acyclic.

Definition 5.1.3. An acyclic resolution of a sheaf $\mathscr{F}$ is an exact sequence

$$0 \to \mathscr{F} \to \mathscr{F}^0 \to \mathscr{F}^1 \to \cdots$$

of sheaves such that each $\mathscr{F}^i$ is acyclic.

A functor between abelian categories, such as Ab or Ab(X), need not take exact sequences to exact sequences, but it will always take complexes to complexes. In particular, given a complex of sheaves $\mathscr{F}^\bullet$, the sequence

$$\Gamma(X,\mathscr{F}^0) \to \Gamma(X,\mathscr{F}^1) \to \cdots$$

is necessarily a complex of abelian groups.

Theorem 5.1.4. *Given an acyclic resolution $\mathscr{F}^\bullet$ of $\mathscr{F}$, we have*

$$H^i(X,\mathscr{F}) \cong \mathscr{H}^i(\Gamma(X,\mathscr{F}^\bullet)).$$

Proof. Let $\mathscr{K}^{-1} = \mathscr{F}$ and $\mathscr{K}^i = \ker(\mathscr{F}^{i+1} \to \mathscr{F}^{i+2})$ for $i \geq 0$. Then there are exact sequences

$$0 \to \mathscr{K}^{i-1} \to \mathscr{F}^i \to \mathscr{K}^i \to 0$$

for $i \geq 0$. Since each $\mathscr{F}^i$ is acyclic, Theorem 4.2.3 implies that

$$0 \to H^0(\mathscr{K}^{i-1}) \to H^0(\mathscr{F}^i) \to H^0(\mathscr{K}^i) \to H^1(\mathscr{K}^{i-1}) \to 0 \tag{5.1.1}$$

is exact, and

$$H^j(\mathscr{K}^i) \cong H^{j+1}(\mathscr{K}^{i-1}) \tag{5.1.2}$$

for $j > 0$. We have a diagram

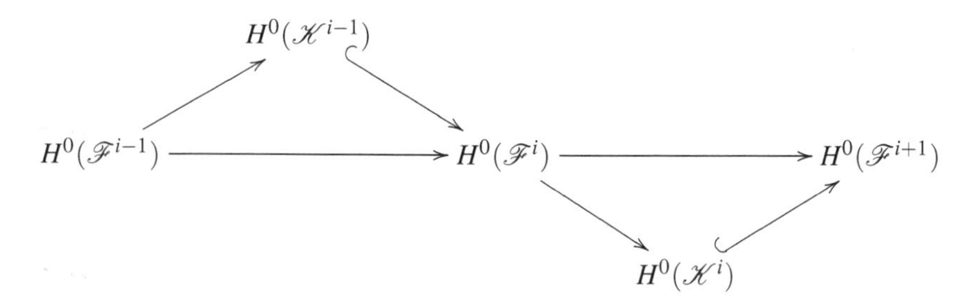

which is commutative, since the morphism $\mathscr{F}^{i-1} \to \mathscr{F}^i$ factors through $\mathscr{K}^{i-1}$ and so on. The oblique line in the diagram is part of (5.1.1), so it is exact. In particular, the first hooked arrow is injective. The injectivity of the second hooked arrow follows for similar reasons. Thus

$$\mathrm{im}[H^0(\mathscr{F}^i) \to H^0(\mathscr{K}^i)] = \mathrm{im}[H^0(\mathscr{F}^i) \to H^0(\mathscr{F}^{i+1})]. \tag{5.1.3}$$

Suppose that $\alpha \in H^0(\mathscr{F}^i)$ maps to 0 in $H^0(\mathscr{F}^{i+1})$. Then it maps to 0 in $H^0(\mathscr{K}^i)$. Therefore α lies in the image of $H^0(\mathscr{K}^{i-1})$. Thus

$$H^0(\mathscr{K}^{i-1}) = \ker[H^0(\mathscr{F}^i) \to H^0(\mathscr{F}^{i+1})]. \tag{5.1.4}$$

This already implies the theorem when $i = 0$. Replacing i by $i+1$ in (5.1.4) and combining it with (5.1.1) and (5.1.3) shows that

$$H^1(\mathscr{K}^{i-1}) \cong \frac{H^0(\mathscr{K}^i)}{\mathrm{im}[H^0(\mathscr{F}^i) \to H^0(\mathscr{K}^i)]} = \frac{\ker[H^0(\mathscr{F}^{i+1}) \to H^0(\mathscr{F}^{i+2})]}{\mathrm{im}[H^0(\mathscr{F}^i) \to H^0(\mathscr{F}^{i+1})]}.$$

Combining this with the isomorphisms

$$H^{i+1}(\mathscr{F}) = H^{i+1}(\mathscr{K}^{-1}) \cong H^i(\mathscr{K}^0) \cong \cdots \cong H^1(\mathscr{K}^{i-1})$$

of (5.1.2) proves the theorem for positive exponents. □

Example 5.1.5. Let $\mathscr{F}$ be a sheaf. Using the notation from Section 4.2, define $\mathbf{G}^i(\mathscr{F}) = \mathbf{G}(\mathbf{C}^i(\mathscr{F}))$. We define $d : \mathbf{G}^i(\mathscr{F}) \to \mathbf{G}^{i+1}(\mathscr{F})$ as the composition of the natural maps $\mathbf{G}^i(\mathscr{F}) \to \mathbf{C}^{i+1}(\mathscr{F}) \to \mathbf{G}^{i+1}(\mathscr{F})$. This can be seen to give an acyclic resolution of $\mathscr{F}$.

Exercises

5.1.6. Check that $\mathbf{G}^\bullet(\mathscr{F})$ gives an acyclic resolution of $\mathscr{F}$.

5.1.7. A sheaf $\mathscr{I}$ is called *injective* if given a monomorphism of sheaves $\mathscr{A} \to \mathscr{B}$, any morphism $\mathscr{A} \to \mathscr{I}$ extends to a morphism of $\mathscr{B} \to \mathscr{I}$. Show that an injective module is flasque and hence acyclic. (Hint: given an open set $U \subseteq X$, let $\mathbb{Z}_U = \ker[\mathbb{Z}_X \to \mathbb{Z}_{X-U}]$; check that $\mathrm{Hom}(\mathbb{Z}_U, \mathscr{F}) = \mathscr{F}(U)$.) Conclude that if $0 \to \mathscr{F} \to \mathscr{I}^0 \to \mathscr{I}^1 \to \cdots$ is an injective resolution, then $H^i(X, \mathscr{F}) = \mathscr{H}^i(\Gamma(X, \mathscr{I}^\bullet))$; this is usually taken as the definition of H^i.

5.1.8. A morphism of complexes is a collection of maps $\mathscr{F}^i \to \mathscr{G}^i$ commuting with the differentials d. This would induce a map on cohomology. Suppose that $\mathscr{F} \to \mathscr{F}^0 \to \cdots$ and $\mathscr{G} \to \mathscr{G}^0 \to \cdots$ are acyclic resolutions of sheaves $\mathscr{F}$ and $\mathscr{G}$, and suppose that we have a morphism $\mathscr{F} \to \mathscr{G}$ that extends to a morphism of the resolutions. Show that we can choose the isomorphisms so that the diagram

$$\begin{array}{ccc} H^i(\mathscr{F}) & \cong & \mathscr{H}^i(\Gamma(\mathscr{F}^\bullet)) \\ \downarrow & & \downarrow \\ H^i(\mathscr{G}) & \cong & \mathscr{H}^i(\Gamma(\mathscr{G}^\bullet)) \end{array}$$

commutes.

5.2 De Rham's Theorem

Let X be a C^∞ manifold and $\mathscr{E}^k = \mathscr{E}^k_X$ the sheaf of k-forms on it. Note that $\mathscr{E}^0_X = C^\infty_X$. If $U \subset X$ is a coordinate neighborhood with coordinates $x_1, \dots, x_n$, then $\mathscr{E}^k(U)$ is a free $C^\infty(U)$-module with basis

$$\{dx_{i_1} \wedge \cdots \wedge dx_{i_k} \mid i_1 < \cdots < i_k\}.$$

Theorem 5.2.1. *There exist canonical $\mathbb{R}$-linear maps $d : \mathscr{E}^k_X \to \mathscr{E}^{k+1}_X$, called exterior derivatives, satisfying the following:*

(a) $d : \mathscr{E}^0_X \to \mathscr{E}^1_X$ is the operation introduced in Section 2.6.
(b) $d^2 = 0$.
(c) $d(\alpha \wedge \beta) = d\alpha \wedge \beta + (-1)^i \alpha \wedge d\beta$ for all $\alpha \in \mathscr{E}^i(X)$, $\beta \in \mathscr{E}^j(X)$.
(d) If $g : Y \to X$ is a C^∞ map, $g^* \circ d = d \circ g^*$.

Proof. A complete proof can be found in almost any book on manifolds (e.g., [110, 117]). We will only sketch the construction. When $U \subset X$ is a coordinate neighborhood with coordinates x_i, we can see that there is a unique operation satisfying the above rules (a) and (c) given by

$$d\left(\sum_{i_1<\cdots<i_k} f_{i_1\dots i_k} dx_{i_1} \wedge \cdots \wedge dx_{i_k}\right) = \sum_{i_1<\cdots<i_k} \sum_j \frac{\partial f_{i_1\dots i_k}}{\partial x_j} dx_j \wedge dx_{i_1} \wedge \cdots \wedge dx_{i_k}.$$

By uniqueness, these local d's patch to define an operator on X. Taking the derivative again yields

$$\begin{aligned}
&\sum_{i_1\dots} \sum_{j,\ell} \frac{\partial^2 f_{i_1\dots i_k}}{\partial x_j \partial x_\ell} dx_j \wedge dx_\ell \wedge dx_{i_1} \wedge \cdots dx_{i_k} \\
&= \sum_{i_1\dots} \sum_{j<\ell} \left(\frac{\partial^2 f_{i_1\dots i_k}}{\partial x_j \partial x_\ell} - \frac{\partial^2 f_{i_1\dots i_k}}{\partial x_\ell \partial x_j}\right) dx_j \wedge dx_\ell \wedge dx_{i_1} \wedge \cdots dx_{i_k}, \\
&= 0
\end{aligned}$$

which proves (b). □

When $X = \mathbb{R}^3$, d can be realized as the div, grad, curl of vector calculus. The theorem tells us that $\mathscr{E}^\bullet(X)$ forms a complex, called the *de Rham complex*.

Definition 5.2.2. The de Rham cohomology groups (actually vector spaces) of X are defined by

$$H^k_{dR}(X) = \mathscr{H}^k(\mathscr{E}^\bullet(X)).$$

A differential form α is called *closed* if $d\alpha = 0$ and *exact* if $\alpha = d\beta$ for some β. Elements of de Rham cohomology are equivalence classes $[\alpha]$ represented by closed forms, where two closed forms are equivalent if they differ by an exact form.

Given a C^∞ map of manifolds $g: Y \to X$, we get a map $g^*: \mathscr{E}^*(X) \to \mathscr{E}^*(Y)$ of the de Rham complexes that induces a map g^* on cohomology. We easily have the following lemma:

Lemma 5.2.3. *$X \mapsto H^i_{dR}(X)$ is a contravariant functor from manifolds to real vector spaces.*

We compute the de Rham cohomology of Euclidean space.

Theorem 5.2.4 (Poincaré's lemma). *For all n and $k > 0$,*

$$H^k_{dR}(\mathbb{R}^n) = 0.$$

Proof. Assume, for the inductive hypothesis, that the theorem holds for $n-1$. Consider the maps $p: \mathbb{R}^n \to \mathbb{R}^{n-1}$ and $\iota: \mathbb{R}^{n-1} \to \mathbb{R}^n$ defined by $p(x_1, x_2, \dots, x_n) = (x_2, \dots, x_n)$ and $\iota(x_2, \dots, x_n) = (0, x_2, \dots, x_n)$. Let $R = (\iota \circ p)^*$. More explicitly, $R: \mathscr{E}^k(\mathbb{R}^n) \to \mathscr{E}^k(\mathbb{R}^n)$ is the $\mathbb{R}$-linear operator defined by

$$\begin{aligned} &R(f(x_1,\dots,x_n)dx_{i_1} \wedge \dots \wedge dx_{i_k}) \\ &\quad = \begin{cases} f(0,x_2,\dots,x_n)dx_{i_1} \wedge \dots \wedge dx_{i_k} & \text{if } 1 \notin \{i_1, i_2, \dots\}, \\ 0 & \text{otherwise}, \end{cases} \end{aligned}$$

where we always choose $i_1 < i_2 < \cdots$. The image of R can be identified with $p^*\mathscr{E}^k(\mathbb{R}^{n-1})$. Note that R commutes with d. Therefore if $\alpha \in \mathscr{E}^k(\mathbb{R}^n)$ is closed, $dR\alpha = Rd\alpha = 0$. By the induction assumption, $R\alpha$ is exact.

For each k, define a linear map $h: \mathscr{E}^k(\mathbb{R}^n) \to \mathscr{E}^{k-1}(\mathbb{R}^n)$ by

$$h(f(x_1,\dots,x_n)dx_{i_1} \wedge \dots \wedge dx_{i_k}) = \begin{cases} (\int_0^{x_1} f dx_1)dx_{i_2} \wedge \dots \wedge dx_{i_k} & \text{if } i_1 = 1, \\ 0 & \text{otherwise.} \end{cases}$$

Then the fundamental theorem of calculus shows that $dh + hd = I - R$, where I is the identity. (In other words, h is homotopy from I to R.) Given $\alpha \in \mathscr{E}^k(\mathbb{R}^n)$ satisfying $d\alpha = 0$, we have $\alpha = dh\alpha + R\alpha$, which is exact. □

We have an inclusion of the sheaf of locally constant functions $\mathbb{R}_X \subset \mathscr{E}^0_X$. This is precisely the kernel of $d: \mathscr{E}^0_X \to \mathscr{E}^1_X$.

Theorem 5.2.5. *The sequence*

$$0 \to \mathbb{R}_X \to \mathscr{E}^0_X \to \mathscr{E}^1_X \to \cdots$$

is an acyclic resolution of $\mathbb{R}_X$.

Proof. Any ball is diffeomorphic to Euclidean space, and any point on a manifold has a fundamental system of such neighborhoods. Therefore, Poincaré's lemma implies that the above sequence is exact on stalks, and hence exact.

By Corollary 4.4.5, the sheaves $\mathscr{E}^k$ are soft, hence acyclic. □

Corollary 5.2.6 (De Rham's theorem). *There is an isomorphism*

$$H^k_{dR}(X) \cong H^k(X,\mathbb{R}).$$

In particular, de Rham cohomology depends only on the underlying topological space.

Recall that by our convention, $H^k(X,\mathbb{R})$ is $H^k(X,\mathbb{R}_X)$. Later on, we will work with complex-valued differential forms. Essentially the same argument shows that $H^*(X,\mathbb{C})$ can be computed using such forms.

Exercises

5.2.7. We will say that a manifold is of *finite type* if it has a finite open cover $\{U_i\}$ such that any nonempty intersection of the U_i are diffeomorphic to the ball. Compact manifolds are known to have finite type [110, pp. 595–596]. Using Mayer–Vietoris and de Rham's theorem, prove that if X is an n-dimensional manifold of finite type, then $H^k(X,\mathbb{R})$ vanishes for $k > n$, and is finite-dimensional otherwise.

5.2.8. Let X be a manifold, and let t be the coordinate along $\mathbb{R}$ in $\mathbb{R}\times X$. Consider the maps $\iota : X \to \mathbb{R}\times X$ and $p : \mathbb{R}\times X \to X$ given by $x \mapsto (0,x)$ and $(t,x) \mapsto x$. Since $(p\circ\iota) = \mathrm{id}$, conclude that $\iota^* : H^i_{dR}(\mathbb{R}\times X) \to H^i_{dR}(X)$ is surjective.

5.2.9. Continuing the notation from the previous exercise, let $R : \mathcal{E}^k(\mathbb{R}\times X) \to \mathcal{E}^k(\mathbb{R}\times X)$ be the operator $(i\circ p)^*$, and let $h : \mathcal{E}^k(\mathbb{R}\times X) \to \mathcal{E}^{k-1}(\mathbb{R}\times X)$ be the operator that is integration with respect to dt (as in the proof of the Poincaré lemma). Show that $dh + hd = I - R$. Use this to show that R induces the identity on $H^i_{dR}(\mathbb{R}\times X)$. Conclude that $\iota^* : H^i_{dR}(\mathbb{R}\times X) \to H^i_{dR}(X)$ is also injective, and therefore an isomorphism.

5.2.10. Show that $\mathbb{C} - \{0\}$ is diffeomorphic to $\mathbb{R}\times S^1$, and conclude that $H^1_{dR}(\mathbb{C} - \{0\})$ is one-dimensional. Show that $\mathrm{Re}(\frac{dz}{iz}) = \frac{-ydx+xdy}{x^2+y^2}$ generates it.

5.2.11. Let S^n denote the n-dimensional sphere. Use Mayer–Vietoris with respect to the cover $U = S^n - \{\text{north pole}\}$ and $V = S^n - \{\text{south pole}\}$ to compute $H^*(S^n,\mathbb{R})$. (Hint: show that $U\cap V \cong S^{n-1}\times\mathbb{R}$.)

5.3 Künneth's Formula

Suppose that X is a C^∞ manifold. If $\alpha \in \mathcal{E}^i(X)$ and $\beta \in \mathcal{E}^j(X)$ are closed forms, then $\alpha\wedge\beta$ is also closed, by Theorem 5.2.1. The *cup product* of the associated cohomology classes is defined by $[\alpha]\cup[\beta] = [\alpha\wedge\beta]$. This is a well-defined operation that makes de Rham cohomology into a graded ring. An extension of de Rham's theorem shows that this operation is also a topological invariant.

Theorem 5.3.1 (Multiplicative de Rham's theorem). *Under the de Rham isomorphism, the product given above coincides with the cup product in sheaf cohomology constructed in Section 4.6.*

We outline the argument, concentrating on those parts that will be needed later. First, we need a more convenient method for computing cup products. Given complexes of (sheaves of) vector spaces $(A^\bullet, d_A)$ and $(B^\bullet, d_B)$ over a field k, their tensor product is the complex

$$(A^\bullet \otimes B^\bullet)^n = \bigoplus_{i+j=n} A^i \otimes B^j$$

with differential

$$d(a \otimes b) = d_A a \otimes b + (-1)^i a \otimes d_B b, \quad a \in A^i, b \in B^j.$$

The cohomology of this is easily computed by the following result:

Theorem 5.3.2 (Algebraic Künneth formula). *If $A^\bullet$ and $B^\bullet$ are complexes of vector spaces, then*

$$H^n((A^\bullet \otimes B^\bullet)^\bullet) \cong \bigoplus_{i+j=n} H^i(A^\bullet) \otimes H^j(B^\bullet),$$

where the map (from right to left) is induced by the inclusion of $\ker d_A \otimes \ker d_B \subset \ker d$.

Proof. A proof can be found in [108, Chapter 5 §3, Lemma 1; §4] for instance. □

The next lemma is left as an exercise.

Lemma 5.3.3. *The tensor product of two soft sheaves of vector spaces is soft.*

Lemma 5.3.4. *If $\mathscr{F} \to \mathscr{F}^\bullet$ and $\mathscr{G} \to \mathscr{G}^\bullet$ are soft resolutions of sheaves of vector spaces, then $\mathscr{F} \otimes \mathscr{G} \to (\mathscr{F}^\bullet \otimes \mathscr{G}^\bullet)^\bullet$ is again a soft resolution.*

Proof. The previous lemma shows that the sheaves $(\mathscr{F}^\bullet \otimes \mathscr{G}^\bullet)^\bullet$ are soft. To see that it resolves $\mathscr{F} \otimes \mathscr{G}$, use Theorem 5.3.2 to obtain

$$\mathscr{H}^i((\mathscr{F}^\bullet \otimes \mathscr{G}^\bullet)^\bullet_x) = \begin{cases} \mathscr{F}_x \otimes \mathscr{G}_x & \text{if } i = 0, \\ 0 & \text{otherwise.} \end{cases}$$ □

Choosing soft resolutions $\mathscr{F} \to \mathscr{F}^\bullet$ and $\mathscr{G} \to \mathscr{G}^\bullet$, we have a morphism of complexes

$$(\Gamma(\mathscr{F}^\bullet) \otimes \Gamma(\mathscr{G}^\bullet))^\bullet \to \Gamma((\mathscr{F}^\bullet \otimes \mathscr{G}^\bullet))^\bullet,$$

which induces a map on cohomology. The cohomology on the left decomposes into a sum of tensor products of the cohomology of $\mathscr{F}$ and $\mathscr{G}$. Thus on each summand we get a map

$$H^i(X, \mathscr{F}) \otimes H^j(X, \mathscr{G}) \to H^{i+j}(X, \mathscr{F} \otimes \mathscr{G}). \tag{5.3.1}$$

Lemma 5.3.5. *The map in (5.3.1) coincides with the product defined in Section 4.6.*

Proof. This hinges on the fact that the product given in (5.3.1) is well defined and satisfies the axioms of Theorem 4.6.1 by [45, pp. 255–259]. □

We can now sketch the proof of Theorem 5.3.1.

Proof. By the previous lemmas and [45, Chapter II, Theorem 6.6.1], it suffices to observe that the diagram

$$\begin{array}{ccc} \mathbb{R}\otimes\mathbb{R} & \xrightarrow{\sim} & (\mathscr{E}_X^\bullet \otimes \mathscr{E}_X^\bullet)^\bullet \\ \downarrow = & & \downarrow \wedge \\ \mathbb{R} & \xrightarrow{\sim} & \mathscr{E}_X^\bullet \end{array}$$

commutes. □

We can adapt these arguments to deduce a more geometric version of Künneth's formula.

Theorem 5.3.6 (Künneth formula). *Let X and Y be C^∞ manifolds. Then the product $Z = X \times Y$ is also a C^∞ manifold. Let $p : Z \to X$ and $q : Z \to Y$ denote the projections. Then the map*

$$\sum \alpha_i \otimes \beta_j \mapsto \sum p^*\alpha_i \cup q^*\beta_j$$

induces an isomorphism

$$\bigoplus_{i+j=k} H^i_{dR}(X) \otimes_{\mathbb{R}} H^j_{dR}(Y) \cong H^k_{dR}(Z).$$

Proof. The proof involves the sheaves $p^*\mathscr{E}^i_X \otimes_{\mathbb{R}} q^*\mathscr{E}^j_Y$. Their sections on basic opens are

$$p^*\mathscr{E}^i_X \otimes q^*\mathscr{E}^j_Y(U \times V) = \mathscr{E}^i_X(U) \otimes \mathscr{E}^j_Y(V).$$

These map to $\mathscr{E}^{i+j}_Z(U \times V)$ under $\kappa(\alpha \otimes \beta) = p^*\alpha \wedge q^*\beta$. Locally constant functions lie in $p^*\mathscr{E}^0_X \otimes q^*\mathscr{E}^0_Y$. Thus we have a commutative triangle

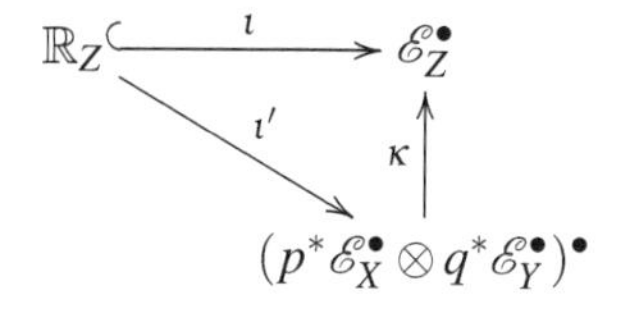

The map ι is a soft resolution, as we saw earlier. An argument similar to the proof of Lemma 5.3.4 shows that ι' is also a soft resolution. Therefore the map κ induces an isomorphism in cohomology. □

Exercises

5.3.7. Check that $[\alpha] \cup [\beta] = [\alpha \wedge \beta]$ yields a well-defined product on $H^*_{dR}(X)$.

5.3.8. Prove Lemma 5.3.3.

5.3.9. Let $e(X) = \sum(-1)^i \dim H^i(X,\mathbb{R})$ denote the Euler characteristic. Prove that $e(X \times Y) = e(X)e(Y)$.

5.3.10. Show that the cohomology ring of a torus $T = (\mathbb{R}/\mathbb{Z})^n$ is isomorphic to the exterior algebra on $\mathbb{R}^n$.

5.4 Poincaré Duality

Let X be a C^∞ manifold. Let $\mathcal{E}^k_c(X)$ denote the set of C^∞ k-forms with compact support. Clearly $d\mathcal{E}^k_c(X) \subset \mathcal{E}^{k+1}_c(X)$, so these form a complex.

Definition 5.4.1. Compactly supported de Rham cohomology is defined by $H^k_{cdR}(X) = \mathcal{H}^k(\mathcal{E}^\bullet_c(X))$.

Lemma 5.4.2. *For all n,*

$$H^k_{cdR}(\mathbb{R}^n) = \begin{cases} \mathbb{R} \text{ if } k = n, \\ 0 \text{ otherwise.} \end{cases}$$

Proof. [14, Corollary 4.7.1]. □

This computation suggests that these groups are roughly opposite to the usual de Rham groups. There is another piece of evidence, which is that H_{cdR} behaves covariantly in certain cases. For example, given an open set $U \subset X$, a form in $\mathcal{E}^k_c(U)$ can be extended by zero to $\mathcal{E}^k_c(X)$. This induces a map $H^k_{cdR}(U) \to H^k_{cdR}(X)$.

The precise statement of duality requires the notion of orientation. An *orientation* on an n-dimensional real vector space V is a connected component of $\wedge^n V - \{0\}$ (there are two). An ordered basis $v_1, \ldots, v_n$ is positively oriented if $v_1 \wedge \cdots \wedge v_n$ lies in the given component. If V were to vary, there is no guarantee that we could choose an orientation consistently. So we make a definition:

Definition 5.4.3. An n-dimensional manifold X is called orientable if $\wedge^n T_X$ minus its zero section has two components. If this is the case, an orientation is a choice of one of these components.

Theorem 5.4.4 (Poincaré duality, version I). *Let X be a connected oriented n-dimensional manifold. Then*

$$H^k_{cdR}(X) \cong H^{n-k}(X,\mathbb{R})^*.$$

There is a standard proof of this using currents, which are to forms what distributions are to functions. However, we can get by with something much weaker. We define the space of *pseudocurrents* of degree k on an open set $U \subset X$ to be

$$\mathscr{C}^k(U) = \mathscr{E}_c^{n-k}(U)^* := \mathrm{Hom}(\mathscr{E}_c^{n-k}(U), \mathbb{R}).$$

This is "pseudo" because we are using the ordinary (as opposed to topological) dual. We make this into a presheaf as follows. Given $V \subseteq U$, $\alpha \in \mathscr{C}_X^k(U)$, $\beta \in \mathscr{E}_c^{n-k}(V)$, define $\alpha|_V(\beta) = \alpha(\tilde{\beta})$, where $\tilde{\beta}$ is the extension of β by 0.

Lemma 5.4.5. *$\mathscr{C}_X^k$ is a sheaf.*

Proof. Let $\{U_i\}$ be an open cover of U, which we may assume is locally finite. Suppose that $\alpha_i \in \mathscr{C}_X^k(U_i)$ is a collection of sections such that $\alpha_i|_{U_i \cap U_j} = \alpha_j|_{U_i \cap U_j}$. This means that $\alpha_i(\beta) = \alpha_j(\beta)$ if β has support in $U_i \cap U_j$. Let $\{\rho_i\}$ be a C^∞ partition of unity subordinate to $\{U_i\}$ (see §4.3). Then define $\alpha \in \mathscr{C}_X^k(U)$ by

$$\alpha(\beta) = \sum_i \alpha_i(\rho_i \beta|_{U_i}).$$

We have to show that $\alpha(\tilde{\beta}) = \alpha_j(\beta)$ for any $\beta \in \mathscr{E}_c^{n-k}(U_j)$ with $\tilde{\beta}$ its extension to U by 0. The support of $\rho_i \tilde{\beta}$ lies in $U_i \cap \mathrm{supp}(\beta) \subset U_i \cap U_j$, so only finitely many of these are nonzero. Therefore

$$\alpha(\tilde{\beta}) = \sum_i \alpha_i(\rho_i \tilde{\beta}) = \sum_i \alpha_j(\rho_i \tilde{\beta}) = \alpha_j(\beta),$$

as required. We leave it to the reader to check that α is the unique current with this property. □

Define a map $\delta : \mathscr{C}_X^k(U) \to \mathscr{C}_X^{k+1}(U)$ by $\delta(\alpha)(\beta) = (-1)^{k+1}\alpha(d\beta)$. One automatically has $\delta^2 = 0$. Thus we have a complex of sheaves.

Let X be an oriented n-dimensional manifold. Then we will recall [109] that one can define an integral $\int_X \alpha$ for any n-form $\alpha \in \mathscr{E}_c^n(X)$. Using a partition of unity, the definition can be reduced to the case that α is supported in a coordinate neighborhood U. Then we can write $\alpha = f(x_1, \ldots, x_n) dx_1 \wedge \cdots \wedge dx_n$, where the order of the coordinates is chosen so that $\partial/\partial x_1, \ldots, \partial/\partial x_n$ gives a positive orientation of T_X. Then

$$\int_X \alpha = \int_{\mathbb{R}^n} f(x_1, \ldots, x_n) dx_1 \cdots dx_n.$$

The functional $\int_X$ defines a canonical global section of $\mathscr{C}_X^0$.

Theorem 5.4.6 (Stokes's theorem). *Let X be an oriented n-dimensional manifold; then $\int_X d\beta = 0$.*

Proof. See [109]. □

Corollary 5.4.7. $\int_X \in \ker[\delta]$.

We define a map $\mathbb{R}_X \to \mathscr{C}_X^0$ by sending r to $r\int_X$. The key lemma to establish Theorem 5.4.4 is the following:

Lemma 5.4.8.
$$0 \to \mathbb{R}_X \to \mathscr{C}_X^0 \to \mathscr{C}_X^1 \to \cdots$$
is an acyclic resolution.

Proof. Lemma 5.4.2 implies that this complex is exact. Given $f \in C^\infty(U)$ and $\alpha \in \mathscr{C}^k(U)$, define
$$f\alpha(\beta) = \alpha(f\beta).$$
This makes $\mathscr{C}^k$ into a C^∞-module, and it follows that it is soft and therefore acyclic. □

We can now prove Theorem 5.4.4.

Proof. We can use the complex $\mathscr{C}_X^\bullet$ to compute the cohomology of $\mathbb{R}_X$ to obtain
$$H^i(X,\mathbb{R}) \cong \mathscr{H}^i(\mathscr{C}_X^\bullet(X)) = \mathscr{H}^i(\mathscr{E}_c^{n-\bullet}(X)^*).$$
The right-hand space is isomorphic to $H^i_{cdR}(X,\mathbb{R})^*$. This completes the proof of the theorem. □

Corollary 5.4.9. *If X is a compact oriented n-dimensional manifold, then*
$$H^k(X,\mathbb{R}) \cong H^{n-k}(X,\mathbb{R})^*.$$

The following is really a corollary of the proof.

Corollary 5.4.10. *If X is a connected oriented n-dimensional manifold, then the map* $\alpha \mapsto \int_X \alpha$ *induces an isomorphism*
$$\int_X : H^n_{cdR}(X,\mathbb{R}) \cong \mathbb{R}.$$

We can make the Poincaré duality isomorphism more explicit:

Theorem 5.4.11 (Poincaré duality, version II). *If* $f \in H^{n-k}_{cdR}(X)^*$, *then there exists a closed form* $\alpha \in \mathscr{E}^k(X)$ *such that* $f([\beta]) = \int_X \alpha \wedge \beta$. *Moreover, the class* $[\alpha] \in H^k_{dR}(X)$ *is unique.*

Proof. Define
$$P : \mathscr{E}_X^k \to \mathscr{C}_X^k$$
by
$$P(\alpha)(\beta) = \int_U \alpha \wedge \beta$$
for $\alpha \in \mathscr{E}^k(U)$ and $\beta \in \mathscr{E}_c^{n-k}(U)$. With the help of Stokes's theorem, we see that $\delta P(\alpha) = P(d\alpha)$. Therefore, P gives a morphism of complexes of sheaves. Note also that $P(1) = \int_X$. Thus we have a morphism of resolutions

$$\begin{array}{ccc} \mathbb{R}_X & \to & \mathscr{E}_X^\bullet \\ || & & \downarrow \\ \mathbb{R}_X & \to & \mathscr{C}_X^\bullet \end{array}$$

So the theorem follows from Exercise 5.1.8. □

Corollary 5.4.12. *The cup product (induced by $\wedge$) followed by integration gives a nondegenerate pairing*

$$H^k_{dR}(X) \times H^{n-k}_{cdR}(X) \to H^n_{cdR}(X) \cong \mathbb{R}.$$

Here is a simple example to illustrate this.

Example 5.4.13. Consider the torus $T = \mathbb{R}^n/\mathbb{Z}^n$. We will show later, in Section 8.2, that every de Rham cohomology class on T contains a unique form with constant coefficients. This will imply that there is an algebra isomorphism $H^*(T,\mathbb{R}) \cong \wedge^*\mathbb{R}^n$. Poincaré duality becomes the standard isomorphism

$$\wedge^k\mathbb{R}^n \cong \wedge^{n-k}\mathbb{R}^n.$$

Exercises

5.4.14. Prove that the Euler characteristic $\sum(-1)^i \dim H^i(X,\mathbb{R})$ is zero when X is an odd-dimensional compact orientable manifold.

5.4.15. If X is a connected oriented n-dimensional manifold, show that

$$H^n(X,\mathbb{R}) \cong \begin{cases} \mathbb{R} & \text{if } X \text{ is compact,} \\ 0 & \text{otherwise.} \end{cases}$$

5.4.16.(a) Let $S^2 \subset \mathbb{R}^3$ denote the unit sphere. Show that

$$\alpha = xdy \wedge dz + ydz \wedge dx + zdx \wedge dy$$

generates $H^2_{dR}(S^2)$.

(b) The real projective plane is defined by $\mathbb{RP}^2 = S^2/i$, where $i(x,y,z) = -(x,y,z)$. This is a compact manifold. Show that $H^2_{dR}(\mathbb{RP}^2) = 0$ by identifying it with the i^*-invariant part of $H^2_{dR}(S^2)$, and conclude that it cannot be orientable.

5.4.17. Assuming the exercises of §4.3, prove that $H^i_c(X,\mathbb{R}) \cong H^i_{cdR}(X)$.

5.5 Gysin Maps

Let $f : Y \to X$ be a C^∞ map of compact oriented manifolds of dimension m and n respectively. Then we have a natural map

$$f^* : H^k_{dR}(X) \to H^k_{dR}(Y)$$

given by pulling back forms. By Poincaré duality, we can identify this with a map

$$H^{n-k}_{dR}(X)^* \to H^{m-k}_{dR}(Y)^*.$$

Dualizing and changing variables yields a map in the opposite direction,

$$f_! : H^k_{dR}(Y) \to H^{k+n-m}_{dR}(X),$$

called the Gysin homomorphism. This is characterized by

$$\int_X f_!(\alpha) \cup \beta = \int_Y \alpha \cup f^*(\beta). \tag{5.5.1}$$

Our goal is to give a more explicit description of this map. Notice that we can factor f as the inclusion of the graph $Y \to Y \times X$ given by $y \mapsto (y, f(y))$, followed by a projection $Y \times X \to X$. Therefore we only need to study what happens in these two special cases.

5.5.1 Projections

Suppose that $Y = X \times Z$ is a product of compact connected oriented manifolds. Let $p : Y \to X$ and $q : Y \to Z$ be the projections. Let $r = m - n = \dim Z$. Choose local coordinates $x_1, \ldots, x_n$ on X and $z_1, \ldots, z_r$ on Z. *Integration along the fiber* $\int_p : \mathcal{E}^k(Y) \to \mathcal{E}^{k-r}(X)$ is defined in local coordinates by

$$\sum f_{i_1,\ldots,i_{k-n}}(x_1, \ldots, x_n, z_1, \ldots, z_r) dz_1 \wedge \cdots \wedge dz_n \wedge dx_{i_1} \wedge \cdots \wedge dx_{i_{k-r}} \mapsto$$
$$\sum \left(\int f_{i_1,\ldots,i_{k-n}}(x_1, \ldots, x_n, z_1, \ldots, z_r) dz_1 \cdots dz_n \right) dx_{i_1} \wedge \cdots \wedge dx_{i_{k-r}}.$$

Note that $\int_p \alpha = 0$ if none of its terms contains $dz_1 \wedge \cdots \wedge dz_n$.

Lemma 5.5.1. *$p_! \alpha$ is represented by $\int_p \alpha$.*

Proof. Fubini's theorem in calculus gives

$$\int_Y \alpha \wedge p^* \beta = \int_X \left(\int_p \alpha \right) \wedge \beta, \tag{5.5.2}$$

so that $\int_p$ satisfies (5.5.1). □

The cohomology of Y is the tensor product of the cohomology of X and Z by the Künneth formula, Theorem 5.3.6. The Gysin map $p_!$ is simply the projection onto one of the Künneth factors,

$$H^k_{dR}(Y) \to H^{k+n-m}_{dR}(X) \otimes H^{m-n}_{dR}(Z) \cong H^{k+n-m}_{dR}(X).$$

5.5.2 Inclusions

Now suppose that $i : Y \hookrightarrow X$ is an inclusion of a closed submanifold. We need the following:

Theorem 5.5.2. *There exists an open neighborhood* Tube_Y *of* Y *in* X*, called a tubular neighbourhood. This possesses a* C^∞ *map* $\pi : \mathrm{Tube}_Y \to Y$ *that makes* Tube_Y *a locally trivial bundle over* Y*, with fibers diffeomorphic to* $\mathbb{R}^{n-m}$.

Proof. The details can be found in [110, Chapter 9, addendum]. However, we give a brief outline, since we will need to understand a bit about the construction later on. We choose a Riemannian metric on X. This amounts to a family of inner products on the tangent spaces of X; among other things, this allows one to define the length of a curve. The Riemannian distance between two points is the infimum of the lengths of curves joining the points. This is a metric in the sense of point set topology. Tube_Y is given by the set of points with Riemannian distance less than ε from Y for $0 < \varepsilon \ll 1$. In order to see the bundle structure, we give an alternative description. We can take the normal bundle N to be the fiberwise orthogonal complement to the tangent bundle T_Y in $T_X|_Y$. N inherits a Riemannian metric, and we let $\mathrm{Tube}'_Y \subset N$ be the set of points of distance less than ε from the zero section. Given a point $(y,v) \in N$, let $\gamma_{y,v}(t)$ be the geodesic emanating from y with velocity v. Then the map $(y,v) \mapsto \gamma_{y,v}(1)$ defines a diffeomorphism from Tube'_Y to Tube_Y. □

Then the map i^* can be factored as

$$H^{m-k}_{dR}(X) \to H^{m-k}_{dR}(\mathrm{Tube}_Y) \xrightarrow{\sim} H^{m-k}_{dR}(Y).$$

The second map is an isomorphism, since the fibers of π are contractible. Dualizing, we see that $i_!$ is a composition of

$$H^{k}_{dR}(Y) \xrightarrow{\sim} H^{k+n-m}_{cdR}(\mathrm{Tube}_Y) \to H^{k+n-m}_{dR}(X).$$

The first map is called the *Thom isomorphism.* The second map can be seen to be extension by zero. To get more insight into this, let $k = 0$. Then $H^0_{dR}(Y)$ has a natural generator, which is the constant function 1_Y with value 1. Under the Thom isomorphism, this maps to a class $\tau_Y \in H^{n-m}_{cdR}(\mathrm{Tube}_Y)$, called the *Thom class*. This can be represented by (any) differential form with compact support in Tube_Y, which integrates to 1 on the fibers of π. The Thom isomorphism is given explicitly by $\alpha \mapsto \pi^*\alpha \cup \tau_Y$. So to summarize:

Lemma 5.5.3. $i_!\alpha$ *is the extension of* $\pi^*\alpha \cup \tau_Y$ *to* X *by zero.*

Exercises

5.5.4. Let $j : U \to X$ be the inclusion of an open set in a connected oriented manifold. Check that the Poincaré dual of the restriction map $j^* : H^*_{dR}(X) \to H^*_{dR}(U)$ is given by extension by zero.

5.5.5. Let $\pi : \mathrm{Tube}_Y \to Y$ be a tubular neighborhood for $i : Y \hookrightarrow X$ as above. Prove that $i^*\beta = \int_\pi \tau_Y \cup \beta$ for $\beta \in \mathscr{E}^\bullet(X)$, where $\int_\pi$ is defined as above.

5.5.6. With the help of the previous exercise and (the appropriate extension of) (5.5.2), finish the proof of Lemma 5.5.3.

5.5.7. Prove the projection formula $f_!(f^*(\alpha) \cup \beta) = \alpha \cup f_!\beta$.

5.6 Fundamental Class

We can use Gysin maps to construct interesting cohomology classes. Let $i : Y \hookrightarrow X$ be a closed connected oriented m-dimensional submanifold of an n-dimensional oriented manifold.

Definition 5.6.1. The fundamental class of Y in X is $[Y] = i_! 1_Y \in H^{n-m}_{dR}(X)$.

Equivalently, $[Y]$ is the extension of τ_Y by zero. Under the duality isomorphism, $H^0_{dR}(Y) \cong H^m_{dR}(Y)^*$, 1 goes to the functional

$$\beta \mapsto \int_Y \beta,$$

and this maps to

$$\alpha \mapsto \int_Y i^*\alpha$$

in $H^m(X)^*$. Composing this with the isomorphism $H^m(X)^* \cong H^{n-m}(X)$ yields the basic relation

$$\int_Y i^*\alpha = \int_X [Y] \cup \alpha. \tag{5.6.1}$$

Let $Y, Z \subset X$ be oriented submanifolds such that $\dim Y + \dim Z = n$. Then under the duality isomorphism, $[Y] \cup [Z] \in H^n(X, \mathbb{R}) \cong \mathbb{R}$ corresponds to a number $Y \cdot Z$, called the *intersection number*. This has a geometric interpretation that we give under an extra transversality assumption that holds "most of the time." We say that Y and Z are *transverse* if $Y \cap Z$ is finite and if $T_{Y,p} \oplus T_{Z,p} = T_{X,p}$ for each p in the intersection.

Definition 5.6.2. Let Y and Z be transverse, and let $p \in Y \cap Z$. Choose ordered bases $v_1(p), \dots, v_m(p) \in T_{Y,p}$ and $v_{m+1}(p), \dots, v_n(p) \in T_{Z,p}$ that are positively oriented with respect to the orientations of Y and Z. The local intersection number at p is

$$i_p(Y,Z) = \begin{cases} +1 & \text{if } v_1(p), \dots, v_m(p), v_{m+1}(p), \dots, v_n(p) \\ & \quad \text{is a positively oriented basis of } T_{X,p}, \\ -1 & \text{otherwise.} \end{cases}$$

(This is easily seen to be independent of the choice of bases.)

Proposition 5.6.3. *If Y and Z are transverse, then $Y\cdot Z = \sum_{p\in Y\cap Z} i_p(Y,Z)$.*

Proof. Let $m = \dim Y$. Let U_p be a collection of disjoint coordinate neighborhoods for each $p \in Y\cap Z$. (Note that these U_p will be replaced by smaller neighborhoods whenever necessary.) Choose coordinates $x_1,\dots,x_n$ around each p such that Y is given by $x_{m+1} = \cdots = x_n = 0$ and Z by $x_1 = \cdots = x_m = 0$. Next construct suitable tubular neighborhoods $\pi : T \to Y$ of Y and $\pi' : T' \to Z$ of Z. Recall that these neighborhoods depend on a choice of Riemannian metric and radii ε,ε'. By choosing the radii small enough, we can guarantee that $T\cap T'$ lies in the union $\bigcup U_p$. Also, by modifying the metric to be Euclidean near each p, we can assume that π is locally the projection $(x_1,\dots,x_n)\mapsto(x_1,\dots,x_m)$, and likewise for π'.

Then with the above assumptions,

$$Y\cdot Z = \int_X \tau_Y\wedge\tau_Z = \sum_p\int_{U_p}\tau_Y\wedge\tau_Z,$$

where τ_Y and τ_Z are forms representing the Thom classes of T and T'. We will assume that U_p is a ball and hence diffeomorphic to $\mathbb{R}^n$. We can view $\tau_Y|_{U_p}$ as defining a class in

$$H^0_{dR}(\mathbb{R}^m)\otimes H^{n-m}_{cdR}(\mathbb{R}^{n-m}) \cong H^{n-m}_{cdR}(\mathbb{R}^{n-m})$$

and similarly for $\tau_Z|_{U_p}$. Thus we can see that

$$\begin{aligned}\tau_Y|_{U_p} &= f(x_{m+1},\dots,x_n)dx_{m+1}\wedge\cdots\wedge dx_n + d\eta,\\ \tau_Z|_{U_p} &= g(x_1,\dots,x_m)dx_1\wedge\cdots\wedge dx_m + d\xi,\end{aligned}$$

with f and g compactly supported such that

$$\int_{\mathbb{R}^{n-m}} f(x_{m+1},\dots,x_n)dx_{m+1}\cdots dx_n = \int_{\mathbb{R}^m} g(x_1,\dots,x_m)dx_1\cdots dx_m = 1.$$

Fubini's theorem and Stokes's theorem then give

$$\int_{U_p}\tau_Y\wedge\tau_Z = i_p(Y,Z).$$ □

The proposition implies that the intersection number is an integer for transverse intersections. In fact, this is always true. There are a couple of ways to see this. One is by proving that intersections can always be made transverse without altering the intersection numbers. A simpler explanation is that the fundamental classes can actually be defined to take values in integral cohomology $H^*(X,\mathbb{Z})$. Moreover, we have a cup product pairing as indicated:

$$H^k(X,\mathbb{Z})\times H^{n-k}(X,\mathbb{Z}) \to H^n(X,\mathbb{Z}) \cong \mathbb{Z}.$$

The classes that we have defined are images under the natural map $H^*(X,\mathbb{Z}) \to H^*(X,\mathbb{R})$.

Example 5.6.4. Let $T = \mathbb{R}^n/\mathbb{Z}^n$, let $\{e_i\}$ be the standard basis of $\mathbb{R}^n$, and let x_i be coordinates on $\mathbb{R}^n$. If $V_I \subset \mathbb{R}^n$ is the span of $\{e_i \mid i \in I\}$, then $T_I = V_I/(\mathbb{Z}^n \cap V_I)$ is a submanifold of T. Its fundamental class is $dx_{i_1} \wedge \cdots \wedge dx_{i_d}$, where $i_1 < \cdots < i_d$ are the elements of I in increasing order. If J is the complement of I, then T_I and T_J will meet transversally at one point. Therefore $T_I \cdot T_J = \pm 1$.

Example 5.6.5. Consider complex projective space $\mathbb{P}^n_{\mathbb{C}}$. This is the basic example for us, and it will be studied further in Section 7.2. For now we just state the main results. We have

$$H^i_{dR}(\mathbb{P}^n_{\mathbb{C}}) = \begin{cases} \mathbb{R} \text{ if } 0 \le i \le 2n \text{ is even,} \\ 0 \text{ otherwise.} \end{cases}$$

Given a complex subspace $V \subset \mathbb{C}^{n+1}$, the subset $\mathbb{P}(V) \subset \mathbb{P}^n$ consisting of lines lying in V forms a submanifold, which can be identified with another projective space. If W is another subspace with $\dim W = n - \dim V + 2$ and $\dim(V \cap W) = 1$, then $\mathbb{P}(V)$ and $\mathbb{P}(W)$ will meet transversally at one point. In this case, $\mathbb{P}(V) \cdot \mathbb{P}(W)$ is necessarily $+1$ (see the exercises).

Exercises

5.6.6. Show that if $Y, Z \subset X$ are transverse complex submanifolds of a complex manifold, then $i_p(Y,Z) = 1$ for each p in the intersection. Thus $Y \cdot Z$ is the number of points of intersection.

5.6.7. Check that fundamental classes of subtori of $\mathbb{R}^n/\mathbb{Z}^n$ are described as above.

5.6.8. Let $T = \mathbb{R}^2/\mathbb{Z}^2$ and let $V, W \subset \mathbb{R}^2$ be distinct lines with rational slope. Show that the images of V and W in T are transverse. Find an interpretation for their intersection number.

5.7 Lefschetz Trace Formula

Let X be a compact n-dimensional oriented manifold with a C^∞ map $f : X \to X$. The Lefschetz formula is a formula for the number of fixed points counted appropriately. This needs to be explained. Let

$$\begin{aligned} \Gamma_f &= \{(x, f(x)) \mid x \in X\}, \\ \Delta &= \{(x,x) \mid x \in X\}, \end{aligned}$$

be the graph of f and the diagonal respectively. These are both n-dimensional submanifolds of $X \times X$ that intersect precisely at points (x,x) with $x = f(x)$. We define the "number of fixed points" as $\Gamma_f \cdot \Delta$. Since this number could be negative, we need to take this with a grain of salt. If these manifolds are transverse, we see that this can be evaluated as the sum of local intersection numbers over fixed points,

$$\sum_x i_x(\Gamma_f, \Delta),$$

by Proposition 5.6.3. In particular, $\Gamma_f \cdot \Delta$ is the true number of fixed points if each local intersection number is $+1$.

Theorem 5.7.1. *The number $\Gamma_f \cdot \Delta$ is given by*

$$L(f) = \sum_p (-1)^p \operatorname{trace}[f^* : H^p(X,\mathbb{R}) \to H^p(X,\mathbb{R})].$$

Proof. The proof will be based on the elementary observation that if F is an endomorphism of a finite-dimensional vector space with basis $\{v_i\}$ and dual basis $\{v_i^*\}$, then the matrix is given by $(v_i^*(F(v_j)))$. Therefore

$$\operatorname{trace}(F) = \sum_i v_i^*(F(v_i)).$$

For each p, choose a basis $\alpha_{p,i}$ of $H^p(X)$, and let $\alpha_{p,i}^*$ denote the dual basis transported to $H^{n-p}(X)$ under the Poincaré duality isomorphism $H^{n-p}(X) \cong H^p(X)^*$, so that

$$\int_X \alpha_{p,i} \cup \alpha_{p,j}^* = \delta_{ij}.$$

Let $\pi_i : X \times X \to X$ denote the projections. Then by Künneth's formula, $\{A_{p,i,j} = \pi_1^*\alpha_{p,i} \cup \pi_2^*\alpha_{p,j}^*\}_{p,i,j}$ and $\{A_{p,i,j}^* = (-1)^{n-p}\pi_1^*\alpha_{p,i}^* \cup \pi_2^*\alpha_{p,j}\}$ both give bases for $H^n(X \times X)$, which are dual to this in the sense that

$$\int_{X\times X} A_{p,i,j} \cup A_{p',i',j'}^* = \delta_{(p,i,j),(p',i',j')}.$$

Thus we can express

$$[\Delta] = \sum c_{p,i,j} A_{p,i,j}.$$

The coefficients can be computed by integrating against the dual basis:

$$c_{p,i,j} = \int_{X\times X} [\Delta] \cup A_{p,i,j}^* = \int_\Delta A_{p,i,j}^* = (-1)^{n-p} \int_X \alpha_{p,i} \cup \alpha_{p,j}^* = (-1)^{n-p}\delta_{ij}.$$

Therefore

$$[\Delta] = \sum_{i,p} (-1)^{n-p} \pi_1^*\alpha_{p,i} \cup \pi_2^*\alpha_{p,i}^*. \tag{5.7.1}$$

Consequently,

$$\begin{aligned}
\Gamma_f \cdot \Delta &= \int_{\Gamma_f} [\Delta] \\
&= \sum_p (-1)^{n-p} \sum_i \int_{\Gamma_f} \pi_1^* \alpha_{p,i} \cup \pi_2^* \alpha_{p,i}^* \\
&= \sum_p (-1)^{n-p} \sum_i \int_X \alpha_{p,i} \cup f^* \alpha_{p,i}^* \\
&= \sum_p (-1)^{n-p} \operatorname{trace}[f^* : H^{n-p}(X,\mathbb{R}) \to H^{n-p}(X,\mathbb{R})] \\
&= L(f).
\end{aligned}$$

□

Corollary 5.7.2. *If $L(f) \neq 0$, then f has a fixed point.*

Proof. If $\Gamma_f \cap \Delta = \emptyset$, then $\Gamma_f \cdot \Delta = 0$. □

Corollary 5.7.3. *$\Delta \cdot \Delta$ is the Euler characteristic $e(X)$.*

Exercises

5.7.4. We say that two C^∞ maps $f,g : X \to Y$ between manifolds are homotopic if there exists a C^∞ map $h : X \times \mathbb{R} \to Y$ such that $f(x) = h(x,0)$ and $g(x) = h(x,1)$. Using Exercise 5.2.9, show that $f^* = g^*$ if f and g are homotopic. Conclude that g has a fixed point if $L(f) \neq 0$.

5.7.5. Let $v(x)$ be C^∞ vector field on a compact manifold X. By the existence and uniqueness theorem for ordinary differential equations, there is an $\varepsilon > 0$ such that for each x there is unique curve $\gamma_x : [0,\varepsilon] \to X$ with $\gamma_x(0) = x$ and $d\gamma_x(t) = v(\gamma_x(t))$. Moreover, the map $x \mapsto \gamma_x(\delta)$ is a diffeomorphism from X to itself for every $\delta \leq \varepsilon$. Use this to show that v must have a zero if $e(X) \neq 0$.

5.7.6. Let A be a nonsingular $n \times n$ matrix. Then it acts on $\mathbb{P}^{n-1}$ by $[v] \mapsto [Av]$, and the fixed points correspond to eigenvectors. Show that A is homotopic to the identity. Use this to show that $L(A) \neq 0$, and therefore that A has an eigenvector. Deduce the fundamental theorem of algebra from this.

Chapter 6
Riemann Surfaces

Riemann surfaces are the same thing as one-dimensional complex manifolds or nonsingular complex curves. Although the basic concept goes back to Riemann, the rigorous definition appears to be due to Weyl [122]. We already considered these briefly in the first chapter. But now we are in a position to make a more thorough study using the geometric and homological tools introduced in the intervening chapters. Nevertheless, we will only scratch the surface of this rich subject. Further details can be found in [5, 20, 49, 38] in addition to Weyl's book.

6.1 Genus

The coarsest classification is topological. A Riemann surface can be regarded as a manifold of real dimension 2. It has a canonical orientation: if we identify the real tangent space at any point with the complex tangent space, then for any nonzero vector v, we declare the ordered basis (v, iv) to be positively oriented. Let us now forget the complex structure and consider the purely topological problem of classifying these surfaces up to homeomorphism.

Given two connected 2-dimensional topological manifolds X and Y with points $x \in X$ and $y \in Y$, we can form a new topological manifold $X\#Y$, called the connected sum. To construct this, choose open disks $D_1 \subset X$ and $D_2 \subset Y$. Then $X\#Y$ is obtained by gluing $X - D_1 \cup S^1 \times [0,1] \cup Y - D_2$ appropriately. The homeomorphism class is independent of the choices made. Figure 6.1 depicts the connected sum of two tori.

Theorem 6.1.1. *A compact connected orientable 2-dimensional topological manifold is classified, up to homeomorphism, by a nonnegative integer called the genus. A genus-0 surface is homeomorphic to the 2-sphere S^2. A manifold of genus $g > 0$ is homeomorphic to a connected sum of the 2-torus and a surface of genus $g - 1$.*

Proof. An equivalent formulation is that every such manifold is homeomorphic to a sphere with g handles. A classical reference for this is [99, p. 145]. □

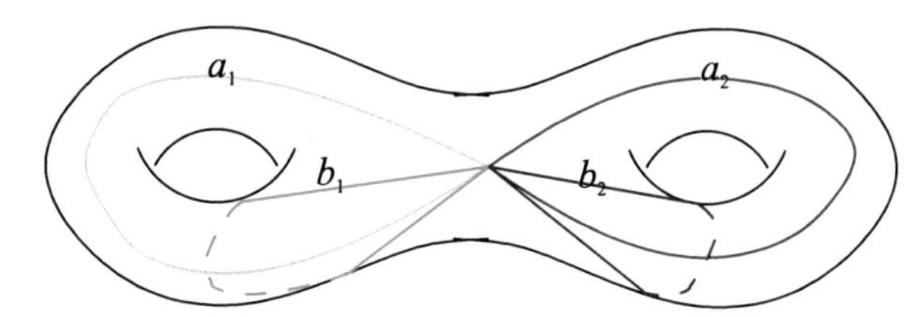

Fig. 6.1 Genus-2 surface.

There is another standard model for these surfaces [99] that is also quite useful (for instance for computing the fundamental group). A genus-g surface can be constructed by gluing the sides of a $2g$-gon. It is probably easier to visualize this in reverse. After cutting the genus-2 surface of Figure 6.1 along the indicated curves, it can be opened up to an octagon (see Figure 6.2).

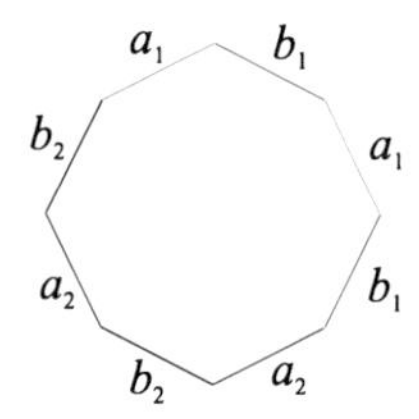

Fig. 6.2 Genus-2 surface cut open.

The topological Euler characteristic of the space X is

$$e(X) = \sum(-1)^i \dim H^i(X, \mathbb{R}).$$

From Exercise 4.5.5, we have the following lemma:

Lemma 6.1.2. *If X is a union of two open sets U and V, then $e(X) = e(U) + e(V) - e(U \cap V)$.*

Corollary 6.1.3. *If X is a manifold of genus g, then $e(X) = 2 - 2g$, and the first Betti number is given by $\dim H^1(X, \mathbb{R}) = 2g$.*

Proof. This will be left for the exercises. □

When $g = 2$, this gives $\dim H^1(X, \mathbb{R}) = 4$. We can find explicit generators by taking the fundamental classes of the curves a_1, a_2, b_1, b_2 in Figure 6.1, after choosing orientations. To see that these generate, $H^1(X, \mathbb{R})$, it suffices to prove that they are linearly independent. For this, consider the pairing

$$(\alpha, \beta) \mapsto \int_X \alpha \wedge \beta$$

on $H^1(X,\mathbb{R})$. This restricts to the intersection pairing on fundamental classes. After orienting the curves suitably, their intersection matrix is

$$\begin{pmatrix} 0 & 0 & 1 & 0 \\ 0 & 0 & 0 & 1 \\ -1 & 0 & 0 & 0 \\ 0 & -1 & 0 & 0 \end{pmatrix},$$

and this shows independence. A similar basis with intersection matrix

$$\begin{pmatrix} 0 & I \\ -I & 0 \end{pmatrix}$$

can be found for any g. These basis vectors will generate a lattice inside $H^1(X,\mathbb{R})$ that can be identified with $H^1(X,\mathbb{Z})$.

To answer the obvious question, we show that every genus occurs by direct construction. A much harder question is, what does a typical genus g look like? Many examples of compact Riemann surfaces can be given explicitly as nonsingular projective algebraic curves, and we start with this approach. Later, we will see that all compact Riemann surfaces arise in this way.

Example 6.1.4. Let $f(x,y,z)$ be a homogeneous polynomial of degree d. Suppose that the partials of f have no common zeros in $\mathbb{C}^3$ except $(0,0,0)$. Then the curve $V(f) = \{f(x,y,z) = 0\}$ in $\mathbb{P}^2_{\mathbb{C}}$ is nonsingular. We will see later that the genus is $\frac{(d-1)(d-2)}{2}$. In particular, not every genus occurs for these examples.

Example 6.1.5. Given a collection of homogeneous polynomials $f_i \in \mathbb{C}[x_0,\dots,x_n]$ such that $X = V(f_1, f_2, \dots) \subset \mathbb{P}^n_{\mathbb{C}}$ is a nonsingular algebraic curve, then X will be a complex submanifold of $\mathbb{P}^n_{\mathbb{C}}$ and hence a Riemann surface. By a generic projection argument, $n = 3$ suffices to give all such examples.

Example 6.1.6. Choose $2g+2$ distinct points in $a_i \in \mathbb{C}$. Consider the affine curve $C_1 \subset \mathbb{C}^2$ defined by

$$y^2 = \prod (x - a_i).$$

We can compactify this to obtain a nonsingular projective curve C by taking the projective closure and normalizing. Here is an alternative description of the same curve. Consider the affine curve C_2:

$$Y^2 = \prod (1 - a_i X).$$

Glue this to C_1 by identifying $X = x^{-1}$ and $Y = yx^{-g-1}$. Since C is nonsingular, it can be viewed as a Riemann surface. By construction, C comes equipped with a morphism $f : C \to \mathbb{P}^1_{\mathbb{C}}$ that is 2-to-1 except at the branch points $\{a_i\}$. In the exercises, it will be shown that the genus of X is g. Curves of this form are called hyperelliptic (if $g > 1$), and they are very nice to work with. However, "most" curves are not hyperelliptic.

From a more analytic point of view, we can construct many examples as quotients of $\mathbb{C}$ or the upper half-plane. In fact, the *uniformization theorem* [38] tells us that all examples other than $\mathbb{P}^1$ arise in this way.

Example 6.1.7. Let $L \subset \mathbb{C}$ be a lattice, i.e., an abelian subgroup generated by two $\mathbb{R}$-linearly independent numbers. The quotient $E = \mathbb{C}/L$ can be made into a Riemann surface (Exercise 2.2.20) called an elliptic curve. Since this is topologically a torus, the genus is 1. Conversely, any genus-1 curve is of this form.

The surface $E = \mathbb{C}/L$ can be realized as an algebraic curve in a couple of ways. As we saw in the first chapter, we can embed it into $\mathbb{P}^2$ as the cubic curve

$$zy^2 = 4x^3 - g_2(L)xz^2 - g_3(L)z^3 \tag{6.1.1}$$

by $z \mapsto [\wp(z), \wp'(z), 1]$. Alternatively, $\wp : E \to \mathbb{P}^1$ realizes this directly as a two-sheeted cover branched at four points. We can assume without loss of generality that these are $0, 1, \infty, t$. Then E is given by Legendre's equation

$$y^2 z = x(x-z)(x-tz). \tag{6.1.2}$$

Let us now consider quotients of the upper half-plane $H = \{z \,|\, Im(z) > 0\}$. The group $\mathrm{SL}_2(\mathbb{R})$ acts on H by fractional linear transformations:

$$z \mapsto \frac{az+b}{cz+d}.$$

Note that $-I$ acts trivially, so the action factors through $\mathrm{PSL}_2(\mathbb{R}) = \mathrm{SL}_2(\mathbb{R})/\{\pm I\}$. The action of subgroup $\Gamma \subset \mathrm{SL}_2(\mathbb{R})$ on H is *properly discontinuous* if every point has a neighborhood V such that $\gamma V \cap V \neq \emptyset$ for all but finitely many $\gamma \in \Gamma$. A point $x \in V$ is called a *fixed point* if its stabilizer in $\Gamma/\{\pm I\}$ is nontrivial. The action is *free* if it has no fixed points.

Proposition 6.1.8. *If Γ acts properly discontinuously on H, the quotient $X = H/\Gamma$ becomes a Riemann surface. If $\pi : H \to X$ denotes the projection, the structure sheaf is defined by $f \in \mathscr{O}_X(U)$ if and only if $f \circ \pi \in \mathscr{O}_H(\pi^{-1}U)$.*

When Γ acts freely with compact quotient, then it has genus $g > 1$. The quickest way to see this is by applying the Gauss–Bonnet theorem [56], which says that $e(H/\Gamma) = 2 - 2g$ can be computed by integrating the Gaussian curvature for any Riemannian metric. The hyperbolic metric on H, which descends to the quotient, has negative curvature. A *fundamental domain* for this action is a region $R \subset H$ such that $\cup_\gamma \gamma \bar{R} = H$ and such that two translates of $\bar{R}$ can meet only at their boundaries. The fundamental domain in this case can be chosen to be the interior of a geodesic $2g$-gon.

The modular group $\mathrm{SL}_2(\mathbb{Z})$ is a particularly important example where the action is not free.

Theorem 6.1.9. *The quotient $H/\mathrm{SL}_2(\mathbb{Z})$ is isomorphic to $\mathbb{C}$, and its points correspond to isomorphism classes of elliptic curves (which are genus-one curves with a distinguished point).*

Proof. We sketch the idea. The details can be found in [32, 106] or [103]. A fundamental domain for the action is given in Figure 6.3. The marked points in the diagram are $\rho = e^{2\pi\imath/3}, i$, and $\rho + 1$.

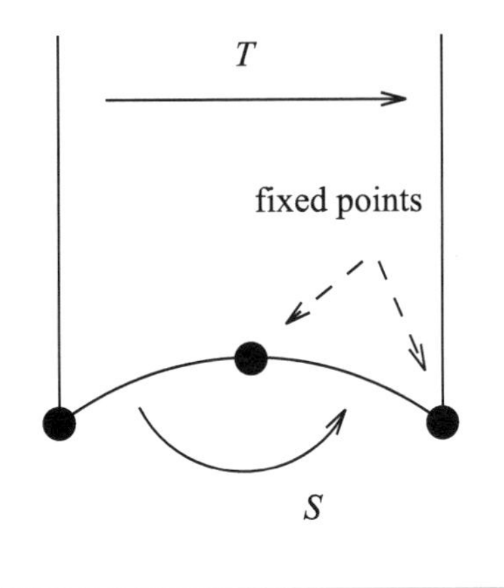

Fig. 6.3 Fundamental domain of $\mathrm{SL}_2(\mathbb{Z})$.

After identifying the sides of the domain by the transformations

$$S = \begin{pmatrix} 0 & -1 \\ 1 & 0 \end{pmatrix}, \quad T = \begin{pmatrix} 1 & 1 \\ 0 & 1 \end{pmatrix},$$

as indicated in the diagram, one sees that that the quotient $H/\mathrm{SL}_2(\mathbb{Z})$ is homeomorphic to $\mathbb{C}$. By the Riemann mapping theorem, $H/\mathrm{SL}_2(\mathbb{Z})$ is isomorphic to either $\mathbb{C}$ or H. To see that it is $\mathbb{C}$, we show that the one-point compactification $H/\mathrm{SL}_2(\mathbb{Z}) \cup \{*\}$ can also be made into a Riemann surface, necessarily isomorphic to $\mathbb{P}^1$. Any continuous function f defined in a neighborhood of $*$ can be pulled back to an $\mathrm{SL}_2(\mathbb{Z})$-invariant function $\tilde{f}$ on H. Invariance under T implies that $\tilde{f}$ is periodic; hence it can be expanded in a Fourier series

$$\tilde{f}(z) = \sum_{n=-\infty}^{\infty} a_n e^{2\pi inz}.$$

We declare f to be holomorphic if all the a_n are zero for $n < 0$. To put this another way, $q = e^{2\pi iz}$ is taken as a local analytic coordinate at $*$.

The relation to elliptic curves is given in the exercises. □

Exercises

6.1.10. Prove Corollary 6.1.3.

6.1.11. Check that the genus of the hyperelliptic curve constructed above is g, either by triangulating in such way that the $\{a_i\}$ are included in the set of vertices or by applying Lemma 6.1.2 directly to a suitable open cover.

6.1.12. Let $g(x,y,z)$ be a homogeneous polynomial of degree d such that its partials have no common zeros. Let $f_0(x,y,z)$ be a product of d linear polynomials, and let $f_t(x,y,z) = f_0 + tg$. The variety $V(f_0)$ is a union of lines L_i, and let $X = V(f_t)$ for small t. Choose small closed balls B_{ij} around each crossing point. Show that $B_{ij} \cap X$ is diffeomorphic to the cylinder $S^1 \times [0,1]$.

6.1.13. Continuing the notation from the previous exercise, the complements $X - \cup B_{ij}$ and $V(f_0) - \cup B_{ij}$ can be seen to be diffeomorphic using a version of Ehressman's Theorem 13.1.3. Granting this, we obtain the following topological model for X. Take d copies $L_1, \ldots, L_d$ of S^2, remove $d-1$ disjoint disks $D_{i,j}$ from each L_i, and join L_i to L_j by gluing a cylinder from $\partial D_{i,j}$ to $\partial D_{j,i}$ to get X. Use this to verify that its genus is $(d-1)(d-2)/2$.

6.1.14. Fix $\tau, \tau' \in H$. Check that the following are equivalent:

(a) The elliptic curves $\mathbb{C}/\mathbb{Z}+\mathbb{Z}\tau$ and $\mathbb{C}/\mathbb{Z}+\mathbb{Z}\tau'$ are isomorphic, i.e., there exists a holomorphic bijection preserving 0 and taking $\mathbb{Z}+\mathbb{Z}\tau$ to $\mathbb{Z}+\mathbb{Z}\tau'$.
(b) There exists $\lambda \in \mathbb{C}^*$ such that $\lambda(\mathbb{Z}+\mathbb{Z}\tau) = \mathbb{Z}+\mathbb{Z}\tau'$.
(c)
$$\tau' = \frac{a\tau + b}{c\tau + d}$$
with $\begin{pmatrix} a & b \\ c & d \end{pmatrix} \in \mathrm{SL}_2(\mathbb{Z})$.

6.2 $\bar{\partial}$-Cohomology

Let $U \subset \mathbb{C}$ be an open set. Let x and y be real coordinates on $\mathbb{C}$, and $z = x + \mathrm{i}y$. Given a complex-valued C^∞ function $f : U \to \mathbb{C}$, let

$$\frac{\partial f}{\partial z} = \frac{1}{2}\left(\frac{\partial f}{\partial x} - \mathrm{i}\frac{\partial f}{\partial y}\right),$$

$$\frac{\partial f}{\partial \bar{z}} = \frac{1}{2}\left(\frac{\partial f}{\partial x} + \mathrm{i}\frac{\partial f}{\partial y}\right).$$

With this notation, the Cauchy–Riemann equation is simply $\frac{\partial f}{\partial \bar{z}} = 0$. Define the complex-valued 1-forms $dz = dx + idy$ and $d\bar{z} = dx - idy$. With this notation, we can formulate Cauchy's formula for C^∞ functions.

Theorem 6.2.1. *Let $D \subset \mathbb{C}$ be a disk. If $f \in C^\infty(\bar{D})$, then*

$$f(\zeta) = \frac{1}{2\pi i}\int_{\partial D} \frac{f(z)}{z-\zeta}\,dz + \frac{1}{2\pi i}\int_D \frac{\partial f(z)}{\partial \bar{z}}\,\frac{dz\wedge d\bar{z}}{z-\zeta}.$$

Proof. This follows from Stokes's theorem; see [49, pp. 2–3]. □

The following is an analogue of the Poincaré lemma for $\bar{\partial}$.

Theorem 6.2.2. *Let $D \subset \mathbb{C}$ be an open disk. Given $f \in C^\infty(\bar{D})$, the function $g \in C^\infty(D)$ given by*

$$g(\zeta) = \frac{1}{2\pi i}\int_D \frac{f(z)}{z-\zeta}\,dz\wedge d\bar{z}$$

satisfies $\frac{\partial g}{\partial \bar{z}} = f$.

Proof. Decompose $f(z) = f_1(z) + f_2(z)$ into a sum of C^∞ functions, where $f_1(z) \equiv f(z)$ in a small neighborhood of $z_0 \in D$ and vanishes near the boundary of D. In particular, f_2 is zero in a neighborhood of z_0. Let g_1 and g_2 be the functions obtained by substituting f_1 and f_2 for f in the integral of the theorem. Differentiating under the integral sign yields

$$\frac{\partial g_2(\zeta)}{\partial \bar{\zeta}} = \frac{1}{2\pi i}\int_D \frac{\partial}{\partial \bar{\zeta}}\left(\frac{f_2(z)}{z-\zeta}\right) dz\wedge d\bar{z} = \frac{1}{2\pi i}\int_D d\left(\frac{f_2(z)dz}{z-\zeta}\right) = 0$$

for ζ close to z_0. Since f_1 is compactly supported,

$$g_1(\zeta) = \frac{1}{2\pi i}\int_{\mathbb{C}} \frac{f_1(z)}{z-\zeta}\,dz\wedge d\bar{z}.$$

Then doing a change of variables $w = z-\zeta$ yields

$$g_1(\zeta) = \frac{1}{2\pi i}\int_{\mathbb{C}} \frac{f_1(w+\zeta)}{w}\,dw\wedge d\bar{w}.$$

Thus for ζ close to z_0,

$$\begin{aligned}
\frac{\partial g(\zeta)}{\partial \bar{\zeta}} &= \frac{\partial g_1(\zeta)}{\partial \bar{\zeta}} \\
&= \frac{1}{2\pi i}\int_{\mathbb{C}} \frac{\partial f_1(w+\zeta)}{\partial \bar{\zeta}}\,\frac{dw\wedge d\bar{w}}{w} \\
&= \frac{1}{2\pi i}\int_D \frac{\partial f_1(z)}{\partial \bar{z}}\,\frac{dz\wedge d\bar{z}}{z-\zeta}.
\end{aligned}$$

Since f_1 vanishes on the boundary, the last integral equals $f_1(\zeta) = f(\zeta)$ by Cauchy's formula in Theorem 6.2.1. □

In order to make it easier to globalize the above operators to Riemann surfaces, we reinterpret them in terms of differential forms. Let $C^\infty(U)$ and $\mathscr{E}^\bullet(U)$ denote the space of complex-valued C^∞ functions and forms on $U \subseteq \mathbb{C}$ (we will continue this convention *from now on*). The exterior derivative extends to a $\mathbb{C}$-linear operator between these spaces. Set

$$\begin{aligned} \partial f &= \frac{\partial f}{\partial z} dz, \\ \bar{\partial} f &= \frac{\partial f}{\partial \bar{z}} d\bar{z}. \end{aligned} \tag{6.2.1}$$

We extend this to 1-forms by

$$\begin{aligned} \partial(f d\bar{z}) &= \frac{\partial f}{\partial z} dz \wedge d\bar{z}, \\ \partial(f dz) &= 0, \\ \bar{\partial}(f dz) &= \frac{\partial f}{\partial \bar{z}} d\bar{z} \wedge dz, \\ \bar{\partial}(f d\bar{z}) &= 0. \end{aligned} \tag{6.2.2}$$

A 1-form $\alpha = f dz$ with f holomorphic is holomorphic. This is equivalent to the condition $\bar{\partial}\alpha = 0$. The following identities can be easily verified:

$$\begin{aligned} d &= \partial + \bar{\partial}, \\ \partial^2 &= \bar{\partial}^2 = 0, \\ \partial\bar{\partial} + \bar{\partial}\partial &= 0. \end{aligned} \tag{6.2.3}$$

Let X be a Riemann surface with $\mathscr{O}_X$ and Ω_X^1 its sheaves of holomorphic functions and holomorphic 1-forms respectively. We write C_X^∞ and $\mathscr{E}_X^n$ for the sheaves of complex-valued C^∞ functions and n-forms for $n = 1, 2$. We define a C_X^∞-submodule $\mathscr{E}_X^{(1,0)} \subset \mathscr{E}_X^1$ (respectively $\mathscr{E}_X^{(0,1)} \subset \mathscr{E}_X^1$), by $\mathscr{E}_X^{(1,0)}(U) = C^\infty(U)dz$ (respectively $\mathscr{E}_X^{(0,1)}(U) = C^\infty(U)d\bar{z}$) for any coordinate neighborhood U with holomorphic coordinate z. We have a decomposition

$$\mathscr{E}_X^1 = \mathscr{E}_X^{(1,0)} \oplus \mathscr{E}_X^{(0,1)}.$$

We set $\mathscr{E}_X^{(1,1)} = \mathscr{E}_X^2$, since this is locally generated by $dz \wedge d\bar{z}$.

Lemma 6.2.3. *There exist $\mathbb{C}$-linear maps $\partial, \bar{\partial}$ on the sheaves $\mathscr{E}_X^\bullet$ that coincide with the previous expressions* (6.2.1) *and* (6.2.2) *in local coordinates.*

It follows that the identities (6.2.3) hold globally. We have inclusions $\mathscr{O}_X \subset C_X^\infty$ and $\Omega_X^1 \subset \mathscr{E}_X^{(1,0)}$.

Lemma 6.2.4. *The sequences of sheaves*

$$0 \to \mathscr{O}_X \to C_X^\infty \xrightarrow{\bar{\partial}} \mathscr{E}_X^{(0,1)} \to 0,$$

$$0 \to \Omega_X^1 \to \mathscr{E}_X^{(1,0)} \xrightarrow{\bar{\partial}} \mathscr{E}_X^{(1,1)} \to 0$$

are acyclic resolutions.

Proof. The exactness can be checked on a disk, where it follows from Theorem 6.2.2. The sheaves $C_X^\infty, \mathscr{E}_X^\bullet$ are C^∞-modules and hence soft. □

Corollary 6.2.5.

$$H^1(X, \mathscr{O}_X) = \frac{\mathscr{E}^{(0,1)}(X)}{\bar{\partial} C^\infty(X)},$$

$$H^1(X, \Omega_X^1) = \frac{\mathscr{E}^{(1,1)}(X)}{\bar{\partial}\mathscr{E}_X^{(1,0)}(X)},$$

and

$$H^i(X, \mathscr{O}_X) = H^i(X, \Omega_X^1) = 0$$

if $i > 1$.

Next, we give a holomorphic analogue of the de Rham complex.

Proposition 6.2.6. *There is an exact sequence of sheaves*

$$0 \to \mathbb{C}_X \to \mathscr{O}_X \xrightarrow{d} \Omega_X^1 \to 0.$$

Proof. The only nontrivial part of the assertion is that $\mathscr{O}_X \to \Omega_X^1$ is an epimorphism. We can check this by replacing X by a disk D. A holomorphic 1-form α on D is automatically closed; therefore $\alpha = df$ by the usual Poincaré lemma. Since df is holomorphic, $\bar{\partial} f = 0$. Therefore f is holomorphic. □

Corollary 6.2.7. *There is a long exact sequence*

$$0 \to H^0(X, \mathbb{C}) \to H^0(X, \mathscr{O}_X) \to H^0(X, \Omega_X^1) \to H^1(X, \mathbb{C}) \to \cdots.$$

Holomorphic 1-forms are closed, and

$$H^0(X, \Omega_X^1) \to H^1(X, \mathbb{C}) \tag{6.2.4}$$

is the map that sends a holomorphic form to its class in (complex-valued) de Rham cohomology.

Lemma 6.2.8. *When X is compact and connected, the map* (6.2.4) *is an injection.*

Proof. Since global holomorphic functions on X are constant (Proposition 2.7.2),

$$H^0(X,\mathbb{C}) \to H^0(X,\mathcal{O}_X)$$

is surjective. □

We will postpone the proof of the following proposition to Section 9.1.

Proposition 6.2.9. *The dimensions of $H^0(X,\Omega^1_X)$ and $H^1(X,\mathcal{O}_X)$ both coincide with the genus.*

Corollary 6.2.10. $H^1(X,\Omega^1_X) \cong \mathbb{C}$.

Proof. The proposition implies that the map

$$H^1(X,\mathbb{C}) \to H^1(X,\mathcal{O}_X)$$

is surjective, and therefore that $H^1(X,\Omega^1_X)$ is isomorphic to $H^2(X,\mathbb{C}) = \mathbb{C}$. □

Exercises

6.2.11. Check the identities (6.2.3). Show that a form $f(z)dz$ is closed if and only if it is holomorphic.

6.2.12. Let V be a vector space with a nondegenerate skew-symmetric pairing $\langle , \rangle$. A subspace $W \subset V$ is called isotropic if $\langle w, w' \rangle = 0$ for all $w, w' \in W$. Prove that $\dim W \le \dim V/2$ if W is isotropic. Let X be a compact Riemann surface of genus g. Show that $H^0(X,\Omega^1_X)$ is isotropic for the pairing $\int \alpha \wedge \beta$. Use this to conclude that dim $H^0(X,\Omega^1_X) \le g$.

6.2.13. Show that the differentials $x^i dx/y$, with $0 \le i < g$, are holomorphic on the hyperelliptic curve in Example 6.1.6. Conclude directly that $H^0(X,\Omega^1_X) = g$ in this case.

6.3 Projective Embeddings

Fix a compact Riemann surface X. We introduce some standard shorthand: $h^i =$ dim H^i and $\omega_X = \Omega^1_X$. We will follow standard practice of referring to locally free $\mathcal{O}_X$-modules of rank one as "line bundles." (As we will see, there is a one-to-one correspondence, so this not unreasonable.)

Definition 6.3.1. A divisor D on X is a finite integer linear combination $\sum n_i p_i$, where $p_i \in X$. It is effective if all the coefficients are nonnegative. The degree deg D is equal to $\sum n_i$.

Divisors form an abelian group $\mathrm{Div}(X)$ in the obvious way. For every meromorphic function defined in a neighborhood of $p \in X$, let $\mathrm{ord}_p(f)$ be the order of vanishing (or minus the order of the pole) of f at p. If D is a divisor, define $\mathrm{ord}_p(D)$ to be the coefficient of p in D (or 0 if p is absent). Define the sheaf $\mathscr{O}_X(D)$ by

$$\mathscr{O}_X(D)(U) = \{f : U \to \mathbb{C} \cup \{\infty\} \text{ meromorphic} \mid \mathrm{ord}_p(f) + \mathrm{ord}_p(D) \geq 0, \forall p \in U\}.$$

Lemma 6.3.2.

(a) $\mathscr{O}_X(D)$ *is a line bundle.*
(b) $\mathscr{O}_X(D+D') \cong \mathscr{O}_X(D) \otimes \mathscr{O}_X(D')$.

Proof. Let z be a local coordinate defined in a neighborhood U. Let $D = \sum n_i p_i + E$, where $p_i \in U$ and E is a sum of points not in U. Then it can be checked that

$$\mathscr{O}_X(D)(U) = \mathscr{O}_X(U) \frac{1}{(z-p_1)^{n_1}(z-p_2)^{n_2}\cdots},$$

and this is free of rank one. If $D' = \sum n_i' p_i + E'$ is a second divisor, then

$$[\mathscr{O}(D) \otimes \mathscr{O}(D')](U) \cong \mathscr{O}_X(U) \frac{1}{(z-p_1)^{n_1}\cdots} \frac{1}{(z-p_1)^{n_1'}\cdots} = \mathscr{O}(D+D')(U) \quad \square$$

In terminology to be discussed later, this says that $D \mapsto \mathscr{O}_X(D)$ is a homomorphism from $\mathrm{Div}(X)$ to $\mathrm{Pic}(X)$. If D is effective, $O(-D)$ is a sheaf of ideals. In particular, $\mathscr{O}_X(-p)$ is exactly the maximal ideal sheaf at p. We have an exact sequence

$$0 \to \mathscr{O}_X(-p) \to \mathscr{O}_X \to \mathbb{C}_p \to 0, \tag{6.3.1}$$

where

$$\mathbb{C}_p(U) = \begin{cases} \mathbb{C} & \text{if } p \in U, \\ 0 & \text{otherwise,} \end{cases}$$

is a so-called *skyscraper sheaf*. Observe that $\mathbb{C}_p \otimes \mathscr{L} \cong \mathbb{C}_p$ for any line bundle $\mathscr{L}$. Therefore tensoring (6.3.1) by $\mathscr{O}_X(D)$ yields

$$0 \to \mathscr{O}_X(D-p) \to \mathscr{O}_X(D) \to \mathbb{C}_p \to 0. \tag{6.3.2}$$

In the same way, we get a sequence

$$0 \to \omega_X(D-p) \to \omega_X(D) \to \mathbb{C}_p \to 0, \tag{6.3.3}$$

where $\omega_X(D) = \omega_X \otimes \mathscr{O}_X(D)$.

Lemma 6.3.3. *For all D, $H^i(X, \mathscr{O}_X(D))$ and $H^i(X, \omega_X(D))$ are finite-dimensional for $i = 0, 1$, and zero if $i > 1$.*

Proof. We prove this for $\mathscr{O}_X(D)$, since the argument for $\omega_X(D)$ is the same. Let $D=\sum n_p p$. The proof goes by induction on $d=\sum |n_p|$. The initial case $D=0$ follows from Propositions 2.7.2, 6.2.9 and Corollary 6.2.5.

Observe that $\mathbb{C}_p$ is flasque, and thus has no higher cohomology. The exact sequence (6.3.2) yields

$$0 \to H^0(\mathscr{O}_X(D-p)) \to H^0(\mathscr{O}_X(D)) \to \mathbb{C} \to H^1(\mathscr{O}_X(D-p)) \to H^1(\mathscr{O}_X(D)) \to 0$$

and isomorphisms

$$H^i(\mathscr{O}_X(D-p)) \cong H^i(\mathscr{O}_X(D))$$

for $i > 1$. By adding or subtracting a point, we can drop d by one, and use induction. □

A meromorphic 1-form is a holomorphic form on the complement of a finite set S such that it has finite-order poles at the points of S. The residue of a meromorphic 1-form α at p is

$$\mathrm{res}_p(\alpha) = \frac{1}{2\pi \mathrm{i}} \int_C \alpha,$$

where C is any loop "going once counterclockwise" around p and containing no singularities other than p. Alternatively, if $\alpha = f(z)dz$ locally for some local coordinate z at p, $\mathrm{res}_p(\alpha)$ is the coefficient of $\frac{1}{z}$ in the Laurent expansion of $f(z)$.

Lemma 6.3.4 (Residue theorem). *If α is a meromorphic 1-form, the sum of its residues is 0.*

Proof. Let $\{p_1,\dots,p_n\}$ denote the set of singular points of α. For each i, choose an open disk D_i containing p_i and no other singularity. Then by Stokes's theorem,

$$\sum \mathrm{res}_{p_i}\alpha = \frac{1}{2\pi \mathrm{i}} \int_{X-\cup D_i} d\alpha = 0.$$

□

Theorem 6.3.5. *Suppose that D is a nonzero effective divisor. Then*

(a) (Kodaira vanishing) $H^1(\omega_X(D)) = 0$.
(b) (Weak Riemann–Roch) $h^0(\omega_X(D)) = \deg D + g - 1$.

Proof. This can be proved by induction on the degree of D. We carry out the initial step here; the rest is left as an exercise. Suppose $D=p$. Then $H^0(\omega_X(p))$ consists of the space of meromorphic 1-forms α with at worst a simple pole at p and no other singularities. The residue theorem implies that such an α must be holomorphic. Therefore $H^0(\omega_X(p)) = H^0(\omega_X)$. Therefore $h^0(\omega_X(p)) = g$ as predicted by (b). By the long exact sequence of cohomology groups associated to (6.3.3), we have

$$0 \to \mathbb{C} \to H^1(\omega_X) \to H^1(\omega_X(p)) \to 0.$$

Since the space in the middle is one-dimensional by Corollary 6.2.10, $H^1(\omega_X(p)) = 0$. □

Corollary 6.3.6. *There exists a divisor K (called a canonical divisor) such that $\omega_X \cong \mathscr{O}_X(K)$. If $g > 0$, then K can be chosen to be effective.*

Proof. Given a meromorphic 1-form α, locally $\alpha = f dz$, we define $\mathrm{ord}_p(\alpha) = \mathrm{ord}_p(f)$ (this is independent of the coordinate z). A section of $H^0(\omega_X(D))$ can be interpreted as a meromorphic form α, so that $\mathrm{ord}_p\alpha + \mathrm{ord}_p D \geq 0$ for all $p \in X$. Therefore, we may choose D so that $H^0(\omega_X(D))$ possesses a nonzero section α. Then

$$K = (\alpha) - D = \sum \mathrm{ord}_p(\alpha)p - D$$

satisfies the required properties. □

The degree of K is given by the following proposition:

Proposition 6.3.7. *For any canonical divisor,* $\deg K = 2g - 2$.

The proof will be given in the exercises.

We say that a line bundle $\mathscr{L}$ on X is *globally generated* if for any point $x \in X$, there exists a section $f \in H^0(X, \mathscr{L})$ such that $f(x) \neq 0$. Suppose that this is the case. Choose a basis $f_0, \ldots, f_N$ for $H^0(X, \mathscr{L})$. If we fix an isomorphism $\tau : \mathscr{L}|_U \cong \mathscr{O}_U$, then $\tau(f_i)$ are holomorphic functions on U. Thus we get a holomorphic map $U \to \mathbb{C}^{N+1}$ given by $x \mapsto (\tau(f_i(x)))$. By our assumption, the image lies in the complement of 0, and thus descends to a map to projective space. The image is independent of τ, and hence we get a well-defined holomorphic map

$$\phi_{\mathscr{L}} : X \to \mathbb{P}^N.$$

This map has the property that $\phi_{\mathscr{L}}^* \mathscr{O}_{\mathbb{P}^N}(1) = L$. The line bundle $\mathscr{L}$ is called *very ample* if $\phi_{\mathscr{L}}$ is a closed immersion, that is, if it is an isomorphism onto $\phi_{\mathscr{L}}(X)$.

Proposition 6.3.8. *A sufficient condition for $\mathscr{L}$ to be globally generated is that $H^1(X, \mathscr{L}(-p)) = 0$ for all $p \in X$. A sufficient condition for $\mathscr{L}$ to be very ample is that $H^1(X, \mathscr{L}(-p-q)) = 0$ for all $p, q \in X$.*

Proof. We have an exact sequence

$$0 \to \mathscr{L}(-p) \to \mathscr{L} \to \mathbb{C}_p \to 0.$$

Then $H^1(\mathscr{L}(-p)) = 0$ implies that $H^0(\mathscr{L})$ surjects onto $H^0(\mathbb{C}_p)$. This implies that $\mathscr{L}$ has a global section that is nonzero at p.

To show that $\phi_{\mathscr{L}}(X)$ is a closed immersion, we need to check that it is injective and injective on tangent spaces. When $p \neq q$, the assumption $H^1(\mathscr{L}(-p-q)) = 0$ would imply that $H^0(\mathscr{L})$ surjects onto $H^0(\mathbb{C}_p \oplus \mathbb{C}_q)$. Therefore $\mathscr{L}$ has sections vanishing at p but not q and at q but not p. This shows that $\phi_{\mathscr{L}}$ is injective. If $p = q$, then we see that $H^0(\mathscr{L})$ surjects onto $H^0(\mathscr{O}_p/m_p^2)$. Therefore there is a section f_i with nonzero image in $m_p/m_p^2 = T_p^* \cong \mathbb{C}$. After identifying f_i with a function, this just means that $f_i'(p) \neq 0$. Thus the derivative of $\phi_{\mathscr{L}}$ at p is nonzero, and hence injective. □

Corollary 6.3.9. *$\omega_X(D)$ is very ample if D is nonzero and effective with $\deg D > 2$. In particular, any compact Riemann surface can be embedded into a projective space.*

We will use this much later on (Corollary 15.4.5) to prove that every compact Riemann surface is a nonsingular projective algebraic curve.

Exercises

6.3.10. Finish the proof of Theorem 6.3.5 by writing $D = p + D'$ and applying (6.3.3).

6.3.11. A divisor is called principal if it is given by $(f) = \sum \mathrm{ord}_p(f)p$ for some nonzero meromorphic function f. Prove that a principal divisor has degree 0. Conclude that any two canonical divisors have the same degree.

6.3.12. Prove Proposition 6.3.7 by checking it for $\mathbb{P}^1$, and then showing that it behaves like Riemann-Hurwitz for a branched covering $f : X \to \mathbb{P}^1$.

6.3.13. Suppose that X is a compact Riemann surface. Prove that $H^1(X, \mathscr{O}_X(-p)) \cong H^1(X, \mathscr{O}_X)$ for any $p \in X$.

6.3.14. Using the previous exercise, show that $H^1(X, \mathscr{O}(q-p)) = 0$ for any p, q if X has genus 0. Then conclude that $\mathscr{O}_X(q)$ is globally generated. Show that the corresponding map $X \to \mathbb{P}^1$ is an isomorphism.

6.3.15. The Euler characteristic is given by $\chi(\mathscr{O}_X(D)) = \dim H^0(X, \mathscr{O}_X(D)) - \dim H^1(X, \mathscr{O}_X(D))$. Show that $\chi(\mathscr{O}(D+p)) = \chi(\mathscr{O}(D)) + 1$.

6.3.16. We stated a version of the Riemann–Roch theorem above. The actual theorem says that for any divisor D on a genus-g compact Riemann surface, $\chi(\mathscr{O}_X(D)) = \deg D + 1 - g$. Prove this by induction.

6.4 Function Fields and Automorphisms

Fix a smooth projective curve X over $\mathbb{C}$. By the results of the previous section, all compact Riemann surfaces are of this form. The function field $\mathbb{C}(X)$ is the field of rational functions on X, or equivalently, by Exercise 15.4.7, the field of meromorphic functions on it. Since $\mathbb{C}(X) = \mathbb{C}(U)$ for any affine open set, we can see that this field is a finitely generated extension of $\mathbb{C}$ with transcendence degree one. Let us refer to such fields as function fields. We claim, conversely, that any such field F arises in this way. We can write F as an algebraic extension of $\mathbb{C}(x)$. By the primitive element theorem [76], it is necessarily obtained by adjoining one element, say y, satisfying a polynomial $f \in \mathbb{C}(x)[t]$. After clearing denominators, we have $f(x,y) = 0$

with $f \in \mathbb{C}[x,t]$. Thus F is the function field of the curve $V(f)$. After replacing this by the normalization of its projective closure, we see that F is the function field of a smooth projective curve. A more refined argument yields the following equivalence:

Theorem 6.4.1. *There is an antiequivalence between the category of smooth projective curves and surjective morphisms, and the category of function fields and inclusions.*

Recall that two categories are (anti-)equivalent if there is a (contravariant) functor inducing an isomorphism of Hom's and if every object in the target is isomorphic to an object in the image.

Proof. The details can be found in [60, I, 6.12], but it is worth understanding some key steps. A point of a smooth projective curve X gives a discrete valuation $v = \mathrm{ord}_p : \mathbb{C}(X)^* \to \mathbb{Z}$, which means that it satisfies

1. $v(fg) = v(f) + v(g)$.
2. $v(f+g) \geq \min(v(f), v(g))$.

In fact, any surjective discrete valuation is of this form for a unique p. Thus we can recover the points from the function field. To see that an inclusion $\mathbb{C}(Y) \subseteq \mathbb{C}(X)$ of fields corresponds to a map $f : X \to Y$, we can proceed as follows. A surjective discrete valuation v on $\mathbb{C}(X)$ yields a discrete valuation on the subfield by restriction. This can be made surjective by rescaling it to $v' = \frac{1}{e} v|_{\mathbb{C}(Y)^*}$ for a unique integer $e > 0$. Then $v \mapsto v'$ determines a set-theoretic map $X \to Y$, which can be checked to be a morphism. □

Corollary 6.4.2 (Lüroth's theorem). *Any subfield of* $\mathbb{C}(x)$ *properly containing* $\mathbb{C}$ *is abstractly isomorphic to* $\mathbb{C}(x)$.

Proof. Let $L \subseteq \mathbb{C}(x)$ be such a field. It is necessarily a function field. Therefore L is the function field of some curve Y, and the inclusion corresponds to a surjective morphism $f : \mathbb{P}^1 \to Y$. The following lemma implies that Y has genus 0, and is therefore isomorphic to $\mathbb{P}^1$ by Exercise 6.3.14. □

Lemma 6.4.3. *If* $f : X \to Y$ *is a surjective holomorphic map between compact Riemann surfaces, then the genus of* Y *is less than or equal to the genus of* X.

Proof. If α is a nonzero holomorphic 1-form on Y, $f^*\alpha$ is easily seen to be nonzero. Therefore, there is an injection $H^0(Y, \Omega^1_Y) \hookrightarrow H^0(X, \Omega^1_X)$. □

The set of holomorphic automorphisms of a compact Riemann surface X forms a group $\mathrm{Aut}(X)$, which can be identified with the group of automorphisms of $\mathbb{C}(X)$ fixing $\mathbb{C}$. For $X = \mathbb{P}^1$, this group is infinite, since it contains (and is in fact equal to) $\mathrm{PGL}_2(\mathbb{C})$. Likewise, for an elliptic curve $X = \mathbb{C}/L$, $\mathrm{Aut}(X)$ is infinite, since it contains translations. For curves of larger genus, we have the following result.

Theorem 6.4.4 (Hurwitz). *If* X *is a compact Riemann surface of genus* $g \geq 2$, *Then* $\mathrm{Aut}(X)$ *is a finite group of cardinality at most* $84(g-1)$.

We want to sketch a proof of the finiteness part, since it gives a nice application of the ideas developed so far. Given an automorphism $f : X \to X$, let $\Gamma_f \subset X \times X$ be its graph. The Lefschetz trace formula, Theorem 5.7.1, computes the intersection number of Γ_f and the diagonal as an alternating sum of traces of the induced maps $f^* : H^i(X,\mathbb{R}) \to H^i(X,\mathbb{R})$. For $i = 0,2$, these maps are easily seen to be the identity; thus

$$\Gamma_f \cdot \Delta = 2 - \mathrm{trace}[f^* : H^1(X,\mathbb{R}) \to H^1(X,\mathbb{R})].$$

Lemma 6.4.5. *If f is different from the identity, then $\Gamma_f \cdot \Delta \geq 0$.*

Proof. The point here is that $\Gamma_f \cdot \Delta$ can be computed as a sum of intersection numbers. These are necessarily nonnegative, as we will see in Section 11.2. For the special case of transversal intersections, this was discussed already in Section 5.6 and its exercises. □

Corollary 6.4.6. *If X has genus $g \geq 2$, then f is the identity if and only if $f^* : H^1(X,\mathbb{R}) \to H^1(X,\mathbb{R})$ is the identity.*

Proof. If f^* is the identity, then $\Gamma_f \cdot \Delta < 0$. The other direction is clear. □

We are now ready to sketch the proof of the finiteness statement of Theorem 6.4.4.

Proof. The previous corollary implies that there is an injective homomorphism $\mathrm{Aut}(X) \to \mathrm{Aut}(H^1(X,\mathbb{R}))$, and it suffices to prove that the image is finite. We do this by proving that the group is both compact and discrete. For discreteness we note that for any automorphism, f^* preserves the lattice $H^1(X,\mathbb{Z})$. After choosing a basis lying in $H^1(X,\mathbb{Z})$, this amounts to asserting that $\mathrm{Aut}(X)$ lies in $\mathrm{GL}_{2g}(\mathbb{Z})$, which is certainly a discrete subgroup of $\mathrm{GL}_{2g}(\mathbb{R})$.

The automorphisms f^* preserves the Hermitian pairing

$$\langle \alpha, \beta \rangle = \int_X \alpha \wedge \overline{\beta}$$

on $H^1(X,\mathbb{R})$. This pairing will be shown to be positive definite in the exercises of Section 9.1. Therefore $\mathrm{Aut}(X)$ lies in the corresponding unitary group, which is compact. □

Exercises

6.4.7. Prove that $\mathrm{Aut}(\mathbb{P}^1) = \mathrm{Aut}_{\mathbb{C}}(\mathbb{C}(x)) = \mathrm{PGL}_2(\mathbb{C})$ acting by Möbius transformations $x \mapsto \frac{ax+b}{cx+d}$.

6.4.8. Classify the discrete valuations on $\mathbb{C}(x)$ and show that they are all of the form $e \cdot \mathrm{ord}_p$ for a unique $e \in \{1,2,3,\ldots\}$ and $p \in \mathbb{P}^1$. (Hint: first show that any discrete valuation on $\mathbb{C}$ is zero.)

6.4.9. If X is as in Hurwitz's theorem with $G = \mathrm{Aut}(X)$, then $Y = X/G$ can be made into a compact Riemann surface of genus say h. At the level of function fields, $\mathbb{C}(Y) = \mathbb{C}(X)^G$. A special case of Riemann–Hurwitz's formula gives

$$2g - 2 = |G|(2h-2) + \sum_x (|\mathrm{stab}(x)| - 1),$$

where $\mathrm{stab}(x)$ is the stabilizer of x. Assuming this, show that $(2g-2)/|G| \geq 1/42$, and thus complete the proof of the theorem.

6.4.10. Given a finite group G, construct a compact Riemann surface X with $G \subseteq \mathrm{Aut}(X)$. (It is enough to do this for the symmetric group.)

6.5 Modular Forms and Curves

Let $\Gamma \subset \mathrm{SL}_2(\mathbb{R})$ be a subgroup acting properly discontinuously on H such that H/Γ is compact. Let k be a positive integer. An *automorphic form* of weight $2k$ is a holomorphic function $f : H \to \mathbb{C}$ on the upper half-plane satisfying

$$f(z) = (cz+d)^{-2k} f\left(\frac{az+b}{cz+d}\right) \tag{6.5.1}$$

for each

$$\begin{pmatrix} a & b \\ c & d \end{pmatrix} \in \Gamma.$$

Choose a weight-$2k$ automorphic form f. Then $f(z)(dz)^{\otimes k}$ is invariant under the group precisely when f is automorphic of weight $2k$. Let us suppose that the group $\bar{\Gamma} = \Gamma/\{\pm I\}$ acts freely. Then the quotient $X = H/\Gamma$ is a Riemann surface, and an automorphic form of weight $2k$ descends to a section of the sheaf $\omega_X^{\otimes k}$. We can apply Theorem 6.3.5 to calculate the dimensions of these spaces.

Proposition 6.5.1. *Suppose that $\bar{\Gamma}$ acts freely on H and that the quotient $X = H/\Gamma$ is compact of genus g. Then the dimension of the space of automorphic forms of weight $2k$ is*

$$\begin{cases} g & \text{if } k = 1, \\ (g-1)(2k-1) & \text{if } k > 1. \end{cases}$$

Proof. When $k = 1$, this is clear. When $k > 1$, we have

$$h^0(\omega^{\otimes k}) = h^0(\omega((k-1)K)) = (k-1)(\deg K) + g - 1 = (2k-1)(g-1). \qquad \square$$

The above conditions are a bit too stringent, since they exclude some of the most interesting examples such as the modular group $\mathrm{SL}_2(\mathbb{Z})$ and its finite-index subgroups such as the nth principal congruence group

$$\Gamma(n) = \left\{ \begin{pmatrix} a & b \\ c & d \end{pmatrix} \in \mathrm{SL}_2(\mathbb{Z}) \mid a-1 \equiv d-1 \equiv b \equiv c \equiv 0 \mod n \right\}.$$

The quotient $H/\mathrm{SL}_2(\mathbb{Z})$ can be identified with $\mathbb{C}$, as we saw. The natural compactification $\mathbb{P}^1_{\mathbb{C}}$ can be constructed as a quotient as follows: H corresponds to the upper hemisphere of $\mathbb{P}^1_{\mathbb{C}} = S^2$, and $\mathbb{R} \cup \{\infty\}$ corresponds to the equator. We take H and add rational points on the boundary $H^* = H \cup \mathbb{Q} \cup \{\infty\}$. These are called cusps. Then $\mathrm{SL}_2(\mathbb{Z})$ acts on H^*. The cusps form a single orbit, so we may identify $H^*/\mathrm{SL}_2(\mathbb{Z}) \cong \mathbb{P}^1$ by sending this orbit to the point at infinity in $\mathbb{P}^1$. We can define a topology on H^* by pulling back the open sets under the quotient map $\pi : H^* \to \mathbb{P}^1$. This makes π continuous. It follows that H^*/Γ is a topological space with a continuous projection to $\mathbb{P}^1$ for any subgroup $\Gamma \subseteq \mathrm{SL}_2(\mathbb{Z})$. In fact, somewhat more is true.

Theorem 6.5.2. *Given a finite-index subgroup $\Gamma \subseteq \mathrm{SL}_2(\mathbb{Z})$, H^*/Γ can be made into a compact Riemann surface, called a modular curve, such that $H^*/\Gamma \to H^*/\mathrm{SL}_2(\mathbb{Z}) \cong \mathbb{P}^1$ is holomorphic.*

Proof. See [32, §2.4]. □

Modular curves are of fundamental importance in number theory. A more immediate application is the construction of Riemann surfaces with large automorphism groups. When Γ is normal, then $\mathrm{PSL}_2(\mathbb{Z})/\bar{\Gamma}$ lies in $\mathrm{Aut}(H^*/\Gamma)$ and $H^*/\Gamma \to H^*/\mathrm{SL}_2(\mathbb{Z})$ is the quotient map. This is discussed further in the exercises.

A *modular form* of weight $2k$ for $\Gamma \subseteq \mathrm{SL}_2(\mathbb{Z})$ is a holomorphic function $f : H \to \mathbb{C}$ satisfying (6.5.1) and certain growth conditions that force it to correspond to a holomorphic (or possibly meromorphic) object on H^*/Γ. For example, the Eisenstein series

$$G_{2k}(z) = \sum_{(m,n)\in\mathbb{Z}^2-\{(0,0)\}} \frac{1}{(m+nz)^{2k}}$$

are modular of weight $2k$ for $\mathrm{SL}_2(\mathbb{Z})$ if $k > 1$. Their significance for elliptic curves is that $g_2 = 60G_4(\tau)$, $g_3 = 140G_6(\tau)$ give coefficients of the Weierstrass equation (6.1.1) for the lattice $\mathbb{Z} + \mathbb{Z}\tau$.

As we saw, the points of the quotient $H/\mathrm{SL}_2(\mathbb{Z})$ correspond to isomorphism classes of elliptic curves. More generally, the points of the quotient $H/\Gamma(n)$ correspond to elliptic curves with some extra structure. Let us spell this out for $n = 2$. An elliptic curve is an abelian group, and a level-two structure is a minimal set of generators for its subgroup of 2-torsion elements.

Proposition 6.5.3. *There is a bijection between*

(a) The points of $H/\Gamma(2)$.
(b) The set of elliptic curves with level-two structure.
(c) $\mathbb{P}^1 - \{0, 1, \infty\}$.

Proof. Given $\tau \in H$, we get an elliptic curve $E_\tau = \mathbb{C}/(\mathbb{Z} + \mathbb{Z}\tau)$ with the level-two structure $(1/2, \tau/2) \mod \mathbb{Z} + \mathbb{Z}\tau$. If τ' lies in the orbit of τ under $\Gamma(2)$, then there is an isomorphism $E_\tau \cong E_{\tau'}$ taking $(1/2, \tau/2) \mod \mathbb{Z} + \mathbb{Z}\tau$ to $(1/2, \tau'/2) \mod \mathbb{Z} + \mathbb{Z}\tau'$. Furthermore, any elliptic curve is isomorphic to an E_τ, and the isomorphism can be chosen so that a given level-two structure goes over to the standard one. Thus $H/\Gamma(2)$ classifies elliptic curves with level-two structure as claimed. We can describe the set of such curves in another way. Given an elliptic curve E defined by Legendre's equation (6.1.2), with $t \in \mathbb{P}^1 - \{0, 1, \infty\}$, the ramification points (the points on E lying over $0, 1, \infty, t$) are precisely the 2-torsion points. We can take the ramification point at ∞ to be the origin, and then any other pair of branch points determines a level-two structure. Conversely, given an elliptic curve E, with origin o and a level-two structure (p, q), we have $h^0(\mathscr{O}_E(2o)) = 2$. This means that there is a meromorphic function $f : E \to \mathbb{P}^1_{\mathbb{C}}$ with a double pole at o. It is not hard to see that f is ramified precisely at the 2-torsion points $o, p, q, p+q$ ($+$ refers to the group law on E). By composing f with a (unique) automorphism of $\mathbb{P}^1_{\mathbb{C}}$, we can put E in Legendre form such that f is projection to the x-axis, and $f(o) = \infty$, $f(p) = t \in \mathbb{P}^1 - \{0, 1, \infty\}$, $f(q) = 0$, $f(p+q) = 1$. Thus $H/\Gamma(2)$ is isomorphic to $\mathbb{P}^1 - \{0, 1, \infty\}$. □

We note that $\Gamma(2)/\{\pm I\}$ acts freely on H, and the quotient is isomorphic to $\mathbb{P}^1 - \{0, 1, \infty\}$ as a Riemann surface. A nice application of this to complex analysis is given in the following theorem:

Theorem 6.5.4 (Picard's little theorem). *An entire function omitting two or more points must be constant.*

Proof. The universal cover of $\mathbb{P}^1 - \{0, 1, \infty\}$ is H, which is isomorphic to the unit disk D. Let f be an entire function omitting two points, which we can assume are 0 and 1. Then f lifts to a holomorphic map $\mathbb{C} \to D$ that is bounded and therefore constant by Liouville's theorem [1]. □

Exercises

Let p be an odd prime in the following exercises.

6.5.5. Show that the stabilizers of the action of $\mathrm{PSL}_2(\mathbb{Z})$ on points of H are trivial unless they lie in the orbit of i or $\rho = e^{2\pi i/3}$, in which case they are cyclic of order 2 and 3 respectively. (Hint: refer back to the fundamental domain in Figure 6.3.)

6.5.6. If $\Gamma \subset \mathrm{PSL}_2(\mathbb{Z})$ is a finite-index normal subgroup, show that $\pi : H^*/\Gamma \to H^*/\mathrm{SL}_2(\mathbb{Z}) \overset{j}{=} \mathbb{P}^1$ is a Galois cover with Galois group $G = \mathrm{PSL}_2(\mathbb{Z})/\Gamma$. Show that the branch points of π are the images $j(i), j(\rho), \infty$ in $\mathbb{P}^1$.

6.5.7. Show that $|\mathrm{PSL}_2(\mathbb{Z}/p\mathbb{Z})| = \frac{p^3-p}{2} = d_p$. Check that the order of the stabilizers of the $\mathrm{PSL}_2(\mathbb{Z}/p\mathbb{Z})$-action on $H^*/\Gamma(p)$ at points lying over $j(i), j(\rho), \infty$ are $2, 3$, and p respectively.

6.5.8. Use the Riemann–Hurwitz formula (as stated in Exercise 6.4.9) to calculate that the genus of $H^*/\Gamma(p)$ is

$$g = 1 + \frac{d_p(p-6)}{12p}.$$

Show that Hurwitz's bound is attained with this example when $p = 7$.

Chapter 7
Simplicial Methods

In this chapter, we introduce simplicial methods that can be used to give concrete realizations for many of the cohomology theories encountered so far. We start with simplicial (co)homology, which is historically the first such theory. It is highly computable, at least in principle, but it suffers the disadvantage of depending on a triangulation. This can be overcome by working with singular (co)homology, which we briefly discuss, since it is the standard approach in topology. Fortunately, for reasonable spaces singular cohomology coincides with sheaf cohomology with constant coefficients. So all is well.

Simplicial cohomology also provides a model for Čech cohomology for arbitrary sheaves, which is dealt with later in this chapter. Čech cohomology coincides with sheaf cohomology in good cases, and for many problems is a more convenient theory to work with. In particular, it will be used for many of the computations done later on. Since it is quite concrete, Čech theory is often given a primary role in many expositions of sheaf theory. Nevertheless, as we shall see, it is not without its drawbacks.

7.1 Simplicial and Singular Cohomology

A systematic development of the ideas in this section can be found in Hatcher [61] or Spanier [108]. The standard n-simplex is

$$\Delta^n = \left\{(t_1,\dots,t_{n+1}) \in \mathbb{R}^{n+1} \mid \sum t_i = 1,\, t_i \geq 0\right\}.$$

The ith face Δ_i^n is the intersection of Δ^n with the hyperplane $t_i = 0$ (see Figure 7.1). Each face is homeomorphic to an $(n-1)$-simplex, and this can be realized by an explicit affine map $\delta_i : \Delta^{n-1} \to \Delta_i^n$. In general, we refer to the intersection of Δ with the linear space $t_{i_1+1} = \cdots = t_{i_k+1} = 0$ as a face. The faces are labeled by nonempty subsets of $\{0,\dots,n\}$.

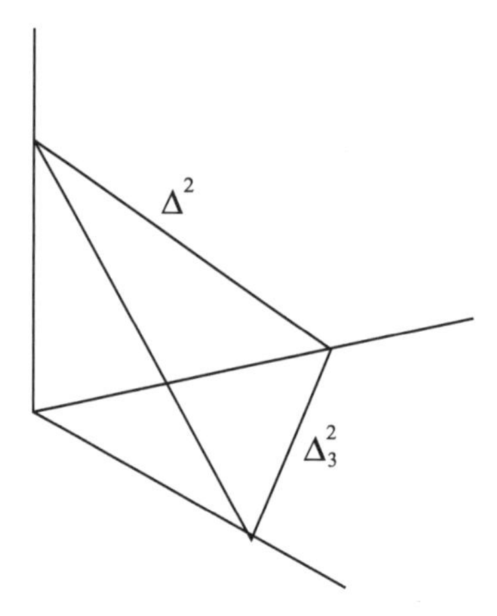

Fig. 7.1 2-simplex.

Some fairly complicated topological spaces, called *polyhedra or triangulable* spaces, can be built up by gluing simplices. It is known, although by no means obvious, that manifolds and algebraic varieties (with classical topology) can be triangulated. The combinatorics of the gluing is governed by a simplicial complex.

Definition 7.1.1. A simplicial complex (V,Σ) consists of a set V, called the set of vertices, and a collection of finite nonempty subsets Σ of V containing all the singletons and closed under taking nonempty subsets.

We can construct a topological space $|(V,\Sigma)|$ out of a simplicial complex roughly as follows. To each maximal element $S\in\Sigma$, choose an n-simplex $\Delta(S)$, where $n+1$ is the cardinality of S. Glue $\Delta(S)$ to $\Delta(S')$ along the face labeled by $S\cap S'$ whenever this is nonempty. (When V is infinite, this gluing process requires some care; see [108, Chapter 3].)

Let $K=(V,\Sigma)$ be a simplicial complex, and assume that V is linearly ordered. We will refer to an element of Σ as a n-simplex if it has cardinality $n+1$. We define an *n-chain* on a simplicial complex to be a finite formal integer linear combination $\sum_i n_i\Delta_i$ where the Δ_i are n-simplices. In other words, the set of n-chains $C_n(K)$ is the free abelian group generated by the set of n-simplices of K. Given an abelian group, let $C_n(K,A)=C_n(K)\otimes_{\mathbb{Z}}A$. Dually, the set of *$n$-cochains* with values in A is $C^n(K,A)=\mathrm{Hom}(C_n(K),A)$. In other words, an n-cochain is a function that assigns an element of A to every n-simplex. One can think of an n-cochain as some sort of combinatorial analogue of an n-form. As in integration theory, we need to worry about orientations, and this is where the ordering comes in. An alternative, which is probably more standard, is to use oriented simplices; the complexes one gets in this way are bigger, but the resulting cohomology theory is the same.

We define the ith face of a simplex

$$\delta_i(\{v_0,\ldots,v_n\})=\{v_0,\ldots,\widehat{v_i},\ldots,v_n\},$$

$v_0 < v_1 < \cdots < v_n$. (The notation $\widehat{x}$ means omit x.) Given an n-chain $C = \sum_j a_j \Delta_j$, we define an $(n-1)$-chain $\partial(C)$, called the boundary of C, by

$$\delta(C) = \sum\sum(-1)^i a_j \delta_i(\Delta_j).$$

This operation can be extended by scalars to $C_\bullet(K,A)$, and it induces a dual operation $\partial : C^n(K,A) \to C^{n+1}(K,A)$ by $\partial(F)(C) = F(\delta C)$. The key relation is as follows:

Lemma 7.1.2. $\delta_j \delta_i = \delta_i \delta_{j-1}$ *for* $i < j$.

Corollary 7.1.3. $\delta^2 = 0$ *and* $\partial^2 = 0$.

Thus we have a complex. The *simplicial homology* of K is defined by

$$H_n(K,A) = \frac{\ker[\delta : C_n(K,A) \to C_{n-1}(K,A)]}{\operatorname{im}[\delta : C_{n+1}(K,A) \to C_n(K,A)]}.$$

Elements of the numerator are called cycles, and elements of the denominator are called boundaries. The simplicial cohomologies are defined likewise by

$$H^n(K,A) = \mathscr{H}^n(C^\bullet(K,A)) = \frac{\ker[\partial : C^n(K,A) \to C^{n+1}(K,A)]}{\operatorname{im}[\partial : C^{n-1}(K,A) \to C^n(K,A)]}.$$

Note that when V is finite, these groups are automatically finitely generated and computable. The choices of A of interest to us are $\mathbb{Z}, \mathbb{R}$, and $\mathbb{C}$. The relationships are given by the following theorem:

Theorem 7.1.4 (Universal coefficient theorem). *If A is torsion-free, then there are isomorphisms*

$$\begin{aligned} H_i(K,A) &\cong H_i(K,\mathbb{Z}) \otimes A, \\ H^i(K,A) &\cong H^i(K,\mathbb{Z}) \otimes A \cong Hom(H_i(K,\mathbb{Z}),A). \end{aligned}$$

Proof. [108]. □

One advantage of cohomology over homology is that it has a multiplicative structure. When A is replaced by a commutative ring R, there is a product on cohomology analogous to the product in de Rham induced by wedging forms. Given two cochains $\alpha \in C^n(K,R)$, $\beta \in C^m(K,R)$, their *cup product* $\alpha \cup \beta \in C^{n+m}(K,R)$ is given by

$$\alpha \cup \beta(\{v_0,\ldots,v_{n+m}\}) = \alpha(\{v_0,\ldots,v_n\})\beta(\{v_n,\ldots,v_{n+m}\}), \qquad (7.1.1)$$

where $v_0 < v_1 < \cdots$.

Lemma 7.1.5. $\partial(\alpha \cup \beta) = \partial(\alpha) \cup \beta + (-1)^n \alpha \cup \partial(\beta)$.

Corollary 7.1.6. $\cup$ *induces an operation on cohomology that makes $H^*(K,R)$ into a graded ring.*

Singular (co)homology was introduced partly in order to give conceptual proof of the fact that $H_i(K)$ and $H^i(K)$ depend only on $|K|$, i.e., that these are independent of the triangulation. Here we concentrate on singular cohomology. A singular n-simplex on a topological space X is simply a continuous map from $f : \Delta^n$ to X. When X is a manifold, we can require the maps to be C^∞. We define a singular n-cochain on X to be a map that assigns an element of A to any n-simplex on X. Let $S^n(X,A)$ (or $S^n_\infty(X,A)$) denote the group of (C^∞) n-cochains with values in A. When F is an n-cochain, its coboundary is the $(n+1)$-cochain

$$\partial(F)(f) = \sum(-1)^i F(f \circ \delta_i).$$

The following has more or less the same proof as Corollary 7.1.3.

Lemma 7.1.7. $\partial^2 = 0$.

The *singular cohomology* groups of X are

$$H^i_{sing}(X,A) = \mathscr{H}^i(S^*(X,A)).$$

Singular cohomology is clearly contravariant in X. A basic property of this cohomology theory is its homotopy invariance. We state this in the form that we will need. A subspace $Y \subset X$ is called a *deformation retract* if there exists a a continuous map $F : [0,1] \times X \to X$ such that $F(0,x) = x$, $F(1,X) = Y$, and $F(1,y) = y$ for $y \in Y$; X is called contractible if it deformation retracts to a point.

Proposition 7.1.8. *If $Y \subset X$ is a deformation retract, then*

$$H^i_{sing}(X,A) \to H^i_{sing}(Y,A)$$

is an isomorphism for any A.

Corollary 7.1.9. *In particular, the higher cohomology vanishes on a contractible space.*

The corollary is an analogue of Poincaré's lemma. We call a space *locally contractible* if every point has a contractible neighborhood. Manifolds and varieties with classical topology are examples of such spaces.

Theorem 7.1.10. *If X is a paracompact Hausdorff space (e.g., a metric space) that is locally contractible, then $H^i(X,A_X) \cong H^i_{sing}(X,A)$ for any abelian group A. Moreover, the cup product in sheaf cohomology coincides with the cup product in singular cohomology.*

A complete proof, excluding the last statement, can be found in [108, Chapter 6 §§7–§9] (note that Spanier uses the Čech approach discussed in the next section). A proof that is more natural from our point of view can be found in [117], although it is stated only for manifolds. The key step is to consider the sheaves $\mathscr{S}^n$ associated to the presheaves $U \mapsto S^n(U,A)$. These sheaves are soft. When $A = \mathbb{R}$ this is clear, since they are modules over the sheaf of real-valued continuous functions. The local contractability guarantees that

$$0 \to A_X \to \mathscr{S}^0 \to \mathscr{S}^1 \to \cdots$$

is a soft resolution. Thus one gets

$$H^i(X, A_X) \cong \mathscr{H}^i(\mathscr{S}^*(X)).$$

The product (7.1.1) also extends to these sheaves, and thus we can conclude the coincidence of cup products using [45, Theorem 6.6.1]. It remains to check that the natural map

$$S^*(X,A) \to \mathscr{S}^*(X)$$

induces an isomorphism on cohomology. We refer the reader to [117, pp. 196–197] for this.

As a corollary of this (and the universal coefficient theorem), we obtain the form of de Rham's theorem that most people think of.

Corollary 7.1.11 (De Rham's theorem, version 2). *If X is a manifold, then*

$$H^*_{dR}(X,\mathbb{R}) \cong H^*_{sing}(X,\mathbb{R}) \cong H^*_{sing}(X,\mathbb{Z}) \otimes \mathbb{R}$$

as graded algebras.

The theorem and its corollary also hold when C^∞ cochains are used. In this case, the map can be defined directly on the level of complexes by

$$\alpha \mapsto \left(f \mapsto \int_\Delta f^* \alpha \right).$$

Singular cohomology carries a cup product given by formula (7.1.1). A stronger form of de Rham's theorem shows that the above map is a ring isomorphism [117]. Fundamental classes of oriented submanifolds can be constructed in $H^*(X,\mathbb{Z})$. This can be used to show that the intersection numbers $Y \cdot Z$ are always integers. This can also be deduced from Proposition 5.6.3 and transervsality theory.

Exercises

7.1.12. Prove Lemma 7.1.2 and it corollary.

7.1.13. Calculate the simplicial cohomology with $\mathbb{Z}$ coefficients for the "tetrahedron," which is the power set of $V = \{1,2,3,4\}$ with $\emptyset$ and V removed.

7.1.14. Let S^n be the n-sphere realized as the unit sphere in $\mathbb{R}^{n+1}$. Let $U_0 = S^n - \{(0,\ldots,0,1)\}$ and $U_1 = S^n - \{(0,\ldots,0,-1)\}$. Prove that U_i are contractible, and that $U_0 \cap U_1$ deformation retracts onto the "equatorial" $(n-1)$-sphere.

7.1.15. Prove that

$$H^i(S^n,\mathbb{Z}) = \begin{cases} \mathbb{Z} & \text{if } i=0,n, \\ 0 & \text{otherwise,} \end{cases}$$

using Mayer–Vietoris.

7.2 Cohomology of Projective Space

Our goal in this section is to calculate the cohomology of projective space. But first we will need to develop a few more tools. Let X be a space satisfying the assumptions of Theorem 7.1.10, and $Y \subset X$ a closed subspace satisfying the same assumptions. We will insert the restriction map

$$H^i(X,\mathbb{Z}) \to H^i(Y,\mathbb{Z})$$

into a long exact sequence. This can be done in a number of ways: by defining cohomology of the pair (X,Y), by using sheaf theory, or by using a mapping cone. We will choose the last option, which is the most geometric.

Proposition 7.2.1. *Let X/Y denote the space obtained by collapsing Y to a point. Then for $i>0$, there is a long exact sequence*

$$\cdots \to H^i(X/Y,\mathbb{Z}) \to H^i(X,\mathbb{Z}) \to H^i(Y,\mathbb{Z}) \to H^{i+1}(X/Y,\mathbb{Z}) \to \cdots. \tag{7.2.1}$$

The sequence is constructed as follows. Let C be obtained by first gluing the base of the cylinder $\{0\}\times Y \subset [0,1]\times Y$ to X along Y, and then collapsing the top $\{1\}\times Y$ to a point P (Figure 7.2).

Let $U_1 = C - P$, and let $U_2 \subset C$ be the open cone $(0,1]\times Y/\{1\}\times Y$. One sees that U_1 deformation retracts to X, U_2 is contractible, and $U_1 \cap U_2$ deformation retracts to Y. The Mayer–Vietoris sequence, together with Proposition 7.1.8, yields a long exact sequence

$$\cdots \to H^i(C,\mathbb{Z}) \to H^i(X,\mathbb{Z}) \to H^i(Y,\mathbb{Z}) \to H^{i+1}(C,\mathbb{Z}) \to \cdots$$

when $i>0$. To make this really useful, note that the map $C \to C/\overline{U_2}$ that collapses the closed cone to a point is a homotopy equivalence. Therefore it induces an isomorphism on cohomology. Since we can identify $C/\overline{U_2}$ with X/Y, we obtain a sequence (7.2.1)

We can now carry out a basic computation. Let $\mathbb{P}^n = \mathbb{P}^n_{\mathbb{C}}$ with its classical topology.

Theorem 7.2.2.

$$H^i(\mathbb{P}^n,\mathbb{Z}) = \begin{cases} \mathbb{Z} & \text{if } 0 \le i \le 2n \text{ is even,} \\ 0 & \text{otherwise.} \end{cases}$$

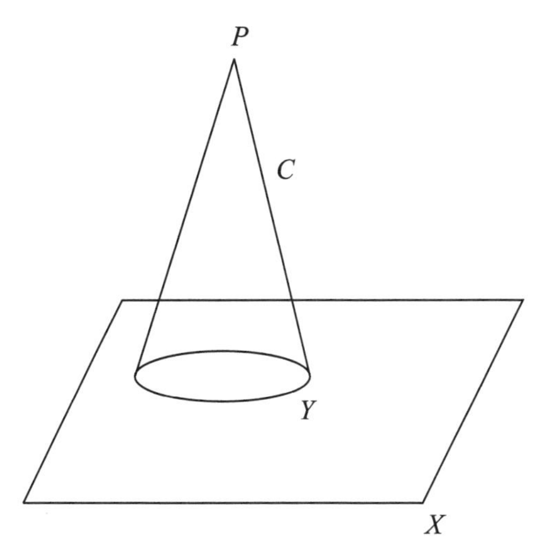

Fig. 7.2 Mapping cone.

Proof. We apply this when $X = \mathbb{P}^n$ and $Y = \mathbb{P}^{n-1}$ embedded as a hyperplane. The complement $X - Y$ is equal to $\mathbb{C}^n$. Collapsing Y to a point amounts to adding a point at infinity to $\mathbb{C}^n$; thus $X/Y = S^{2n}$. Since projective spaces are connected,

$$H^0(\mathbb{P}^n, \mathbb{Z}) \cong H^0(\mathbb{P}^{n-1}, \mathbb{Z}) \cong \mathbb{Z}.$$

For $i > 0$, (7.2.1) and the previous exercise yield isomorphisms

$$H^i(\mathbb{P}^n, \mathbb{Z}) \cong H^i(\mathbb{P}^{n-1}, \mathbb{Z}), \text{ when } i < 2n, \tag{7.2.2}$$
$$H^{2n}(\mathbb{P}^n, \mathbb{Z}) \cong \mathbb{Z}.$$

Theorem 7.2.2 now follows by induction. □

Exercises

7.2.3. Let $L \subset \mathbb{P}^n$ be a linear subspace of codimension i. Prove that its fundamental class $[L]$ is generated by $H^{2i}(\mathbb{P}^n, \mathbb{Z})$.

7.2.4. Let $X \subset \mathbb{P}^n$ be a smooth projective variety. Then $[X] = d[L]$ for some d, where L is a linear subspace of the same dimension; d is called the degree of X. Bertini's theorem, which you can assume, implies that there exists a linear space L' of complementary dimension transverse to X. Check that $X \cdot L' = \#(X \cap L') = d$.

7.2.5. An algebraic variety X admits an affine cell decomposition if there is a sequence of Zariski closed sets $X = X_n \supset X_{n-1} \supset \cdots$ such that each difference

$X_i - X_{i-1}$ is an affine space. Generalize the above procedure to compute $H^*(X,\mathbb{Z})$ assuming that X possesses such a decomposition.

7.2.6. Let $G = \mathbb{G}(2,4)$ be the Grassmannian of lines in $\mathbb{P}^3$. We define a sequence of subsets by imposing so-called Schubert conditions. Fix a plane $P \subset \mathbb{P}^3$, a line $L \subset P$, and a point $Q \in L$. Let G_3 be the set of lines meeting L, G_2 the set of lines containing Q or contained in P, $G_1 \subset G$ the set of lines containing Q and contained in P, and let $G_0 = \{L\}$. Show that this gives an affine cell decomposition. Use this to calculate the cohomology of G.

7.3 Čech Cohomology

We return to sheaf theory proper. We will introduce the Čech approach to cohomology, which has the advantage of being quite explicit and computable (and the disadvantage of not always giving the "right" answer). Roughly speaking, Čech bears the same relation to sheaf cohomology as simplicial does to singular cohomology.

One starts with an open cover $\{U_i \,|\, i \in I\}$ of a space X indexed by a linearly ordered set I. If $J \subseteq I$, let U_J be the intersection of U_j with $j \in J$. Let $\mathscr{F}$ be a sheaf of abelian groups on X. The group of *Čech n-cochains* is

$$C^n = C^n(\{U_i\},\mathscr{F}) = \prod_{i_0<\cdots<i_n} \mathscr{F}(U_{i_0\ldots i_n}).$$

The coboundary map $\partial : C^n \to C^{n+1}$is defined by

$$\partial(f)_{i_0\ldots i_{n+1}} = \sum_k (-1)^k f_{i_0\ldots \hat{i}_k \ldots i_{n+1}}|_{U_{i_0\ldots i_{n+1}}}. \tag{7.3.1}$$

By an argument similar to the proof of corollary 7.1.3, we have the following:

Lemma 7.3.1. $\partial^2 = 0$.

Definition 7.3.2. The nth Čech cohomology group is

$$\check{H}^n(\{U_i\},\mathscr{F}) = \mathscr{H}^n(C^\bullet(\{U_i\},\mathscr{F})) = \frac{\ker(\partial : C^n \to C^{n+1})}{\mathrm{im}(\partial : C^{n-1} \to C^n)}.$$

To get a feeling for this, let us write out the first couple of groups explicitly:

$$\begin{aligned} \check{H}^0(\{U_i\},\mathscr{F}) &= \{(f_i) \in \prod \mathscr{F}(U_i) \,|\, f_i = f_j \text{ on } U_{ij}\} = \mathscr{F}(X), \\ \check{H}^1(\{U_i\},\mathscr{F}) &= \frac{\{(f_{ij}) \in \prod \mathscr{F}(U_{ij}) \,|\, f_{ik} = f_{ij} + f_{jk} \text{ on } U_{ijk}\}}{\{(f_{ij}) \,|\, \exists(\phi_i),\, f_{ij} = \phi_i - \phi_j\}}. \end{aligned} \tag{7.3.2}$$

There is a strong similarity with simplicial cohomology. This can be made precise by introducing a simplicial complex called the *nerve* of the cover. For the set of vertices, we take the index set I. The set of simplices is given by

$$\Sigma = \{\{i_0, \dots i_n\} \,|\, U_{i_0,\dots i_n} \neq \emptyset\}.$$

If we assume that each $U_{i_0,\dots i_n}$ is connected, then we see that the Čech complex $C^n(\{U_i\}, A_X)$ coincides with the simplicial complex of the nerve with coefficients in A.

Even though we are primarily interested in sheaves of abelian groups, it will be convenient to extend (7.3.2) to a sheaf of arbitrary groups $\mathscr{G}$,

$$\check{H}^1(\{U_i\}, \mathscr{G}) = \left\{ (g_{ij}) \in \prod_{i<j} \mathscr{G}(U_{ij}) \,|\, g_{ik} = g_{ij} g_{jk} \text{ on } U_{ijk} \right\} \Big/ \sim,$$

where $(g_{ij}) \sim (\bar{g}_{ij})$ if there exists $(\gamma_i) \in \prod \mathscr{G}(U_i)$ such that $g_{ij} = \gamma_i \bar{g}_{ij} \gamma_j^{-1}$. Note that in general this is not a group, but just a set with distinguished element $g_{ij} = 1$. The (g_{ij}) are called 1-cocycles with values in $\mathscr{G}$. It will be useful to drop the requirement $i < j$ by setting $g_{ji} = g_{ij}^{-1}$ and $g_{ii} = 1$. This will not affect the outcome. See Exercise 7.3.8 for a generalization of this.

As an example of a sheaf of nonabelian groups, take $U \mapsto \mathrm{GL}_n(\mathscr{R}(U))$, where $(X, \mathscr{R})$ is a ringed space (i.e., a space with a sheaf of commutative rings).

Theorem 7.3.3. *Let $(X, \mathscr{R})$ be a manifold or a variety over k, an $\{U_i\}$ an open cover of X. There are one-to-one correspondences between the following sets:*

$A =$ *The set of isomorphism classes of rank-n vector bundles over $(X, \mathscr{R})$ trivializable over $\{U_i\}$.*
$B =$ *The set of isomorphism classes of locally free $\mathscr{R}$-modules $\mathscr{M}$ such that $\mathscr{M}|_{U_i} \cong \mathscr{R}|_{U_i}^n$.*
$C = \check{H}^1(\{U_i\}, \mathrm{GL}_n(\mathscr{R}))$.

Proof. We merely describe the correspondences.

$A \to B$: Given a vector bundle V on X, let $\mathscr{V}$ be its sheaf of sections.
$B \to C$: Given a locally free sheaf $\mathscr{V}$, choose isomorphisms $F_i : \mathscr{R}^n_{U_i} \to \mathscr{V}|_{U_i}$. Set $g_{ij} = F_i \circ F_j^{-1}$. This determines a well-defined element of $\check{H}^1$.
$C \to A$: Define an equivalence relation $\equiv$ on the disjoint union $W = \bigcup U_i \times k^n$ as follows. Given $(x_i, v_i) \in U_i \times k^n$ and $(x_j, v_j) \in U_j \times k^n$, $(x_i, v_i) \equiv (x_j, v_j)$ if and only if $x_i = x_j$ and $v_i = g_{ij}(x) v_j$. Let $V = (W/\equiv)$ with the quotient topology. Given an open set $(U'/\equiv) = U \subset V$, define $f : U \to k$ to be regular, C^∞, or holomorphic (as the case may be) if its pullback to U' has this property. □

Implicit above is a construction that associates to a 1-cocycle $\gamma = (g_{ij})$, the locally free sheaf

$$\mathscr{M}_\gamma(U) = \left\{ (v_i) \in \prod \mathscr{R}(U \cap U_i)^n \,|\, v_i = g_{ij} v_j \right\}.$$

Example 7.3.4. Consider the case of projective space $\mathbb{P} = \mathbb{P}^n_k$. Suppose $x_0, \dots, x_n$ are homogeneous coordinates. Let U_i be the complement of the hyperplane $x_i = 0$. Then U_i is isomorphic to $\mathbb{A}^n_k$ by

$$[x_0,\ldots,x_n] \to \left(\frac{x_0}{x_i},\ldots,\frac{\widehat{x_i}}{x_i},\ldots\right).$$

Define $g_{ij} = x_j/x_i \in \mathscr{O}_{\mathbb{P}}(U_{ij})^*$. This is a 1-cocycle, and $M_{g_{ij}} \cong \mathscr{O}_{\mathbb{P}}(1)$. Likewise, $(x_j/x_i)^d$ is the 1-cocyle for $\mathscr{O}_{\mathbb{P}}(d)$.

We get rid of the dependence on covers by taking direct limits. If $\{V_j\}$ is a refinement of $\{U_i\}$, in the sense that each V_j lies in some U_i, then there is a natural restriction map

$$\check{H}^i(\{U_i\},\mathscr{F}) \to \check{H}^i(\{V_j\},\mathscr{F}).$$

Definition 7.3.5. The nth Čech cohomology group is the direct limit

$$\check{H}^i(X,\mathscr{F}) = \varinjlim \check{H}^i(\{U_j\},\mathscr{F})$$

over all open covers under refinement.

From Theorem 7.3.3, we can extract the following corollary:

Corollary 7.3.6. *There are one to one correspondences between the following sets:*

$A =$ *The set of isomorphism classes of rank n-vector bundles over* $(X,\mathscr{R})$.
$B =$ *The set of isomorphism classes of locally free* $\mathscr{R}$*-modules* $\mathscr{M}$ *of rank n.*
$C = \check{H}^1(X,\mathrm{GL}_n(\mathscr{R}))$.

A *line bundle* is a rank-one vector bundle. We will not distinguish between line bundles and rank-one locally free sheaves. The set of isomorphism classes of line bundles carries the structure of a group, namely $\check{H}^1(X,\mathscr{R}^*)$. This group is called the *Picard group*, and is denoted by $\mathrm{Pic}(X)$.

Exercises

7.3.7. Check that the Čech coboundary satisfies $\partial^2 = 0$.

7.3.8. It is often convenient to use alternating cochains, where

$$C^n_{alt}(\{U_i\},\mathscr{F}) \subset \prod \mathscr{F}(U_{i_0\ldots i_n})$$

consists of families such that $f_{i_0\ldots i_{n+1}} = 0$ when indices are repeated and

$$f_{\sigma(i_0)\ldots\sigma(i_{n+1})} = \mathrm{sign}(\sigma) f_{i_0\ldots i_{n+1}}$$

for any permutation σ. Show that the coboundary (7.3.1) preserves this condition. Define $\iota : C^n(\{U_i\},\mathscr{F}) \to C^n_{alt}(\{U_i\},\mathscr{F})$ by

$$\iota(f)_{i_0,\ldots i_p} = \begin{cases} \mathrm{sign}(\sigma) f_{\sigma(i_0)\ldots\sigma(i_{n+1})} & \text{where } \sigma(i_0) < \cdots < \sigma(i_{n+1}), \\ 0 & \text{if there are repetitions.} \end{cases}$$

Show that this induces an isomorphism of complexes and therefore an isomorphism on cohomology.

7.3.9. Fill in the details of the proof of Theorem 7.3.3.

7.3.10. Show that multiplication in $\mathrm{Pic}(X)$ can be interpreted as tensor product of line bundles. Show that the determinant map $\det : \mathrm{GL}_n(\mathscr{R}) \to \mathscr{R}^*$ induces a map $\check{H}^1(X, \mathrm{GL}_n(\mathscr{R})) \to \mathrm{Pic}(X)$ corresponding to the operation $\mathscr{M} \mapsto \wedge^n \mathscr{M}$ on locally free sheaves.

7.3.11. Given an exact sequence of sheaves $0 \to \mathscr{A} \to \mathscr{B} \to \mathscr{C} \to 0$ on X and $\gamma \in \mathscr{C}(X)$, define $\delta(\gamma)$ as follows. Choose an open cover $\{U_i\}$ so that $\gamma|_{U_i} = \mathrm{im}(\beta_i)$, with $\beta_i \in \mathscr{B}(U_i)$. Let $\delta(\gamma)$ be the class of $\beta_i - \beta_j \in C^1(\{U_i\}, \mathscr{A})$.

(a) Show that $\delta(\gamma)$ gives a well-defined element of $\check{H}^1(X, \mathscr{A})$.
(b) Show that $\delta(\gamma) = 0$ if and only if γ lies in the image of $\mathscr{B}(X)$.

7.3.12. Let X be an algebraic variety and let K^* be the constant sheaf associated to the nonzero elements of the function field $k(X)^*$. There is a natural inclusion $\mathscr{O}_X^* \subset K^*$. A (principal) Cartier divisor is an element $F \in K^*/\mathscr{O}_X^*(X)$ (respectively $\mathrm{im}\, K^*(X)$). The obstruction to being principal is the class $\delta(F) \in \check{H}^1(X, \mathscr{O}_X^*)$. Describe this in explicit terms.

7.4 Čech Versus Sheaf Cohomology

We define a sheafified version of the Čech complex. Given a sheaf $\mathscr{F}$ on a space X and an open cover $\{U_i\}$, let

$$\mathscr{C}^n(\{U_i\}, \mathscr{F}) = \prod_{i_0 < \ldots < i_n} \iota_* \mathscr{F}|_{U_{i_0 \ldots i_n}},$$

where ι denotes the inclusion $U_{i_0 \ldots i_n} \subset X$. We construct a differential ∂ using the same formula as before (7.3.1). Taking global sections yields the usual Čech complex.

Lemma 7.4.1. *These sheaves fit into an exact sequence*

$$0 \to \mathscr{F} \to \mathscr{C}^0(\{U_i\}, \mathscr{F}) \to \mathscr{C}^1(\{U_i\}, \mathscr{F}) \to \cdots. \tag{7.4.1}$$

Proof. [60, III, Lemma 4.2]. □

Lemma 7.4.2. *Suppose that $\mathscr{F}$ is flasque. Then $\check{H}^n(\{U_i\}, \mathscr{F}) = 0$ for all $n > 0$.*

Proof. If $\mathscr{F}$ is flasque, then the sheaves $\mathscr{C}^n(\{U_i\}, \mathscr{F})$ are also seen to be flasque. Then (7.4.1) gives an acyclic resolution of $\mathscr{F}$. Therefore

$$\check{H}^n(\{U_i\}, \mathscr{F}) = \mathscr{H}^n(\mathscr{C}^\bullet(\{U_i\}, \mathscr{F})) = H^n(X, \mathscr{F}) = 0.$$ □

Lemma 7.4.3. *Suppose that $H^1(U_J,\mathscr{A})=0$ for all nonempty finite sets J. Then given an exact sequence*

$$0\to\mathscr{A}\to\mathscr{B}\to\mathscr{C}\to 0$$

of sheaves, there is a long exact sequence

$$0\to\check{H}^0(\{U_i\},\mathscr{A})\to\check{H}^0(\{U_i\},\mathscr{B})\to\check{H}^0(\{U_i\},\mathscr{C})\to\check{H}^1(\{U_i\},\mathscr{A})\to\cdots.$$

Proof. The hypothesis guarantees that there are short exact sequences

$$0\to\mathscr{A}(U_J)\to\mathscr{B}(U_J)\to\mathscr{C}(U_J)\to 0.$$

Thus we have an exact sequence of complexes

$$0\to C^\bullet(\{U_i\},\mathscr{A})\to C^\bullet(\{U_i\},\mathscr{B})\to C^\bullet(\{U_i\},\mathscr{C})\to 0.$$

The long exact sequence now follows from a standard result in homological algebra (cf. [118, Theorem 13.1]). □

Definition 7.4.4. An open cover $\{U_i\}$ is called a Leray cover for a sheaf $\mathscr{F}$ if $H^n(U_J,\mathscr{F})=0$ for all nonempty finite sets J and all $n>0$.

Theorem 7.4.5. *If $\{U_i\}$ is a Leray cover for the sheaf $\mathscr{F}$, then*

$$\check{H}^n(\{U_i\},\mathscr{F})\cong H^n(X,\mathscr{F})$$

for all n.

Proof. This is clearly true for $n=0$. With the notation of Section 4.2, we have an exact sequence

$$0\to\mathscr{F}\to\mathbf{G}(\mathscr{F})\to\mathbf{C}^1(\mathscr{F})\to 0$$

with

$$H^1(X,\mathscr{F})=\operatorname{coker}[\Gamma(X,\mathbf{G}(\mathscr{F}))\to\Gamma(X,\mathbf{C}^1(\mathscr{F}))] \tag{7.4.2}$$

and

$$H^{n+1}(\mathscr{F})=H^n(\mathbf{C}^1(\mathscr{F})). \tag{7.4.3}$$

Lemmas 7.4.3 and 7.4.2 imply that

$$\check{H}^1(\mathscr{F})\cong\operatorname{coker}[\Gamma(X,\mathbf{G}(\mathscr{F}))\to\Gamma(X,\mathbf{C}^1(\mathscr{F}))] \tag{7.4.4}$$

and

$$\check{H}^{n+1}(\mathscr{F})=\check{H}^n(\mathbf{C}^1(\mathscr{F})) \tag{7.4.5}$$

for $n>0$. Formulas (7.4.2) and (7.4.4) already imply the theorem when $n=1$. The remaining cases when $n>1$ can be handled by induction, using (7.4.3), (7.4.5), and the fact that the cover is also Leray with respect to $\mathbf{C}^1(\mathscr{F})$:

$$H^i(U_J,\mathbf{C}^1(\mathscr{F}))=H^{i+1}(U_J,\mathscr{F})=0.$$

□

Corollary 7.4.6. *If every cover admits a Leray refinement, then* $\check{H}^n(X,\mathscr{F}) \cong H^n(X,\mathscr{F})$.

We state a few more general results for the record.

Proposition 7.4.7. *For any sheaf* $\mathscr{F}$,

$$\check{H}^1(X,\mathscr{F}) \cong H^1(X,\mathscr{F}).$$

Proof. See [45, Chapter II, Corollary 5.9.1]. □

Corollary 7.4.8. *If* $(X,\mathscr{R})$ *is a ringed space,* $\mathrm{Pic}(X) \cong H^1(X,\mathscr{R}^*)$.

Theorem 7.4.9. *If X is a paracompact Hausdorff space (e.g., a metric space), then for any sheaf and all i,*

$$\check{H}^i(X,\mathscr{F}) \cong H^i(X,\mathscr{F})$$

for all i.

Proof. See [45, Chapter II, Corollary 5.10.1]. □

Exercises

7.4.10. If $\{U_i\}$ is an open cover of a space X such that every nonempty intersection U_I is contractible, show that it is Leray with respect to a constant sheaf A_X. Thus $H^i(X,A_X)$ is the simplicial cohomology of the nerve associated to such a cover.

7.4.11. A finite-dimensional cell complex, which refines the notion of a simplicial complex, is a space constructed as follows. Start with a set of points X^0, then glue a set of intervels (1-cells) by attatching their endpoints to X^0 to obtain X^1. Glue a set of 2-disks (2-cells) by attaching their boundaries to X^1, and so on. $X = X^n$ for some n. Show that X admits an open cover as in the previous exercise.

7.4.12. When $X = U_1 \cup U_2$, Mayer–Vietoris yields a map $\mathscr{F}(U_1 \cap U_2) \to H^1(X,\mathscr{F})$. This induces a map $\check{H}^1(\{U_1,U_2\}\mathscr{F}) \to H^1(X,\mathscr{F})$, which is an isomorphism when the cover is Leray. Show more generally that for any finite cover $\{U_j\}$, there is a natural map $\check{H}^i(\{U_j\},\mathscr{F}) \to H^i(X,\mathscr{F})$ with the same property.

7.4.13. Let X be an algebraic variety. Prove that $\mathrm{Pic}(X)$ is isomorphic to the quotient of the group of Cartier divisors (Exercise 7.3.12) by principal divisors.

7.4.14. Let $A = \mathbb{A}^2_k$, where k is algebraically closed, and let $C \subset A$ be the union of two irreducible curves C_i meeting in two points. Let $K = \ker[\mathbb{Z}_A \to \mathbb{Z}_C]$. Show that $H^2(X,\mathbb{Z}) = \mathbb{Z}$ using Exercise 5 of §4.5. On the other hand, it can be shown that $\check{H}^2(X,\mathbb{Z}) = 0$; cf. [51, pp. 177–179].

7.5 First Chern Class

Let $(X, \mathcal{O}_X)$ be a complex manifold or algebraic variety over $\mathbb{C}$. Then we have isomorphisms

$$\mathrm{Pic}(X) \cong \check{H}^1(X, \mathcal{O}_X^*) \cong H^1(X, \mathcal{O}_X^*).$$

Let $e(x) = \exp(2\pi \mathrm{i} x)$ be the normalized exponential.

Lemma 7.5.1. *When X is a complex manifold, there is an exact sequence*

$$0 \to \mathbb{Z}_X \to \mathcal{O}_X \xrightarrow{e} \mathcal{O}_X^* \to 1 \tag{7.5.1}$$

called the exponential sequence.

Proof. It suffices to check this when X is replaced by a ball. Since a ball is simply connected, given a holomorphic function $f \in \mathcal{O}^*(X)$, we can find a single-valued branch of the logarithm $\log f$. Then $f = e(\log f/2\pi \mathrm{i})$. This proves surjectivity. The remaining steps are straightforward. □

Definition 7.5.2. Given a line bundle L, its first Chern class $c_1(L) \in H^2(X, \mathbb{Z})$ is the image of L under the connecting map $\mathrm{Pic}(X) \to H^2(X, \mathbb{Z})$ associated to (7.5.1).

This can be carried out for C^∞ manifolds as well, provided one interprets $\mathcal{O}_X$ as the sheaf of complex-valued C^∞ functions, and $\mathrm{Pic}(X)$ as the group of C^∞ complex line bundles. In this case, c_1 is an isomorphism. It is clear that the construction is functorial:

Lemma 7.5.3. *If $f : X \to Y$ is a C^∞ map between manifolds, then $c_1(f^*L) = f^*c_1(L)$.*

We want to calculate the first Chern class explicitly for a compact Riemann surface X. Note that in this case, $H^2(X, \mathbb{Z}) \cong \mathbb{Z}$, so we can view c_1 as a number. To make things even more explicit, we note that this isomorphism is given by the inclusion $H^2(X, \mathbb{Z}) \subset H^2(X, \mathbb{C})$ followed by integration. In order to get a de Rham representative, we consider the commutative diagram

$$\begin{array}{ccccccccc}
0 & \longrightarrow & \mathbb{Z}_X & \longrightarrow & \mathcal{O}_X & \xrightarrow{e} & \mathcal{O}_X^* & \longrightarrow & 1 \\
 & & \downarrow & & \downarrow & & \downarrow{\scriptstyle \frac{1}{2\pi \mathrm{i}} d\log} & & \\
0 & \longrightarrow & \mathbb{C}_X & \longrightarrow & \mathcal{E}_X^0 & \xrightarrow{d} & \mathcal{E}_{X,cl}^1 & \longrightarrow & 0
\end{array} \tag{7.5.2}$$

where $\mathcal{E}_{X,cl}^1$ is the sheaf of closed 1-forms. By definition, it fits into an exact sequence

$$0 \to \mathcal{E}_{X,cl}^1 \to \mathcal{E}_X^1 \to \mathcal{E}_X^2 \to 0.$$

From this we obtain

$$\mathcal{E}^1(X) \to \mathcal{E}^2(X) \to H^1(X, \mathcal{E}_{X,cl}) \to 0.$$

Given a 1-cocycle γ_{jk} representing an element of $H^1(X,\mathcal{E}_{X,cl})$, we can find an explicit lift to $\beta \in \mathcal{E}^2(X)$ as follows. Regarding γ_{jk} as a cocycle in $\mathcal{E}^1_X$, we express it as a coboundary

$$\gamma_{jk} = \alpha_j|_{U_{jk}} - \alpha_k|_{U_{jk}},\ \alpha_j \in \mathcal{E}^1_X(U_j)$$

because $\mathcal{E}^1_X$ is soft. Since γ_{jk} is closed, we have $d\alpha_j = d\alpha_k$ on U_{jk}. Thus these forms patch to give a global 2-form β.

From the diagram (7.5.2), we obtain a commutative square

$$\begin{array}{ccc} H^1(X,\mathcal{O}^*_X) & \xrightarrow{c_1} & H^2(X,\mathbb{Z}) \\ \Big\downarrow{\scriptstyle \frac{1}{2\pi i}d\log} & & \Big\downarrow \\ H^1(X,\mathcal{E}_{X,cl}) & \longrightarrow & H^2(X,\mathbb{C}) \end{array}$$

Combining all of this leads to the following recipe for computing c_1. Apply $\frac{d\log}{2\pi i}$ to a 1-cocycle $g_{jk} \in \mathcal{O}^*_X(U_{jk})$. Then write it as a boundary:

$$\frac{d\log g_{jk}}{2\pi i} = \alpha_j|_{U_{jk}} - \alpha_k|_{U_{jk}},\ \alpha_j \in \mathcal{E}^1_X(U_j).$$

Then $d\alpha_j$ patches to give a well-defined global closed 2-form β that represents the image of c_1 in $H^2(X,\mathbb{C})$. Finally, we integrate β to get a number.

Lemma 7.5.4. *On $\mathbb{P}^1$ with the above identification, we have $c_1(\mathcal{O}(1)) = 1$.*

Proof. Set $\mathbb{P} = \mathbb{P}^1_{\mathbb{C}}$. We can use the standard cover $U_i = \{x_i \neq 0\}$. We identify U_1 with $\mathbb{C}$ with the coordinate $z = x_0/x_1$. The 1-cocycle of $\mathcal{O}_X(1)$ is $g_{01} = z^{-1}$ (Example 7.3.4). Carrying out the above procedure yields

$$\frac{d\log g_{01}}{2\pi i} = -\frac{dz}{2\pi i z} = \alpha_0 - \alpha_1$$

and $\beta = d\alpha_j$.

In order to evaluate the integral $\int \beta$, divide the sphere into two hemispheres $H_1 = \{|z| \le 1\}$ and $H_0 = \{|z| \ge 1\}$. Let C be the curve $|z| = 1$ oriented so that the boundary of H_1 is C. Then with the help of Stokes's theorem, we get

$$\int_{\mathbb{P}} \beta = \int_{H_0} d\alpha_0 + \int_{H_1} d\alpha_1 = -\left(\int_C \alpha_0 - \int_C \alpha_1\right) = \frac{1}{2\pi i}\int_C \frac{dz}{z} = 1.$$

Thus $c_1(\mathcal{O}(1))$ is the fundamental class of $H^2(\mathbb{P}^1,\mathbb{Z})$. □

By the same kind of argument, we obtain the following:

Lemma 7.5.5. *If D is a divisor on a compact Riemann surface X, $c_1(\mathcal{O}(D)) = \deg(D)$.*

We are going to generalize this to higher dimensions. A complex submanifold $D \subset X$ of a complex manifold is called a smooth effective divisor if D is locally definable by a single equation. In other words, we have an open cover $\{U_i\}$ of X and functions $f_i \in \mathscr{O}(U_i)$ such that $D \cap U_i$ is given by $f_i = 0$. We define the $\mathscr{O}_X$-module $\mathscr{O}_X(-D)$ to be the ideal sheaf of D, and $\mathscr{O}_X(D)$ to be the dual. By assumption, $\mathscr{O}_X(-D)$ is locally a principal ideal, and hence a line bundle. The line bundle $\mathscr{O}_X(D)$ is determined by the 1-cocycle $f_i/f_j \in \mathscr{O}(U_i \cap U_j)^*$.

Lemma 7.5.6. *If $H \subset \mathbb{P}^n$ is a hyperplane, then $\mathscr{O}_{\mathbb{P}^n}(H) \cong \mathscr{O}_{\mathbb{P}^n}(1)$.*

Proof. Although we basically proved this in Section 3.6, we give a different argument using cocycles. Let H be given by the homogeneous linear form $\ell = \sum_k a_k x_k = 0$. Then for the standard cover $U_i = \{x_i \neq 0\}$, H is defined by

$$\ell_i = \sum a_k \frac{x_k}{x_i} = 0.$$

Thus $\mathscr{O}(H)$ is determined by the 1-cocycle $\ell_i/\ell_j = x_j/x_i$, which is the cocycle for $\mathscr{O}(1)$. □

Lemma 7.5.7. *$c_1(\mathscr{O}_{\mathbb{P}^n}(1)) = [H]$, where $H \subset \mathbb{P}^n$ is a hyperplane.*

Proof. We have already checked this for $\mathbb{P}^1$ in Lemma 7.5.4. Embed $\mathbb{P}^1 \subset \mathbb{P}^n$ as a line. The restriction map induces an isomorphism on second cohomology with $\mathbb{Z}$ coefficients such that $[H]$ maps to the fundamental class of $\mathbb{P}^1$ by (7.2.2). Since c_1 is compatible with restriction, Lemma 7.5.4 implies that $c_1(\mathscr{O}_{\mathbb{P}^n}(1)) = [H]$. □

Theorem 7.5.8. *If D is a smooth effective divisor, then $c_1(\mathscr{O}_X(D)) = [D]$.*

Proof. There are several ways to prove this. We are going to outline the proof given in [63], which reduces it to the previous lemma. We are going to work in the C^∞ category, since most of the constructions used here cannot be made holomorphic. Choose an open tubular neighborhood T of D. This is a complex C^∞ line bundle such that D corresponds to the zero section. So there is a C^∞ classifying map $X \to \mathbb{P}^n$ such that the line bundle associated to $\mathscr{O}_{\mathbb{P}^n}(-1)$ pulls back to T by Theorem 2.6.6. By composing this with complex conjugation $\mathbb{P}^n \to \mathbb{P}^n$, we can arrange that the dual line bundle U associated to $\mathscr{O}_{\mathbb{P}^n}(1)$ pulls back to T (see the exercises).

The *Thom space* $\mathrm{Th}(M)$ of a line bundle $M \to X$ is just the one-point compactification of M. The Thom space $\mathrm{Th}(T)$ can simply be obtained by collapsing the complement of T in X to a point. From Exercise 3.6.5, we can see that $\mathrm{Th}(U)$ can be identified with $\mathbb{P}^{n+1}$. Under this identification, the zero section corresponds to $\mathbb{P}^n$ embedded as a hyperplane. Since $\mathrm{Th}(T)$ will map to $\mathrm{Th}(U)$, we obtain a commutative diagram.

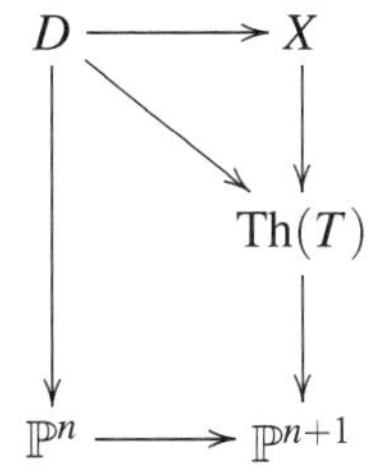

Using this, we can see that the fundamental class of $\mathbb{P}^n$ in $\mathbb{P}^{n+1}$ pulls back to $[D]$. Furthermore the C^∞ line bundle associated to $\mathscr{O}_X(D)$ is isomorphic to the pullback of the C^∞ line bundle associated to $\mathscr{O}_{\mathbb{P}^{n+1}}(1)$. This is because the duals can be identified with the ideal sheaves of D and $\mathbb{P}^n$ respectively. Thus the theorem follows from the previous lemma. □

Exercises

7.5.9. Prove Lemma 7.5.5.

7.5.10. Given a vector bundle V of rank r, define $\det(V) = \wedge^r V$ and $c_1(V) = c_1(\det V)$. Prove that $\det(V_1 \oplus V_2) \cong \det(V_1) \otimes \det(V_2)$. Use this to calculate $c_1(V_1 \oplus V_2)$.

7.5.11. A Hermitian metric $|| \, ||^2$ on a holomorphic line bundle $L \to X$ is a complex-valued C^∞ function on the total space L that restricts to an inner product on the fibers. Every line possesses a metric obtained using a partition of unity. Conclude that L can be represented by a cocycle g_{ij} with values in the unit circle. Use this to show that the pullback $\bar{L}$ of L under complex conjugation is isomorphic to the dual L^*.

7.5.12. Given a Hermitian metric on L and a local trivialization $\phi_i : L|_{U_i} \cong U_i \times \mathbb{C}$, let $h_i = ||\phi_i^{-1}(1)||^2$. This gives a collection of positive C^∞ functions satisfying $h_i = ||g_{ij}||^2 h_j$. Show that the curvature $\omega = \partial\bar{\partial} h_i$ is globally defined, and that $\frac{1}{2\pi\sqrt{-1}}\omega$ represents $c_1(L)$. (See (9.2.1) for a discussion of the $\partial, \bar{\partial}$ operators.)

Part III
Hodge Theory

Chapter 8
The Hodge Theorem for Riemannian Manifolds

Thus far, our approach has been primarily algebraic or topological. We are going to need a basic analytic result, namely the Hodge theorem. This says that on a compact oriented manifold equipped with a metric, every de Rham cohomology class has a unique "smallest" element, called its harmonic representative. When combined with the Kähler identities in later chapters, this will have strong consequences for complex algebraic geometry. The original proof is due to Hodge [64] with a correction by Weyl [121]. Different proofs were given by Bidal and de Rham [12] and Kodaira [72] soon there after. Standard accounts of these results, with the necessary details, can be found in the books of Griffiths and Harris [49], Warner [117], and Wells [120]. These books follow a similar approach of first establishing a weak Hilbert space version of the Hodge theorem, and then applying some regularity results from elliptic PDE theory to deduce the stronger statement. We will depart slightly from these treatments by outlining the heat equation method of Milgram and Rosenbloom [84]. This is an elegant and comparatively elementary approach to the Hodge theorem. As a warmup, we will do a combinatorial version that requires nothing more than linear algebra.

8.1 Hodge Theory on a Simplicial Complex

In order to motivate the general Hodge theorem, we first work out a simple combinatorial analogue. Let $K = (V, \Sigma)$ be a finite simplicial complex. Then the spaces of cochains $C^*(K, \mathbb{R})$ are finite-dimensional. For each simplex S, let

$$\delta_S(S') = \begin{cases} 1 \text{ if } S = S', \\ 0 \text{ otherwise.} \end{cases}$$

These form a basis of $C^*(K, \mathbb{R})$. Now choose inner products on these spaces. A particularly natural choice is determined by making this basis orthonormal. Let

$\partial^* : C^i(K,\mathbb{R}) \to C^{i-1}(K,\mathbb{R})$ be the adjoint to ∂. Then $\Delta = \partial\partial^* + \partial^*\partial$ is the discrete Laplacian.

Lemma 8.1.1. *Let α be a cochain. The following are equivalent:*

(a) $\alpha \in (\operatorname{im}\partial)^\perp \cap \ker\partial$.
(b) $\partial\alpha = \partial^*\alpha = 0$.
(c) $\Delta\alpha = 0$.

Proof. We prove the equivalences of (a) and (b), and of (b) and (c). Suppose that $\alpha \in (\operatorname{im}\partial)^\perp \cap \ker\partial$. Then of course $\partial\alpha = 0$. Furthermore, since

$$\|\partial^*\alpha\|^2 = \langle \alpha, \partial\partial^*\alpha \rangle = 0,$$

it follows that $\partial^*\alpha = 0$. Conversely, assuming (b), we have $\alpha \in \ker\partial$ and

$$\langle \alpha, \partial\beta \rangle = \langle \partial^*\alpha, \beta \rangle = 0.$$

If $\partial\alpha = \partial^*\alpha = 0$, then $\Delta\alpha = 0$. Finally, suppose that $\Delta\alpha = 0$. Then

$$\langle \Delta\alpha, \alpha \rangle = \|\partial\alpha\|^2 + \|\partial^*\alpha\|^2 = 0,$$

which implies (b). □

The cochains satisfying the above conditions are called *harmonic*.

Lemma 8.1.2. *Every simplicial cohomology class has a unique harmonic representative. The harmonic representative minimizes the norm in its cohomology class.*

Proof. To prove the first statement, we check that the map

$$h : (\operatorname{im}\partial)^\perp \cap \ker\partial \to H^n(K,\mathbb{R})$$

sending a harmonic cochain to its cohomology class is an isomorphism. If α lies in the kernel of h, then it is a coboundary. Therefore $\alpha = \partial\beta$, which implies that

$$\|\alpha\|^2 = \langle \alpha, \partial\beta \rangle = 0.$$

This shows that h is injective. Given a cochain α with $\partial\alpha = 0$, we can decompose it as $\alpha = \alpha_1 + \alpha_2$ with $\alpha_1 \in (\operatorname{im}\partial)^\perp \cap \ker\partial$ and $\alpha_2 \in \operatorname{im}\partial$. Thus the cohomology class of α is equal to $h(\alpha_1)$.

Given a harmonic element α,

$$\|\alpha + \partial\beta\|^2 = \|\alpha\|^2 + \|\partial\beta\|^2 > \|\alpha\|^2$$

unless $\partial\beta = 0$. □

Exercises

8.1.3. Prove that the space of cochains can be decomposed into an orthogonal direct sum

$$C^n(K) = \ker \Delta \oplus \operatorname{im} \partial \oplus \operatorname{im} \partial^*.$$

8.1.4. Prove that Δ is a positive semidefinite symmetric operator.

8.1.5. Use the previous exercise to prove that the limit of the "heat kernel"

$$H = \lim_{t \to \infty} e^{-t\Delta}$$

exists and is the orthogonal projection to $\ker \Delta$.

8.1.6. Green's operator G is the endomorphism of $C^n(K)$ that acts by 0 on $\ker \Delta$ and by $\frac{1}{\lambda}$ on the λ-eigenspaces of Δ with $\lambda \neq 0$. Check that $I = H + \Delta G$.

8.2 Harmonic Forms

Let X be an n-dimensional compact oriented manifold. We want to prove an analogue of Lemma 8.1.2 for de Rham cohomology. In order to formulate this, we need inner products. A Riemannian metric $(,)$, is a family of inner products on the tangent spaces that vary in a C^∞ fashion. This means that the inner products are determined by a tensor $g \in \Gamma(X, \mathscr{E}^1_X \otimes \mathscr{E}^1_X)$. The existence of Riemannian metrics can be proved using a standard partition of unity argument [110, 111, 117]. A Riemannian manifold is a C^∞ manifold endowed with a Riemannian metric. A metric determines inner products on exterior powers of the cotangent bundle, which will also be denoted by $(,)$. In particular, the inner product on the top exterior power gives a tensor $\det(g) \in \Gamma(X, \mathscr{E}^n_X \otimes \mathscr{E}^n_X)$. Since X is oriented, it is possible to choose a consistent square root of $\det(g)$, called the volume form $d\mathrm{vol} \in \mathscr{E}^n(X)$. In local coordinates,

$$d\mathrm{vol} = \sqrt{\det(g_{ij})}\, dx_1 \wedge \cdots \wedge dx_n,$$

where $\frac{\partial}{\partial x_1}, \ldots, \frac{\partial}{\partial x_n}$ is positively oriented. The Hodge star operator is a $C^\infty(X)$-linear operator $* : \mathscr{E}^k(X) \to \mathscr{E}^{n-k}(X)$, determined by

$$\alpha \wedge *\beta = (\alpha, \beta) d\mathrm{vol}. \tag{8.2.1}$$

This is easy to calculate explicitly. After choosing a local orthonormal basis or frame e_i for $\mathscr{E}^1_X$ in a neighborhood of a point, we have

$$*(e_{i_1} \wedge \cdots \wedge e_{i_k}) = \varepsilon e_{j_1} \wedge \cdots \wedge e_{j_{n-k}},$$

where $\{j_1,\dots,j_{n-k}\} = \{1,\dots,n\} - \{i_1,\dots,i_k\}$ and ε is given by the sign of the permutation

$$\begin{pmatrix} 1 & \dots & k & k+1 & \dots & n \\ i_1 & \dots & i_k & j_1 & \dots & j_{n-k} \end{pmatrix}.$$

From this, one gets that $** = (-1)^{k(n-k)}$ on k-forms. The spaces $\mathscr{E}^k(X)$ carry inner products,

$$\langle \alpha, \beta \rangle = \int_X (\alpha, \beta) d\mathrm{vol} = \int_X \alpha \wedge *\beta,$$

and hence norms.

Lemma 8.2.1. *For all α, β,*

$$\langle d\alpha, \beta \rangle = \pm \langle \alpha, *d*\beta \rangle.$$

In other words, the adjoint d^* to d is $\pm * d*$.

Proof. The lemma follows by integration by parts. More explicitly, apply Stokes's theorem to the identity

$$d(\alpha \wedge *\beta) = d\alpha \wedge *\beta \pm \alpha \wedge **d*\beta. \qquad \square$$

The key result of this chapter is the following theorem:

Theorem 8.2.2 (The Hodge theorem). *Every de Rham cohomology class has a unique representative that minimizes the norm. This is called the harmonic representative.*

To understand the meaning of the harmonicity condition, we can treat it as a variational problem and find the Euler–Lagrange equation. Let α be a harmonic p-form. Then for any $(p-1)$-form β, we would have to have

$$\frac{d}{dt} ||\alpha + td\beta||^2|_{t=0} = 2\langle \alpha, d\beta \rangle = 2\langle d^*\alpha, \beta \rangle = 0,$$

which forces $d^*\alpha = 0$. Conversely, if $d^*\alpha = 0$, then

$$||\alpha + d\beta||^2 = ||\alpha||^2 + ||d\beta||^2,$$

which implies that α has the minimum norm in its cohomology class. Thus harmonicity can be expressed as a pair of differential equations $d\alpha = 0$ and $d^*\alpha = 0$. It is sometimes more convenient to combine these into a single equation. For this we need the following definition:

Definition 8.2.3. The Hodge Laplacian is $\Delta = d^*d + dd^*$.

Lemma 8.2.4. *The following are equivalent:*

(a) α *is harmonic.*
(b) $d\alpha = 0$ *and* $d^*\alpha = 0$.
(c) $\Delta\alpha = 0$.

Proof. The equivalence of the first two conditions is contained in the previous discussion. The equivalence of the last two conditions follows from the identity $\langle \Delta\alpha, \alpha\rangle = \|d\alpha\|^2 + \|d^*\alpha\|^2$. □

The hard work is contained in the following result, whose proof will be postponed until the next section.

Theorem 8.2.5. *There are linear operators H (harmonic projection) and G (Green's operator) taking* C^∞ *forms to* C^∞ *forms that are characterized by the following properties:*

(a) $H(\alpha)$ *is harmonic,*
(b) $G(\alpha)$ *is orthogonal to the space of harmonic forms,*
(c) $\alpha = H(\alpha) + \Delta G(\alpha)$,

for any C^∞ *form* α.

Corollary 8.2.6. *There is an orthogonal direct sum*

$$\mathscr{E}^i(X) = (\textit{harmonic forms}) \oplus d\mathscr{E}^{i-1}(X) \oplus d^*\mathscr{E}^{i+1}(X).$$

Proof. Exercise. □

We can prove Theorem 8.2.2 modulo Theorem 8.2.5.

Proof. We prove the existence part of the theorem. The uniqueness is straightforward and left as an exercise. Let α be a closed form. It can be written as $\alpha = \beta + dd^*\gamma + d^*d\gamma$ with $\beta = H(\alpha)$ and $\gamma = G(\alpha)$. Since β is harmonic,

$$\langle d^*d\gamma, \beta\rangle = \langle d\gamma, d\beta\rangle = 0.$$

Furthermore,

$$\langle d^*d\gamma, dd^*\gamma\rangle = \langle d^*\gamma, d^2d^*\gamma\rangle = 0.$$

Thus

$$\|d^*d\gamma\|^2 = \langle d^*d\gamma, \alpha\rangle = \langle d\gamma, d\alpha\rangle = 0.$$

Therefore $\alpha = \beta + dd^*\gamma$ is cohomologous to the harmonic form β. □

Theorem 8.2.5 remains, and this will be dealt with later. For now we consider the easy, but instructive, example of the torus.

Example 8.2.7. $X = \mathbb{R}^n/\mathbb{Z}^n$, with the Euclidean metric. A differential form α can be expanded in a Fourier series

$$\alpha = \sum_{\lambda \in \mathbb{Z}^n} \sum_{i_1 < \cdots < i_p} a_{\lambda, i_1 \ldots i_p} e^{2\pi i \lambda \cdot \mathbf{x}} dx_{i_1} \wedge \cdots \wedge dx_{i_p}. \tag{8.2.2}$$

By direct calculation, one finds the Laplacian

$$\Delta = -\sum \frac{\partial^2}{\partial x_i^2},$$

the harmonic projection

$$H(\alpha) = \sum a_{0, i_1 \ldots i_p} dx_{i_1} \wedge \cdots \wedge dx_{i_p},$$

and Green's operator

$$G(\alpha) = \sum_{\lambda \in \mathbb{Z}^n - \{0\}} \sum \frac{a_{\lambda, i_1 \ldots i_p}}{4\pi^2 |\lambda|^2} e^{2\pi i \lambda \cdot \mathbf{x}} dx_{i_1} \wedge \cdots \wedge dx_{i_p}.$$

Since the image of H consists of forms with constant coefficients, this proves the assertion in Example 5.4.13.

As an application of this theorem, we obtain a new proof of Poincaré duality in a strengthened form.

Corollary 8.2.8 (Poincaré duality, reprise). *The pairing*

$$H^i(X, \mathbb{R}) \times H^{n-i}(X, \mathbb{R}) \to \mathbb{R}$$

induced by $(\alpha, \beta) \mapsto \int \alpha \wedge \beta$ *is a perfect pairing.*

Proof. The operator $*$ induces an isomorphism between the space of harmonic i-forms and $(n-i)$-forms. This proves directly that $H^i(X, \mathbb{R})$ and $H^{n-i}(X, \mathbb{R})$ are isomorphic.

Consider the map

$$\lambda : H^i(X, \mathbb{R}) \to H^{n-i}(X, \mathbb{R})^*$$

given by $\lambda(\alpha) = \beta \mapsto \int \alpha \wedge \beta$. We need to prove that λ is an isomorphism. Since these spaces have the same dimension, it is enough to prove that $\ker(\lambda) = 0$. But this is clear, since $\lambda(\alpha)(*\alpha) \neq 0$ whenever α is a nonzero harmonic form. □

Exercises

8.2.9. Give a careful proof of Lemma 8.2.1 and determine the signs.

8.2.10. Show that a harmonic exact form $d\gamma$ satisfies $||d\gamma||^2 = 0$. Use this to prove the uniqueness in Theorem 8.2.2.

8.2.11. Prove Corollary 8.2.6.

8.2.12. Fill in the details of Example 8.2.7.

8.2.13. Let X be a compact Riemannian manifold with a group of isometries G, i.e., G preserves the metric. Then show that if α is harmonic, then $g^*\alpha$ is also harmonic for any $g \in G$.

8.2.14. Suppose that G is a connected Lie group with an action $G \times X \to X$ by isometries on a compact Riemannian manifold X. Then show that any harmonic form is G-invariant.

8.3 The Heat Equation*

We will outline an approach to Theorem 8.2.5 using the heat equation due to Milgram and Rosenbloom [84]. However, we will deviate slightly from their presentation, which is a bit too sketchy in places. The heuristic behind the proof is that if the form α is thought of as an initial temperature, then the temperature should approach a harmonic steady state as the manifold cools. So our task is to solve the heat equation:

$$\frac{\partial A(t)}{\partial t} = -\Delta A(t), \tag{8.3.1}$$

$$A(0) = \alpha, \tag{8.3.2}$$

for all $t > 0$, and study the behavior as $t \to \infty$. For simplicial complexes, this was dealt with in the exercises of the first section.

We start with a few general remarks.

Lemma 8.3.1. *If $A(t)$ is a C^∞ solution of (8.3.1), then $||A(t)||^2$ is (nonstrictly) decreasing.*

Proof. Upon differentiating $||A(t)||^2$, we obtain

$$2\left\langle \frac{\partial A}{\partial t}, A \right\rangle = -2\langle \Delta A, A\rangle = -2(||dA||^2 + ||d^*A||^2) \leq 0. \qquad \square$$

Corollary 8.3.2. *A solution to (8.3.1)–(8.3.2) would be unique.*

Proof. Given two solutions, their difference satisfies $A(0) = 0$, so it remains 0. $\square$

The starting point for the proof of existence is the observation that when X is replaced by Euclidean space and α is a compactly supported function, then one has an explicit solution to (8.3.1)–(8.3.2),

$$A(x,t) = \int_{\mathbb{R}^n} K(x,y,t)\alpha(y)dy = \langle K(x,y,t), \alpha(y)\rangle_y,$$

where the heat kernel is given by

$$K(x,y,t) = (4\pi t)^{-n/2} e^{-\|x-y\|^2/4t}, \tag{8.3.3}$$

and the symbol $\langle , \rangle_y$ is the L^2 inner product with integration carried out with respect to y.

This can be modified to make sense for a p-form on $\mathbb{R}^n$ by replacing K above by

$$K(x,y,t) = (4\pi t)^{-n/2} e^{-\|x-y\|^2/4t} \left(\sum_i dx_i \wedge dy_i \right)^p.$$

When working with forms on $X \times X \times [0,\infty)$, we will abuse notation a bit and write them as local expressions such as $\eta(x,y,t)$. Then $d_x\eta(x,y,t)$ etc. will indicate that the operations $d\ldots$ are preformed with y,t treated as constant (or more correctly, these operations are preformed fiberwise along a projection).

Theorem 8.3.3. *On any compact Riemannian manifold, there exists a C^∞ p-form $K(x,y,t)$ called the heat kernel such that for any p-form α,*

$$A(x,t) = \langle K(x,y,t)\alpha(y)\rangle_y$$

gives a solution to the heat equation (8.3.1) with $A(x,0) = \alpha(x)$.

Proof. We sketch the idea, referring to [11, Chapter 2] for details. See also [19, 94] for alternative presentations. For notational simplicity, assume that $p = 0$. The starting point is the observation that (8.3.3) makes sense on a general Riemannian manifold, provided that $\|x-y\|$ is replaced by the Riemannian distance, which is the infimum of lengths of curves connecting x to y. More precisely, let $\delta(x,y)^2$ be a nonnegative C^∞ function on $X \times X$ that agrees with the square of the Riemannian distance function in a neighborhood of the diagonal and vanishes far away from it. Set

$$K_0(x,y,t) = (4\pi t)^{-n/2} e^{-\delta(x,y)^2/4t}.$$

This is only an approximation to the true heat kernel in general. The idea will be to add successive corrections to K_0, resulting in an infinite series $K = K_0 + \cdots$ called a Volterra series. We note that K_0 satisfies a crucial property, that it approaches the "δ-function of $x-y$" as $t \to 0$ in the sense that

$$\lim_{t\to 0} \langle K_0(x,y,t), \alpha(y)\rangle_y = \alpha(x).$$

We define the convolution of two functions $C(x,y,t)$ and $B(x,y,t)$ by

$$(C * B)(x,y,t) = \int_0^t \langle C(x,z,t-\tau), B(z,y,\tau)\rangle_z d\tau.$$

This is an associative operation. Let $R = \left(\Delta_x + \frac{\partial}{\partial t}\right) K_0$, which is called the remainder. Given a C^∞ function $B(x,y,t)$, we claim that

$$\left(\Delta_x + \frac{\partial}{\partial t}\right)(K_0 * B) = B + R * B. \tag{8.3.4}$$

To see this, let

$$h(x,y,t,\tau) = \langle K_0(x,z,t-\tau), B(z,y,\tau)\rangle_z$$

be the integrand of $K_0 * B$. Then (8.3.4) follows formally by adding

$$\frac{\partial}{\partial t}\int_0^t h(x,y,t,\tau)d\tau = \lim_{\tau\to t} h(x,y,t,\tau) + \int_0^t \frac{\partial}{\partial t} h(x,y,t,\tau)d\tau$$

to

$$\Delta_x \int_0^t h(x,y,t,\tau)d\tau = \int_0^t \Delta_x h(x,y,t,\tau)d\tau.$$

The Volterra series is given by

$$K = K_0 - K_0 * R + K_0 * R * R - \cdots = K_0 + K_0 * (-R + R * R - \cdots). \tag{8.3.5}$$

Applying (8.3.4) formally term by term gives the telescoping series

$$\left(\Delta_x + \frac{\partial}{\partial t}\right) K = R - (R + R * R) + (R * R + R * R * R) - \cdots = 0.$$

It would follow that

$$A(x,t) = \langle K(x,y,t), \alpha(y)\rangle_y$$

satisfies the heat equation and the required initial condition, provided that these formal arguments can be justified and also provided that the "tail" $K_0 * (-R + R * R + \cdots)$ goes to zero with t. In fact, to guarantee uniform convergence of (8.3.5) and its derivatives up to second order on $X^2 \times [0,T]$, K_0 needs to be replaced with an expression K_N with better asymptotics. This step, due to Minakshisundaram and Pleijel, is given by constructing

$$K_N(x,y,t) = K_0(x,y,t)[u_0(x,y) + tu_1(x,y) + \cdots + t^N u_N(x,y)]$$

with $N \gg 0$ and appropriate explicitly chosen functions u_i so that there is an estimate on its remainder

$$|R_N| \le Ct^{N-n/2} \tag{8.3.6}$$

and similarly for its derivatives. This will lead to a bound

$$|R_N^{*\ell}| \le C_\ell t^{\ell(N-n/2)+\ell-1} \tag{8.3.7}$$

and a similar estimate on derivatives, where C_ℓ are constants that go rapidly to 0 as $\ell \to \infty$ (Exercise 8.3.8). This suffices to justify convergence of $-R_N + R_N * R_N - \cdots$ along with the other statements. □

Let

$$T_t(\alpha) = \langle K(x,y,t), \alpha(y)\rangle_y$$

with K as in the last theorem. This is the unique solution to (8.3.1) and (8.3.2). Theorem 8.2.5 can now be deduced from the following:

Theorem 8.3.4.

(a) The semigroup property $T_{t_1+t_2} = T_{t_1}T_{t_2}$ *holds.*
(b) T_t *is formally self-adjoint.*
(c) $T_t\alpha$ *converges to a* C^∞ *harmonic form* $H(\alpha)$.
(d) The integral

$$G(\alpha) = \int_0^\infty (T_t\alpha - H\alpha)dt$$

is well defined, and yields Green's operator.

Proof. We give the main ideas. The semigroup property $T_{t_1+t_2} = T_{t_1}T_{t_2}$ holds because $A(t_1+t_2)$ can be obtained by solving the heat equation with initial condition $A(t_2)$ and then evaluating it at $t = t_1$.

To see that T_t is self-adjoint, calculate

$$\begin{aligned}\frac{\partial}{\partial t}\langle T_t\eta, T_\tau\xi\rangle &= \left\langle \frac{\partial}{\partial t}T_t\eta, T_\tau\xi \right\rangle = -\langle \Delta T_t\eta, T_\tau\xi\rangle \\ &= -\langle T_t\eta, \Delta T_\tau\xi\rangle = \left\langle T_t\eta, \frac{\partial}{\partial \tau}T_\tau\xi \right\rangle = \frac{\partial}{\partial \tau}\langle T_t\eta, T_\tau\xi\rangle,\end{aligned}$$

which implies that $\langle T_t\eta, T_\tau\xi\rangle$ can be written as a function of $t+\tau$, say $g(t+\tau)$. Therefore

$$\langle T_t\eta, \xi\rangle = g(t+0) = g(0+t) = \langle \eta, T_t\xi\rangle.$$

Properties (a) and (b) imply that for $h \geq 0$, we have

$$\begin{aligned}||T_{t+2h}\alpha - T_t\alpha||^2 &= ||T_{t+2h}\alpha||^2 + ||T_t\alpha||^2 - 2\langle T_{t+2h}\alpha, T_t\alpha\rangle \\ &= ||T_{t+2h}\alpha||^2 + ||T_t\alpha||^2 - 2||T_{t+h}\alpha||^2 \\ &= (||T_{t+2h}\alpha|| - ||T_t\alpha||)^2 - 2(||T_{t+h}\alpha||^2 - ||T_{t+2h}\alpha|| \cdot ||T_t\alpha||);\end{aligned}$$

$||T_t\alpha||^2$ converges thanks to Lemma 8.3.1. Therefore $||T_{t+2h}\alpha - T_t\alpha||^2$ can be made arbitrarily small for large t. This implies that $T_t\alpha$ converges in the L^2 sense to an L^2 form $H(\alpha)$, i.e., an element of the Hilbert space completion of $\mathcal{E}^p(X)$. Fix $\tau > 0$. The relations $T_t\alpha = T_\tau T_{t-\tau}\alpha$, imply in the limit that $T_\tau H(\alpha) = H(\alpha)$. Since T_τ is given by an integral transform with smooth kernel, it follows that $H(\alpha)$ is C^∞. The equation $T_\tau H(\alpha) = H(\alpha)$ shows that $H(\alpha)$ is in fact harmonic. We also note that H is formally self-adjoint,

$$\langle H\alpha, \beta\rangle = \lim_{t\to\infty}\langle T_t\alpha, \beta\rangle = \lim_{t\to\infty}\langle \alpha, T_t\beta\rangle = \langle \alpha, H\beta\rangle.$$

The pointwise norms $||T_t\alpha(x) - H\alpha(x)||$ can be shown to decay rapidly enough so that the integral

$$G(\alpha) = \int_0^\infty (T_t\alpha - H\alpha)dt$$

is well defined. We will verify formally that this is Green's operator:

$$\Delta G(\alpha) = \int_0^\infty \Delta T_t \alpha dt = -\int_0^\infty \frac{\partial T_t \alpha}{\partial t} dt = \alpha - H(\alpha),$$

and for β harmonic,

$$\langle G(\alpha), \beta \rangle = \int_0^\infty \langle (T_t - H)\alpha, \beta \rangle dt = \int_0^\infty \langle \alpha, (T_t - H)\beta \rangle dt = 0,$$

as required. □

Let us return to the example of a torus, where things can be calculated explicitly.

Example 8.3.5. Let $X = \mathbb{R}^n/\mathbb{Z}^n$. Given α as in (8.2.2), the solution to the heat equation with initial value α is given by

$$T_t \alpha = \sum_{\lambda \in \mathbb{Z}^n} \sum a_{\lambda, i_1 \dots i_p} e^{(2\pi i \lambda \cdot \mathbf{x} - 4\pi^2 |\lambda|^2 t)} dx_{i_1} \wedge \dots \wedge dx_{i_p},$$

and this converges to the harmonic projection

$$H(\alpha) = \sum a_{0, i_1 \dots i_p} dx_{i_1} \wedge \dots \wedge dx_{i_p};$$

$T_t \alpha - H\alpha$ can be integrated term by term to obtain Green's operator

$$G(\alpha) = \sum_{\lambda \in \mathbb{Z}^n - \{0\}} \sum \frac{a_{\lambda, i_1 \dots i_p}}{4\pi^2 |\lambda|^2} e^{2\pi i \lambda \cdot \mathbf{x}} dx_{i_1} \wedge \dots \wedge dx_{i_p}.$$

Exercises

8.3.6. Suppose that α is an eigenfunction, so that $\Delta \alpha = \lambda \alpha$. Give an explicit formula for $T_t \alpha$.

8.3.7. Check that Example 8.3.5 works as claimed.

8.3.8. Given $A(x,y,t)$, check that

$$A^{*\ell}(x,y,t) = \int_{0 \le t_1 \le \dots \le t_{\ell-1} \le t} \int_{X^{\ell-1}} A(x, z_1, t - t_1) A(z_1, z_2, t_1 - t_2) \cdots$$

(suitably interpreted). Deduce a bound

$$|A^{*\ell}| \le \frac{t^{\ell-1} \mathrm{vol}(X)^{\ell-1} M^\ell}{(\ell - 1)!},$$

where $M = \sup |A(x,y,t)|$ over $X^2 \times (0,t)$.

Chapter 9
Toward Hodge Theory for Complex Manifolds

From now on, we are going to work almost exclusively with complex-valued functions and forms. So we revise our notation accordingly. Given a C^∞ manifold X, let C_X^∞ (respectively $\mathscr{E}_X^k$) now denote the space of complex-valued C^∞ functions (respectively k-forms). We write $C_{X,\mathbb{R}}^\infty$ (or $\mathscr{E}_{X,\mathbb{R}}^k$) for the space of real-valued functions or forms. Let us say that a complex-valued form is exact, closed, or harmonic if its real and imaginary parts both have this property. Then de Rham's theorem and Hodge's theorem carry over almost word for word: $H^k(X,\mathbb{C})$ is isomorphic to the space of complex closed k-forms modulo exact forms, and if X is compact and oriented with a Riemannian metric, then it is also isomorphic to the space of complex harmonic k-forms. This can be checked easily by working with the real and imaginary parts separately.

To go deeper, we should ask how de Rham and Hodge theory interact with the holomorphic structure when X is a complex manifold. This is really a central question in complex algebraic geometry. In this chapter, which is really a warmup for the next, we take the first few steps toward answering this. Here we concentrate on some special cases such as Riemann surfaces and tori, which can be handled without explicitly talking about Kähler metrics. In these cases, we will see that the answer is as nice as one can hope for. We will see, for instance, that the genus of a Riemann surface, which a priori is a topological invariant, can be interpreted as the number of linearly independent holomorphic 1-forms.

9.1 Riemann Surfaces Revisited

Fix a compact Riemann surface X with genus g, which we can define to be one-half of the first Betti number. In this section, we tie up a loose end from Chapter 6, by proving Proposition 6.2.9, that

$$g = \dim H^0(X,\Omega_X^1) = \dim H^1(X,\mathscr{O}_X).$$

This will be an easy application of the Hodge theorem. In order to use it, we need to choose a Riemannian metric that is a C^∞ family of inner products on the tangent spaces. We will also impose a compatibility condition that multiplication by $\mathrm{i} = \sqrt{-1}$ preserves the angles determined by these inner products. To say this more precisely, view X as a two-dimensional real C^∞ manifold. Choosing an analytic local coordinate $z = x + \mathrm{i}y$ in a neighborhood of U, the vectors $v_1 = \frac{\partial}{\partial x}$ and $v_2 = \frac{\partial}{\partial y}$ give a basis (or frame) of the real tangent sheaf $\mathscr{T}_X$ of X restricted to U. The automorphism $J_p : \mathscr{T}_X|_U \to \mathscr{T}_X|_U$ represented by

$$\begin{pmatrix} 0 & 1 \\ -1 & 0 \end{pmatrix}$$

in the basis v_1, v_2 is independent of this basis, and hence globally well defined. A Riemannian metric $(,)$ is said to be compatible with the complex structure, or *Hermitian*, if the transformations J_p are orthogonal. In terms of the basis v_1, v_2 this forces the matrix of the bilinear form $(,)$ to be a positive multiple of I by some function h. In coordinates, the metric would be represented by a tensor $h(x,y)(dx \otimes dx + dy \otimes dy)$. The volume form is represented by $h dx \wedge dy$. It follows that $*dx = dy$ and $*dy = -dx$. In other words, $*$ is the transpose of J, which is independent of h. Once we have $*$, we can define all the operators from the last chapter.

Standard partition of unity arguments show that Hermitian metrics always exist. For our purposes, one metric is as good as any other, so we simply choose one.

Lemma 9.1.1. *A* 1*-form is harmonic if and only if its* $(1,0)$ *and* $(0,1)$ *parts are respectively holomorphic and antiholomorphic.*

Proof. Given a local coordinate $z = x + \mathrm{i}y$,

$$*dz = *(dx + \mathrm{i}dy) = dy - \mathrm{i}dx = -\mathrm{i}dz, \tag{9.1.1}$$

and similarly

$$*d\bar{z} = \mathrm{i}d\bar{z}. \tag{9.1.2}$$

If α is a $(1,0)$-form, then $d\alpha = \bar{\partial}\alpha$. Thus α is holomorphic if and only if it is closed if and only if $d\alpha = d * \alpha = 0$. The last condition is equivalent to harmonicity by Lemma 8.2.4. By a similar argument a $(0,1)$-form is antiholomorphic if and only if it is harmonic. This proves one direction.

By (9.1.1) and (9.1.2) the $(1,0)$ and $(0,1)$ parts of a 1-form are linear combinations of α and $*\alpha$. Thus if α is harmonic, then so are its parts. □

Corollary 9.1.2. $\dim H^0(X, \Omega^1_X)$ *equals the genus of* X.

Proof. By the Hodge theorem, the first Betti number $2g$ is the dimension of the space of harmonic 1-forms, which decomposes into a direct sum of the spaces of holomorphic and antiholomorphic 1-forms. Both these spaces have the same dimension, since conjugation gives a real isomorphism between them. Therefore

$$2g = 2\dim H^0(X, \Omega_X^1).$$ □

Lemma 9.1.3. *The images of Δ and $\partial\bar{\partial}$ on $\mathscr{E}^2(X)$ coincide.*

Proof. On the space of 2-forms, we have $\Delta = -d*d*$. Computing in local coordinates yields

$$\partial\bar{\partial} f = -\frac{\mathrm{i}}{2}\left(\frac{\partial^2 f}{\partial x^2} + \frac{\partial^2 f}{\partial y^2}\right) dx \wedge dy$$

and

$$d*df = \left(\frac{\partial^2 f}{\partial x^2} + \frac{\partial^2 f}{\partial y^2}\right) dx \wedge dy,$$

which implies the lemma. □

Proposition 9.1.4. *The map $H^1(X, \mathscr{O}_X) \to H^1(X, \Omega_X^1)$ induced by d vanishes.*

Proof. We use the descriptions of these spaces as $\bar{\partial}$-cohomology groups provided by Corollary 6.2.5. Given $\alpha \in \mathscr{E}^{01}(X)$, let $\beta = d\alpha$. We have to show that β lies in the image of $\bar{\partial}$. Theorem 8.2.5 shows that we can write $\beta = H(\beta) + \Delta G(\beta)$. Since β is exact, we can conclude that $H(\beta) = 0$ by Corollary 8.2.6. Therefore β lies in the image of $\partial\bar{\partial} = -\bar{\partial}\partial$. □

Corollary 9.1.5. *The map $H^1(X, \mathbb{C}) \to H^1(X, \mathscr{O}_X)$ is surjective, and* $\dim H^1(X, \mathscr{O}_X)$ *coincides with the genus of X.*

Proof. The surjectivity is immediate from the exact sequence

$$H^1(X, \mathbb{C}) \to H^1(X, \mathscr{O}_X) \to H^1(X, \Omega_X^1).$$

The second part follows from the equation

$$\dim H^1(X, \mathscr{O}_X) = \dim H^1(X, \mathbb{C}) - \dim H^0(X, \Omega_X).$$ □

Exercises

9.1.6. Show that $H^1(X, \mathbb{C}) \to H^1(X, \mathscr{O}_X)$ can be identified with the projection of harmonic 1-forms to antiholomorphic $(0,1)$-forms.

9.1.7. Calculate the spaces of harmonic and holomorphic one-forms explicitly for an elliptic curve.

9.1.8. Show that the pairing $(\alpha, \beta) \mapsto \int_X \alpha \wedge \overline{\beta}$ is positive definite.

9.2 Dolbeault's Theorem

We now extend the results from Riemann surfaces to higher dimensions. Given an n-dimensional complex manifold X, let $\mathscr{O}_X$ denote the sheaf of holomorphic functions. We can regard X as a $2n$-dimensional (real) C^∞ manifold as explained in Section 2.2. As explained in the introduction, $\mathscr{E}_X^k$ will now denote the sheaf of C^∞ complex-valued k-forms. We have

$$\mathscr{E}_X^k(U) = \mathbb{C} \otimes_{\mathbb{R}} \mathscr{E}_{X,\mathbb{R}}^k(U).$$

By a *real structure* on a complex vector space V, we mean a real vector space $V_{\mathbb{R}}$ and an isomorphism $\mathbb{C} \otimes V_{\mathbb{R}} \cong V$. This gives rise to a $\mathbb{C}$-antilinear involution $v \mapsto \bar{v}$ given by $\overline{a \otimes v} = \bar{a} \otimes v$. Conversely, such an involution gives rise to the real structure $V_{\mathbb{R}} = \{v \mid \bar{v} = v\}$. In particular, $\mathscr{E}_X^k(U)$ has a natural real structure.

The sheaf of holomorphic p-forms Ω_X^p is a subsheaf of $\mathscr{E}_X^p$ stable under multiplication by $\mathscr{O}_X$. This sheaf is locally free as an $\mathscr{O}_X$-module. If $z_1, \ldots, z_n$ are holomorphic coordinates defined on an open set $U \subset X$, then

$$\{dz_{i_1} \wedge \cdots \wedge dz_{i_p} \mid i_1 < \cdots < i_p\}$$

gives a basis for $\Omega_X^p(U)$. To simplify our formulas, we let $dz_I = dz_{i_1} \wedge \cdots \wedge dz_{i_p}$, where $I = \{i_1, \ldots, i_p\}$.

Definition 9.2.1. Let $\mathscr{E}_X^{(p,0)}$ denote the C^∞ submodule of $\mathscr{E}_X^p$ generated by Ω_X^p. Let $\mathscr{E}_X^{(0,p)} = \overline{\mathscr{E}_X^{(p,0)}}$ and $\mathscr{E}_X^{(p,q)} = \mathscr{E}_X^{(p,0)} \wedge \mathscr{E}_X^{(0,q)}$.

In local coordinates, $\{dz_I \wedge d\bar{z}_J \mid \#I = p, \#J = q\}$ gives a basis of $\mathscr{E}_X^{(p,q)}(U)$. All of the operations of Section 6.2 can be extended to the higher-dimensional case. The operators

$$\partial : \mathscr{E}_X^{(p,q)} \to \mathscr{E}_X^{(p+1,q)}$$

and

$$\bar{\partial} : \mathscr{E}_X^{(p,q)} \to \mathscr{E}_X^{(p,q+1)}$$

are given locally by

$$\partial\left(\sum_{I,J} f_{I,J}\, dz_I \wedge d\bar{z}_J\right) = \sum_{I,J} \sum_{i=1}^{n} \frac{\partial f_{I,J}}{\partial z_i} dz_i \wedge dz_I \wedge d\bar{z}_J,$$

$$\bar{\partial}\left(\sum_{I,J} f_{I,J}\, dz_I \wedge d\bar{z}_J\right) = \sum_{I,J} \sum_{j=1}^{n} \frac{\partial f_{I,J}}{\partial \bar{z}_j} d\bar{z}_j \wedge dz_I \wedge d\bar{z}_J.$$

The identities

$$d = \partial + \bar{\partial}, \quad \partial^2 = \bar{\partial}^2 = 0, \quad \partial\bar{\partial} + \bar{\partial}\partial = 0, \tag{9.2.1}$$

hold.

Theorem 6.2.2 has the following extension.

Theorem 9.2.2. *Let $D \subset \mathbb{C}^n$ be an open polydisk (i.e., a product of disks). Given $\alpha \in \mathscr{E}^{(p,q)}(\bar{D})$ with $\bar{\partial}\alpha = 0$, there exists $\beta \in \mathscr{E}^{(p,q-1)}(D)$ such that $\alpha = \bar{\partial}\beta$.*

Proof. See [49, pp. 25–26]. □

Corollary 9.2.3 (Dolbeault's theorem I). *For any complex manifold X,*

(a)

$$0 \to \Omega_X^p \to \mathscr{E}_X^{(p,0)} \xrightarrow{\bar{\partial}} \mathscr{E}_X^{(p,1)} \xrightarrow{\bar{\partial}} \cdots$$

is a soft resolution.

(b)

$$H^q(X, \Omega_X^p) \cong \frac{\ker[\bar{\partial} : \mathscr{E}^{(p,q)}(X) \to \mathscr{E}^{(p,q+1)}]}{\operatorname{im}[\bar{\partial} : \mathscr{E}^{(p,q-1)}(X) \to \mathscr{E}^{(p,q)}]}.$$

Proof. Since exactness can be checked on the stalks, there is no loss in assuming $X = D$ for (a). The only thing not stated above is that Ω_X^p is the kernel of the $\bar{\partial}$ operator on $\mathscr{E}_X^{(p,0)}$. This is a simple calculation. Given a $(p,0)$-form $\sum_I f_I dz_I$,

$$\bar{\partial}\left(\sum_I f_I dz_I\right) = \sum_I \sum_{j=1}^n \frac{\partial f_I}{\partial \bar{z}_j} d\bar{z}_j \wedge dz_I = 0$$

if and only it is holomorphic. Thus the sheaves $\mathscr{E}_X^{(p,\bullet)}$ give a resolution, which is soft since these are modules over C_X^∞.

(b) is now a consequence of Theorem 5.1.4. □

In the sequel, we will refer to elements of $\ker \bar{\partial}$ (or $\operatorname{im} \bar{\partial}$) as $\bar{\partial}$-exact (or $\bar{\partial}$-closed).

Exercises

9.2.4. Check the identities (9.2.1).

9.2.5. Give an explicit description of the map $H^i(X, \mathbb{C}) \to H^i(X, \mathscr{O}_X)$ induced by inclusion $\mathbb{C}_X \to \mathscr{O}_X$ as a projection from de Rham cohomology to $\bar{\partial}$-cohomology.

9.3 Complex Tori

A *complex torus* is a quotient $X = V/L$ of a finite-dimensional complex vector space by a lattice (i.e., a discrete subgroup of maximal rank). Thus it is both a complex manifold and a torus. After choosing a basis, we may identify V with $\mathbb{C}^n$. Let $z_1, \ldots, z_n$ be the standard complex coordinates on $\mathbb{C}^n$, and let $x_i = \operatorname{Re}(z_i)$, $y_i = \operatorname{Im}(z_i)$.

We give X the flat metric induced by the Euclidean metric on V. Recall that harmonic forms with respect to this are the forms with constant coefficients (Example 8.2.7).

Lemma 9.3.1. *A holomorphic form on X has constant coefficients and is therefore harmonic.*

Proof. The coefficients of a holomorphic form $\sum f_I dz_I$ are holomorphic functions on X. These are constant because X is compact. □

This can be refined.

Proposition 9.3.2. *$H^q(X,\Omega_X^p)$ is isomorphic to the space of (p,q)-forms with constant coefficients.*

Corollary 9.3.3. *Set*

$$H^{(p,q)} = \bigoplus_{\#I=p,\#J=q} \mathbb{C} dz_I \wedge d\bar{z}_J.$$

Then $H^q(X,\Omega_X^p) \cong H^{(p,q)} \cong \wedge^p \mathbb{C}^n \otimes \wedge^q \mathbb{C}^n$.

The isomorphism in the corollary is highly noncanonical. A more natural identification is

$$H^q(X,\Omega_X^p) \cong \wedge^p V^* \otimes \wedge^q \bar{V}^*,$$

where V^* is the usual dual, and $\bar{V}^*$ is the set of antilinear maps from V to $\mathbb{C}$.

The proof of Proposition 9.3.2 hinges on a certain identity between Laplacians that we now define. The space of forms carries inner products as in Section 8.2, where X is equipped with the flat metric. Let ∂^* and $\bar{\partial}^*$ denote the adjoints to ∂ and $\bar{\partial}$ respectively. These will be calculated explicitly below. We can define the ∂- and $\bar{\partial}$-Laplacians by

$$\Delta_\partial = \partial^* \partial + \partial \partial^*,$$
$$\Delta_{\bar{\partial}} = \bar{\partial}^* \bar{\partial} + \bar{\partial} \bar{\partial}^*.$$

Lemma 9.3.4. $\Delta = 2\Delta_{\bar{\partial}} = 2\Delta_\partial$.

We give two proofs. The first is by direct calculation.

Proof. Let

$$\alpha = \sum_{I,J} \alpha_{IJ} dz_I \wedge d\bar{z}_J.$$

Then

$$\begin{aligned}
\Delta_{\bar{\partial}}(\alpha) &= -2 \sum_{I,J,i} \frac{\partial^2 \alpha_{IJ}}{\partial z_i \partial \bar{z}_i} dz_I \wedge d\bar{z}_J \\
&= -\frac{1}{2} \sum_{I,J,i} \left(\frac{\partial^2 \alpha_{IJ}}{\partial x_i^2} + \frac{\partial^2 \alpha_{IJ}}{\partial y_i^2} \right) dz_I \wedge d\bar{z}_J \\
&= \frac{1}{2} \Delta(\alpha).
\end{aligned}$$

A similar calculation holds for $\Delta_\partial(\alpha)$. □

This implies Proposition 9.3.2:

Proof. By Dolbeault's theorem,

$$H^q(X,\Omega_X^p) \cong \frac{\ker[\bar{\partial}:\mathscr{E}^{(p,q)}(X)\to\mathscr{E}^{(p,q+1)}]}{\operatorname{im}[\bar{\partial}:\mathscr{E}^{(p,q-1)}(X)\to\mathscr{E}^{(p,q)}]}.$$

Let α be a $\bar{\partial}$-closed (p,q)-form. Decompose

$$\alpha = \beta + \Delta\gamma = \beta + 2\Delta_{\bar{\partial}}\gamma = \beta + \bar{\partial}\gamma_1 + \bar{\partial}^*\gamma_2$$

with β harmonic, which is possible by Theorem 8.2.5. We have

$$\|\bar{\partial}^*\gamma_2\|^2 = \langle\gamma_2,\bar{\partial}\bar{\partial}^*\gamma_2\rangle = \langle\gamma_2,\bar{\partial}\alpha\rangle = 0.$$

It is left as an exercise to check that β is of type (p,q), and that it is unique. Therefore the $\bar{\partial}$-class of α has a unique representative by a constant (p,q)-form. □

We will sketch a second proof of Lemma 9.3.4. Although it is much more complicated than the first, it has the advantage of generalizing nicely to Kähler manifolds. We introduce a number of auxiliary operators. Let i_k and $\bar{i}_k$ denote contraction with the vector fields $2\frac{\partial}{\partial z_k}$ and $2\frac{\partial}{\partial \bar{z}_k}$. Thus for example, $i_k(dz_k\wedge\alpha) = 2\alpha$. If we choose our Euclidean metric so that monomials in dx_i, dy_j are orthonormal, then the contractions i_k and $\bar{i}_k$ can be checked to be adjoints to $dz_k\wedge$ and $d\bar{z}_k\wedge$. Let

$$\omega = \frac{\sqrt{-1}}{2}\sum dz_k\wedge d\bar{z}_k = \sum dx_k\wedge dy_k,\quad L\alpha = \omega\wedge\alpha,$$

and

$$\Lambda = -\frac{\sqrt{-1}}{2}\sum \bar{i}_k i_k.$$

The operators L and Λ are adjoint. Using integration by parts (see [49, p. 113]), we get explicit formulas

$$\partial^*\alpha = -\sum\frac{\partial}{\partial\bar{z}_k}i_k\alpha,\quad \bar{\partial}^*\alpha = -\sum\frac{\partial}{\partial z_k}\bar{i}_k\alpha,$$

where the derivatives above are taken coefficient-wise.

Let $[A,B] = AB - BA$ denote the commutator. Then we have the following first-order Kähler identities:

Proposition 9.3.5.

(a) $[\Lambda,\bar{\partial}] = -\sqrt{-1}\partial^*$.
(b) $[\Lambda,\partial] = \sqrt{-1}\bar{\partial}^*$.

Proof. We check the second identity on the space of $(1,1)$-forms. The general case is more involved notationally but not essentially harder; see [49, p. 114]. There are two cases. First suppose that $\alpha = f dz_j\wedge d\bar{z}_k$, with $j\neq k$. Then

$$\begin{aligned}[\Lambda,\partial]\alpha &= \Lambda\partial\alpha \\ &= \Lambda\left(\sum_m \frac{\partial f}{\partial z_m} dz_m \wedge dz_j \wedge d\bar{z}_k\right) \\ &= 2\sqrt{-1}\frac{\partial f}{\partial z_k} dz_j \\ &= \sqrt{-1}\bar{\partial}^*\alpha.\end{aligned}$$

Next suppose that $\alpha = f dz_k \wedge d\bar{z}_k$. Then

$$\begin{aligned}[\Lambda,\partial]\alpha &= \Lambda\left(\sum_{m\neq k} \frac{\partial f}{\partial z_m} dz_m \wedge dz_k \wedge d\bar{z}_k\right) - \partial\left(-2\sqrt{-1}f\right) \\ &= -2\sqrt{-1}\left(\sum_{m\neq k} \frac{\partial f}{\partial z_m} dz_m - \sum_m \frac{\partial f}{\partial z_m} dz_m\right) \\ &= 2\sqrt{-1}\frac{\partial f}{\partial z_k} dz_k \\ &= \sqrt{-1}\bar{\partial}^*\alpha.\end{aligned}$$

□

We now give a second proof of Lemma 9.3.4. Upon substituting the first-order identities into the definitions of the various Laplacians, some remarkable cancellations take place:

Proof. We first establish $\partial\bar{\partial}^* + \bar{\partial}^*\partial = 0$,

$$\sqrt{-1}(\partial\bar{\partial}^* + \bar{\partial}^*\partial) = \partial(\Lambda\partial - \partial\Lambda) + (\Lambda\partial - \partial\Lambda)\partial = \partial\Lambda\partial - \partial\Lambda\partial = 0.$$

Similarly, we have $\partial^*\bar{\partial} + \bar{\partial}\partial^* = 0$.
Next expand Δ,

$$\begin{aligned}\Delta &= (\partial + \bar{\partial})(\partial^* + \bar{\partial}^*) + (\partial^* + \bar{\partial}^*)(\partial + \bar{\partial}) \\ &= (\partial\partial^* + \partial^*\partial) + (\bar{\partial}\bar{\partial}^* + \bar{\partial}^*\bar{\partial}) + (\partial\bar{\partial}^* + \bar{\partial}^*\partial) + (\partial^*\bar{\partial} + \bar{\partial}\partial^*) \\ &= (\partial\partial^* + \partial^*\partial) + (\bar{\partial}\bar{\partial}^* + \bar{\partial}^*\bar{\partial}) + (\partial\bar{\partial}^* + \bar{\partial}^*\partial) + (\partial^*\bar{\partial} + \bar{\partial}\partial^*) \\ &= \Delta_\partial + \Delta_{\bar{\partial}}.\end{aligned}$$

Finally, we check $\Delta_\partial = \Delta_{\bar{\partial}}$,

$$\begin{aligned}-\sqrt{-1}\Delta_\partial &= \partial(\Lambda\bar{\partial} - \bar{\partial}\Lambda) + (\Lambda\bar{\partial} - \bar{\partial}\Lambda)\partial \\ &= \partial\Lambda\bar{\partial} - \partial\bar{\partial}\Lambda + \Lambda\bar{\partial}\partial - \bar{\partial}\Lambda\partial \\ &= (\partial\Lambda - \Lambda\partial)\bar{\partial} + \bar{\partial}(\partial\Lambda - \Lambda\partial) \\ &= -\sqrt{-1}\Delta_{\bar{\partial}}.\end{aligned}$$

□

Exercises

9.3.6. Check the first order-Kähler identities (Proposition 9.3.5) on the space of all 2-forms.

9.3.7. Show that $\beta + \bar{\partial}\gamma = 0$ forces $\beta = 0$ if β is harmonic.

9.3.8. Suppose that $\alpha = \beta + \bar{\partial}\gamma$ and that α is of type (p,q) and β harmonic. By decomposing $\beta = \sum \beta^{(p',q')}$ and $\gamma = \sum \gamma^{(p',q')}$ into (p',q') type and using the previous exercise, prove that β is of type (p,q) and unique.

Chapter 10
Kähler Manifolds

We come now to the heart of our story. We saw that for Riemann surfaces and tori, holomorphic forms are harmonic and more. For general compact complex manifolds, the relationship is much more complicated. There is, however, an important class of manifolds, called Kähler manifolds, on which these kinds of results do hold. More precisely, harmonic forms on such manifolds decompose into holomorphic, antiholomorphic, and more generally harmonic (p,q) parts. This is the Hodge decomposition, which is the central theorem in the subject.

Kähler manifolds are complex manifolds that carry a special metric called a Kähler metric. Unlike Riemannian or Hermitian metrics, there are topological obstructions for a manifold to carry a Kähler metric. Fortunately, all projective manifold do possess such metrics, and this is the key. Other references that cover this material in more detail are Griffiths and Harris [49], Huybrechts [67], Morrow and Kodaira [89], Voisin [115], and Wells [120].

10.1 Kähler Metrics

Let X be a compact complex manifold with complex dimension n. A Hermitian metric H on X could be defined as in the previous chapter as a Riemannian metric for which multiplication by $\sqrt{-1}$ is orthogonal. However, it is more convenient for our purposes to view it as a choice of Hermitian inner product on the complex tangent spaces that vary in C^∞ fashion. More precisely, H will be given by a section of $\mathscr{E}_X^{(1,0)} \otimes \mathscr{E}_X^{(0,1)}$ such that in some (any) local coordinate system $z_i = x_i + \sqrt{-1}y_i$ around each point, H is given by

$$H = \sum h_{ij} dz_i \otimes d\bar{z}_j$$

with h_{ij} positive definite Hermitian. The real part of this matrix is positive definite symmetric, and the tensor

$$\sum \mathrm{Re}(h_{ij})(dx_i \otimes dx_i + dy_i \otimes dy_i)$$

gives a globally defined Riemannian structure on X. We also have a $(1,1)$-form, ω called the *Kähler form*, which is the normalized image of H under

$$\mathscr{E}_X^{(1,0)} \otimes \mathscr{E}_X^{(0,1)} \to \mathscr{E}_X^{(1,0)} \wedge \mathscr{E}_X^{(0,1)} = \mathscr{E}^{(1,1)}(X).$$

In coordinates,

$$\omega = \frac{\sqrt{-1}}{2} \sum h_{ij} dz_i \wedge d\bar{z}_j.$$

The normalization makes ω real, i.e., $\bar{\omega} = \omega$. It is clear from this formula that ω determines the metric.

Definition 10.1.1. A Hermitian metric on X is called a Kähler metric for any $p \in X$ if there exist analytic coordinates $z_1, \dots, z_n$ with $z_i = 0$ at p for which the metric becomes Euclidean up to second order:

$$h_{ij} \equiv \delta_{ij} \mod (x_1, y_1, \dots, x_n, y_n)^2.$$

A Kähler manifold is a complex manifold that admits a Kähler metric. (Sometimes the term is used for a manifold with a fixed Kähler metric.)

In such a coordinate system, a Taylor expansion gives

$$\omega = \frac{\sqrt{-1}}{2} \sum dz_i \wedge d\bar{z}_i + \text{ terms of second order and higher.}$$

Therefore $d\omega = 0$ at $z_i = 0$. Since such coordinates can be chosen around any point, $d\omega$ is identically zero. This gives a nontrivial obstruction for a Hermitian metric to be Kähler. In fact, this condition characterizes Kähler metrics and is usually taken as the definition:

Proposition 10.1.2. *Given a Hermitian metric H, the following are equivalent:*

(a) H is Kähler.
(b) The Kähler form is closed: $d\omega = 0$.
(c) The Kähler form is locally expressible as $\omega = \partial\bar{\partial} f$.

Proof. Condition (c) clearly implies (b), and (b) will be shown to imply (c) in the exercises.

That (a) implies (b) was explained above. So as to dispel any mystery, we explain the broad strategy for the reverse implication, and refer to [49, p. 107] for further details. Observe that by a linear change of coordinates it is easy to arrange that $h_{ij}(p) = \delta_{ij}$ for any $p \in X$, so that near p, we have

$$\omega = \frac{\sqrt{-1}}{2} \sum (\delta_{ij} + a^i_{jk} z_i + b^i_{jk} \bar{z}_{jk} + O(2)) dz_j \wedge d\bar{z}_k,$$

where $O(2)$ refers to terms of higher order. We want to eliminate the linear coefficients a^i_{jk}, b^i_{jk}, by setting $z_i = w_i + \sum c^i_{jk} w_j w_k$ and solving for the c^i_{jk}. At first glance, this seems hopeless, since there are many more equations than variables. Fortunately, there are redundancies:

$$b^j_{ik} = \bar{a}^i_{jk}$$

and

$$a^i_{jk} = a^k_{ji},$$

which stem from the Hermitianness of the metric and the closedness of ω. With the help of these relations, one can check that

$$c^j_{ki} = -a^i_{jk}$$

gives a solution. □

We will refer to the cohomology class of ω as the *Kähler class*. The function f such that $\omega = \sqrt{-1}\partial\bar{\partial} f$ is called a Kähler potential. A function f is *plurisubharmonic* if it is a Kähler potential, or equivalently in coordinates, this means that

$$\sum \frac{\partial^2 f}{\partial z_i \partial \bar{z}_j} \xi_i \bar{\xi}_j > 0$$

for any nonzero vector ξ.

The following are basic examples of compact Kähler manifolds:

Example 10.1.3. Any Hermitian metric on a Riemann surface is Kähler, since $d\omega$ vanishes for trivial reasons.

Example 10.1.4. Complex tori are Kähler. Any flat metric will do.

Example 10.1.5. Complex projective space $\mathbb{P}^n$ carries a Kähler metric called the Fubini–Study metric. In homogeneous coordinates the Kähler form is given by

$$\omega = \frac{\sqrt{-1}}{2\pi} \partial\bar{\partial} \log(|z_0|^2 + \cdots + |z_n|^2).$$

This really means that its pullback to $\mathbb{C}^{n+1} - \{0\}$ is given by this formula. In inhomogeneous coordinates we can write ω as

$$\frac{\sqrt{-1}}{2\pi} \partial\bar{\partial} \log\left(\left|\frac{z_0}{z_i}\right|^2 + \cdots + 1 + \cdots + \left|\frac{z_n}{z_i}\right|^2 \right). \tag{10.1.1}$$

We leave it as an exercise to check that this is in fact a Kähler form. The significance of the normalization $\frac{\sqrt{-1}}{2\pi}$ will be clear shortly.

The key examples of Kähler manifolds are provided by the following lemma.

Lemma 10.1.6. *A complex submanifold of a Kähler manifold inherits a Kähler metric such that the Kähler class is the restriction of the Kähler class of the ambient manifold.*

Proof. The Kähler form locally, has a plurisubharmonic potential f. It follows immediately from the definition that f restricts to a plurisubharmonic function on a complex submanifold. Thus the Kähler form will restrict to a Kähler form. □

Corollary 10.1.7. *A smooth projective variety $X \subset \mathbb{P}^n$ is Kähler, with Kähler metric induced from the Fubini–Study metric.*

The Kähler class above has algebrogeometric meaning. Recall that for a line bundle L, we have its first Chern class $c_1(L) \in H^2(X,\mathbb{Z})$. We can take its image in $H^2(X,\mathbb{C})$.

Lemma 10.1.8. *The Kähler class of the Fubini–Study metric on $\mathbb{P}^n$ coincides with $c_1(\mathscr{O}(1))$.*

Proof. Since $H^2(\mathbb{P}^n,\mathbb{C})$ is one-dimensional, the Kähler class $[\omega]$ would have to be a nonzero multiple of $c_1(\mathscr{O}(1))$. We can check that the constant is 1 by integrating (10.1.1) over the embedded line $z_2 = \cdots = z_n = 0$ and setting $z = z_0/z_1$ to obtain

$$\frac{\sqrt{-1}}{2\pi}\int_{\mathbb{P}^1} \partial\bar{\partial}\log(1+|z|^2) = \frac{1}{4\pi}\int_0^{2\pi}\int_0^{\infty} \frac{4r\,dr\,d\theta}{(1+r^2)^2} = 1.$$ □

Definition 10.1.9. A line bundle L on a compact complex manifold X is called very ample if there is an embedding $X \subset \mathbb{P}^n$ such that $L \cong \mathscr{O}_{\mathbb{P}^n}(1)|_X$. L is ample if some positive tensor power $L^{\otimes k} = L \otimes \cdots \otimes L$ is very ample.

Corollary 10.1.10. *If X is a smooth projective variety with an ample line bundle L, then $c_1(L)$ is the Kähler class associated to a Kähler metric.*

Proof. Any positive constant multiple of a Kähler form is another Kähler form. Since $c_1(L^{\otimes k}) = kc_1(L)$, we can assume that L is very ample. In this case, the result follows from the previous lemma and Corollary 10.1.8. □

Kähler classes arising as Chern classes are necessarily rational. The converse is the famous Kodaira embedding theorem.

Theorem 10.1.11 (Kodaira embedding theorem). *Suppose that X is a compact Kähler manifold whose Kähler class lies in the image of $H^2(X,\mathbb{Q})$. Then X can be embedded as a submanifold of projective space.*

Proof. See [49, 120]. □

Using Chow's theorem (which will discussed later on, in Section 15.4) we obtain the following:

Corollary 10.1.12. *A compact complex manifold is a nonsingular projective algebraic variety if and only if it possesses a Kähler metric with rational Kähler class.*

Exercises

10.1.13. Show that the product of two Kähler manifolds is Kähler.

10.1.14. If $\pi : Y \to X$ is a covering space, or equivalently a local biholomorphism, show that a Kähler form on X pulls back to a Kähler form on Y. In the opposite direction, show that if Y is Kähler and $X = Y/\Gamma$, where Γ is a discrete group acting nicely (properly discontinuously, freely, holomorphically, and isometrically), then X is Kähler.

10.1.15. Prove (b) $\Rightarrow$ (c) in Proposition 10.1.2. (Use Poincaré's lemma to locally solve $\omega = d\alpha$, then apply the $\bar{\partial}$-Poincaré lemma (9.2.2) to the $(0,1)$-part of α and the conjugate of the $(1,0)$-part.)

10.1.16. Show that the Fubini–Study metric is Kähler by checking that

$$\left|\frac{z_0}{z_i}\right|^2 + \cdots + 1 + \cdots + \left|\frac{z_n}{z_i}\right|^2$$

is plurisubharmonic.

10.1.17. Given a Kähler manifold, prove that the Kähler form ω is nondegenerate in the sense that for every $x \in X$, the associated bilinear form $T_x \times T_X \to \mathbb{R}$ on the real tangent space is nondegenerate. A manifold with a nondegenerate closed 2-form is called *symplectic*.

10.1.18. Let X be an n-dimensional Kähler manifold. Show that the volume form $d\mathrm{vol}$ for the underlying Riemannian metric is proportional to $\omega^n = \omega \wedge \cdots \wedge \omega$, and determine the constant of proportionality. Conclude, in particular, that $d\mathrm{vol}$ is of type (n,n). (Hint: for the almost Euclidean coordinates around p, in the sense of Definition 10.1.1, $d\mathrm{vol} = dx_1 \wedge dy_1 \wedge \cdots dy_n$ at p.)

10.1.19. Use the previous exercise to show that when X is compact and Kähler, the Kähler class $[\omega]$ is nonzero. Therefore the second Betti number is nonzero. This provides a nontrivial topological obstruction to the existence of a Kähler metric.

10.1.20. Construct a complex torus T with no nonzero forms ω of type $(1,1)$ that are rational in the sense that $\int_{2\text{-tori}} \omega \in \mathbb{Q}$. Conclude that T is not projective.

10.2 The Hodge Decomposition

Fix an n-dimensional Kähler manifold X with a Kähler metric. Since Kähler metrics are Euclidean up to second order, we have the following metatheorem: *Any identity involving geometrically defined first-order differential operators on Euclidean space will automatically extend to Kähler manifolds.* Let us introduce the relevant operators. Since X is a complex manifold, it has a canonical orientation given by declaring $\frac{\partial}{\partial x_1}, \frac{\partial}{\partial y_1}, \frac{\partial}{\partial x_2}, \dots$ to be positively oriented for any choice of analytic local coordinates. Thus the Hodge star operator associated to the Riemannian structure can be defined. The operator $*$ will be extended to a $\mathbb{C}$-linear operator on $\mathscr{E}_X^\bullet$, and set $\bar{*}(\alpha) = \overline{*\alpha} = *\overline{\alpha}$. (Be aware that many authors, notably Griffiths and Harris [49], we define $*$ to be what we call $\bar{*}$.) If we let $(\,,\,)$ be the pointwise Hermitian inner product on $\mathscr{E}^\bullet(X)$, then (8.2.1) becomes

$$\alpha \wedge \bar{*}\beta = (\alpha, \beta)d\mathrm{vol}.$$

It follows easily from this and the fact that $d$vol is of type (n,n) (Exercise 10.1.18) that

$$\bar{*}\mathscr{E}^{(p,q)}(X) \subseteq \mathscr{E}^{(n-p,n-q)}(X),$$
$$*\mathscr{E}^{(p,q)}(X) \subseteq \mathscr{E}^{(n-q,n-p)}(X).$$

Let $\partial^* = -\bar{*}\partial\bar{*} = -*\partial*$ and $\bar{\partial}^* = -\bar{*}\bar{\partial}\bar{*} = -*\bar{\partial}*$. These operators are the adjoints of ∂ and $\bar{\partial}$, and have bidegree $(-1,0)$ and $(0,-1)$ respectively. Then we can define the operators with bidegrees indicated on the right:

$\Delta_\partial = \partial^*\partial + \partial\partial^*$	$(0,0)$
$\Delta_{\bar{\partial}} = \bar{\partial}^*\bar{\partial} + \bar{\partial}\bar{\partial}^*$	$(0,0)$
$L = \omega\wedge$	$(1,1)$
$\Lambda = -*L* = -\bar{*}L\bar{*}$	$(-1,-1)$

A form is called *$\bar{\partial}$-harmonic* if it lies in the kernel of $\Delta_{\bar{\partial}}$. For a general Hermitian manifold there is no relationship between harmonicity and $\bar{\partial}$-harmonicity. However, these notions do coincide for the class of Kähler manifolds, by the following:

Theorem 10.2.1. $\Delta = 2\Delta_{\bar{\partial}} = 2\Delta_\partial$.

Proof. Since Laplacians are of second order, the above metatheorem cannot be applied directly. However, the first-order identities of Proposition 9.3.5 do generalize to X by this principle. The rest of the argument is the same as the proof of Lemma 9.3.4. □

Corollary 10.2.2. *If X is compact, then $H^q(X, \Omega_X^p)$ is isomorphic to the space of harmonic (p,q)-forms.*

Proof. Since harmonic forms are $\bar{\partial}$-closed (Exercise 10.2.9), we have a map from the space of harmonic (p,q)-forms to $H^q(X, \Omega_X^p)$. The rest of the argument is identical to the proof of Proposition 9.3.2. □

We obtain the following special case of Serre duality as a consequence:

Corollary 10.2.3. *When X is compact, $H^p(X, \Omega_X^q) \cong H^{n-p}(X, \Omega_X^{n-q})$.*

Proof. $\bar{*}$ induces an $\mathbb{R}$-linear isomorphism between the corresponding spaces of harmonic forms. □

Theorem 10.2.4 (The Hodge decomposition). *Suppose that X is a compact Kähler manifold. Then a differential form is harmonic if and only if its* (p,q) *components are. Consequently, we have (for the moment) noncanonical isomorphisms*

$$H^i(X,\mathbb{C}) \cong \bigoplus_{p+q=i} H^q(X,\Omega_X^p).$$

Furthermore, complex conjugation induces $\mathbb{R}$*-linear isomorphisms between the space of harmonic* (p,q)*- and* (q,p)*-forms. Therefore*

$$H^q(X,\Omega_X^p) \cong H^p(X,\Omega_X^q).$$

Proof. The operator $\Delta_{\bar{\partial}}$ preserves the decomposition

$$\mathcal{E}^i(X) = \bigoplus_{p+q=i} \mathcal{E}^{(p,q)}(X).$$

Therefore a form is harmonic if and only if its (p,q) components are. Since complex conjugation commutes with Δ, conjugation preserves harmonicity. These arguments, together with Corollary 10.2.2, finish the proof. □

The next two results are true more generally. However, for Kähler manifolds, we get these almost for free.

Corollary 10.2.5. *The Hodge numbers* $h^{p,q}(X) = \dim H^q(X,\Omega_X^p)$ *are finite.*

Corollary 10.2.6. *Suppose that X and Y are compact Kähler. Then there is an isomorphism*

$$H^q(X\times Y,\Omega_{X\times Y}^q) \cong \bigoplus_{\substack{a+b=q\\ c+d=p}} H^a(X,\Omega_Y^c)\otimes H^b(Y,\Omega_Y^d).$$

Proof. We make the previous identifications of $H^a(X,\Omega_X^c) \cong H^{ca}(X)$ with harmonic (c,a)-forms X etc. Let π_i be the projections on $X\times Y$. Recall that we have an isomorphism

$$\kappa : \bigoplus_{i+j=k} H^i(X,\mathbb{C})\otimes H^j(Y,\mathbb{C}) \cong H^k(X\times Y,\mathbb{C})$$

(Theorem 5.3.6), where the map $\kappa(\alpha\otimes\beta)$ equals $\pi_1^*\alpha\cup\pi_2^*\beta$. Decomposing into type, we see that

$$\kappa(H^{ca}(X)\otimes H^{db}(Y)) \subset H^{c+d,a+b}(X\times Y).$$

These restrictions are necessarily isomorphisms, since their sum is an isompor-phism. □

The Hodge decomposition theorem has a number of subtle implications for the topology of compact Kähler manifolds.

Corollary 10.2.7. *If i is odd, then the ith Betti number b_i of X is even.*

Proof.

$$b_i = 2 \sum_{p<i/2} h^{p,i-p}.$$

□

Corollary 10.2.8 (Johnson–Rees). *The fundamental group of X cannot be free (unless it is trivial).*

Proof. The proof is taken from [4], which gives a bit more detail. If $\pi_1(X)$ were free on an odd number of generators, then the first Betti number would be odd, which is ruled out as above. If $\pi_1(X)$ is free on an even number of generators, then it can be checked that it contains a subgroup of finite index that is free on an odd number of generators. This subgroup is the fundamental group of a finite sheeted covering $Y \to X$. Since Y inherits a Kähler structure from X (Exercise 10.1.14), this is again impossible. □

More information about the structure of fundamental groups of compact Kähler manifolds—called Kähler groups—can be found in [2]. The finite-dimensionality of $H^q(X, \Omega_X^p)$ is also true for compact complex non-Kähler manifolds by Theorem 16.3.5, but the other corollaries may fail. The Hodge numbers give an important set of holomorphic invariants for X. We can visualize them by arranging them in a diamond:

$$\begin{array}{ccccc} & & h^{00}=1 & & \\ & h^{10} & & h^{01} & \\ h^{20} & & h^{11} & & h^{02} \\ & & \cdots & & \\ & h^{n,n-1} & & h^{n-1,n} & \\ & & h^{nn}=1 & & \end{array}$$

The previous results imply that this picture has both vertical and lateral symmetry (e.g., $h^{10} = h^{01} = h^{n,n-1} = h^{n-1,n}$).

Exercises

10.2.9. Prove that a form α on a compact Kähler manifold is harmonic if and only if $\bar{\partial}\alpha = \bar{\partial}^*\alpha = 0$. Deduce that a $(p,0)$-form is harmonic if and only if it is holomorphic.

10.2.10. Check that the operators $\partial^*, \bar{\partial}^*$, and Λ defined in Section 9.3 for the flat metric are special cases of the corresponding operators defined here.

10.2.11. The Hopf surface S is a complex manifold obtained as a quotient of $\mathbb{C}^2 - \{0\}$ by $\mathbb{Z}$ acting by $z \mapsto 2^n z$. Show that S is homeomorphic to $S^1 \times S^3$, and conclude that it cannot be Kähler.

10.3 Picard Groups

Recall that the Picard group $\mathrm{Pic}(X) \cong H^1(X, \mathscr{O}_X^*)$ of a complex manifold is the group of isomorphism classes of holomorphic line bundles. Our goal is to analyze the structure of this group when X is compact Kähler. There is a discrete part, which is the image of the first Chern class map $\mathrm{Pic}(X) \to H^2(X, \mathbb{Z})$, and a continuous part, which is the kernel. We start by describing the image. Let us identify $H^2(X, \mathbb{C})$ with the space of harmonic 2-forms with respect to a fixed Kähler metric. Let $H^{11}(X)$ be the subspace of harmonic forms of type $(1,1)$. Let us write $H^2(X, \mathbb{Z}) \cap H^{11}(X)$ for the preimage of $H^{11}(X)$ under the map $H^2(X, \mathbb{Z}) \to H^2(X, \mathbb{C})$. This is the sum of the true intersection $\mathrm{im}[H^2(X, \mathbb{Z}) \to H^2(X, \mathbb{C})] \cap H^{11}(X)$ with the torsion subgroup $H^2(X, \mathbb{Z})_{\mathrm{tors}}$.

Theorem 10.3.1 (The Lefschetz (1, 1) theorem). $c_1(\mathrm{Pic}(X)) = H^2(X, \mathbb{Z}) \cap H^{11}(X)$.

Proof. The map $f : H^2(X, \mathbb{Z}) \to H^2(X, \mathscr{O}_X)$ can be factored as $H^2(X, \mathbb{Z}) \to H^2(X, \mathbb{C})$ followed by projection of the space of harmonic 2-forms to the space of harmonic $(2,0)$-forms. From the exponential sequence, $c_1(\mathrm{Pic}(X)) = \ker(f)$, and this certainly contains $H^2(X, \mathbb{Z}) \cap H^{11}(X)$.

Conversely, suppose that $\alpha \in \ker(f)$. Then its image in $H^2(X, \mathbb{C})$ can be represented by a sum of a harmonic $(0,2)$-form α_1 and a harmonic $(1,1)$-form α_2. Since $\bar{\alpha} = \alpha$, α_1 must be zero. Therefore $\alpha \in H^2(X, \mathbb{Z}) \cap H^{11}(X)$. □

Proposition 10.3.2. $\mathrm{Pic}^0(X) = \ker(c_1)$ *is a complex torus.*

Proof. From the exponential sequence, we obtain

$$H^1(X, \mathbb{Z}) \to H^1(X, \mathscr{O}_X) \to \mathrm{Pic}(X) \to H^2(X, \mathbb{Z}).$$

Pic^0 is the cokernel of the first map, that is, $H^1(X, \mathscr{O}_X)/H^1(X, \mathbb{Z})$. We have to prove that $H^1(X, \mathbb{Z})$ sits inside $H^1(X, \mathscr{O}_X)$ as a lattice. Since

$$H^1(X, \mathbb{Z}) \subset H^1(X, \mathbb{R}) \cong H^1(X, \mathbb{Z}) \otimes \mathbb{R}$$

is a lattice, it suffices to prove that the natural map

$$\pi : H^1(X, \mathbb{R}) \to H^1(X, \mathscr{O}_X)$$

is an isomorphism of real vector spaces. We know that these spaces have the same real dimension $b_1 = 2h^{10}$, so it is enough to check that π is injective. The Hodge decomposition implies that $\alpha \in H^1(X, \mathbb{R})$ can be represented by a sum of a harmonic $(1,0)$-form α_1 and a harmonic $(0,1)$-form α_2. Since $\alpha = \bar{\alpha}$, $\alpha_1 = \bar{\alpha}_2$, it follows that $\pi(\alpha)$ is just α_1. Therefore $\pi(\alpha) = 0$ implies that $\alpha = 0$. □

Combining both results gives the following structure theorem:

Theorem 10.3.3. *Let X be a compact Kähler manifold. Then* $\mathrm{Pic}(X)$ *fits into an exact sequence*

$$0 \to \mathrm{Pic}^0(X) \to \mathrm{Pic}(X) \to H^2(X,\mathbb{Z}) \cap H^{11}(X) \to 0$$

with $\mathrm{Pic}^0(X)$ *a complex torus of dimension* $h^{01}(X)$*:*

$\mathrm{Pic}^0(X)$ is called the *Picard torus* (or variety when X is projective). When X is a compact Riemann surface, $\mathrm{Pic}^0(X)$ is usually called the Jacobian and denoted by $J(X)$.

Example 10.3.4. When $X = V/L$ is a complex torus, $\mathrm{Pic}^0(X)$ is a new torus called the dual torus. Using Corollary 9.3.3, it can be seen to be isomorphic to $\bar{V}^*/L^*$, where $\bar{V}^*$ is the antilinear dual and

$$L^* = \{\lambda \in V^* \mid \mathrm{Im}(\lambda)(L) \subseteq \mathbb{Z}\}.$$

The dual of the Picard torus is called the Albanese torus $\mathrm{Alb}(X)$. Since $H^{01} = \bar{H}^{01}$, $\mathrm{Alb}(X)$ is isomorphic to

$$\frac{H^0(X,\Omega_X^1)^*}{H_1(X,\mathbb{Z})/H_1(X,\mathbb{Z})_{\mathrm{tors}}},$$

where the elements γ of the denominator are identified with the functionals $\alpha \mapsto \int_\gamma \alpha$ in the numerator. In a sense that will be discussed in the exercises, $\mathrm{Alb}(X)$ is the torus that is closest to X.

Exercises

10.3.5. Show that $X \mapsto \mathrm{Pic}^0(X)$ extends to a contravariant functor from compact Kähler manifolds to complex tori. Show that $X \mapsto \mathrm{Alb}(X)$ is covariant.

10.3.6. Let X be a compact Riemann surface. Show that the pairing $\langle \alpha, \beta \rangle = \int_X \alpha \wedge \beta$ induces an isomorphism $J(X) \cong \mathrm{Alb}(X)$; in other words, $J(X)$ is self-dual.

10.3.7. Fix a base point $x_0 \in X$. Show that the so-called Abel–Jacobi map $\alpha : X \to \mathrm{Alb}(X)$ given by $x \mapsto \int_{x_0}^x$ (which is well defined modulo H_1) is holomorphic.

10.3.8. Show that every holomorphic 1-form on X is the pullback of a holomorphic 1-form from $\mathrm{Alb}(X)$. In particular, the Abel–Jacobi map cannot be constant if $h^{01} \neq 0$.

10.3.9. Prove that $\mathrm{Alb}(T) \cong T$ if T is a complex torus. Given a holomorphic map $f : X \to T$ sending x_0 to 0, prove that it factors through $\mathrm{Alb}(X) \to \mathrm{Alb}(T) \cong T$.

10.3.10. Consider the map $X^n \to \mathrm{Alb}(X)$ given by $(x_1, \ldots, x_n) \mapsto \sum \alpha(x_i)$. Show that this is surjective for some n. (Hint: calculate the derivate at a general point.)

Chapter 11
A Little Algebraic Surface Theory

Let us return to geometry armed with what we have learned so far. We have already looked at Riemann surfaces (which will be referred to as complex curves from now on) in some detail. So we consider the next step up. A nonsingular complex surface is a two-dimensional complex manifold. By an algebraic surface, we will mean a two-dimensional nonsingular projective variety. So in particular, they are Kähler manifolds. In this chapter, we will present a somewhat breezy account of surface theory, concentrating on topics that illustrate the general theorems from the previous chapters.

Much more systematic introductions to algebraic surface theory can be found in the books by Barth, Peters, and Van de Ven [9] and Beauville [10].

11.1 Examples

The basic discrete invariant of a curve is its genus. For algebraic surfaces, there are several numbers that play a similar role. We can use the Hodge numbers. By the symmetry properties considered earlier, there are only three that matter: h^{10}, h^{20}, h^{11}. The first two are traditionally called (and denoted by) the *irregularity* ($q = h^{10}$) and *geometric genus* ($p_g = h^{20}$). The basic topological invariants are the Betti numbers, which by Poincaré duality and the Hodge decomposition theorem can be expressed as

$$b_1 = b_3 = 2q, \quad b_2 = 2p_g + h^{11}.$$

It is also convenient to consider the Euler characteristic,

$$e = b_0 - b_1 + b_2 - b_3 + b_4 = 2 - 4q + 2p_g + h^{11}.$$

Let us now start our tour.

Example 11.1.1. The most basic example is the projective plane $X = \mathbb{P}^2 = \mathbb{P}^2_{\mathbb{C}}$. We computed the Betti numbers $b_1 = 0$, $b_2 = 1$ in Section 7.2. Therefore $q = p_g = 0$ and $h^{11} = 1$.

Example 11.1.2. If $X = C_1 \times C_2$ is a product of two nonsingular curves of genus g_1 and g_2, then by Künneth's formula (Theorem 5.3.6 and Corollary 10.2.6),

$$q = h^{10}(X) = h^{10}(C_1)h^{00}(C_2) + h^{00}(C_1)h^{10}(C_2) = g_1 + g_2.$$

Similarly, $p_g = g_1 g_2$ and $h^{11} = 2g_1 g_2 + 2$.

Example 11.1.3. As a special case of the previous example, when we have a product of a curve C with $\mathbb{P}^1$, the invariants are $q = g$, $p_g = 0$, and $h^{11} = 2$. More generally, we can consider ruled surfaces over C, which are $\mathbb{P}^1$-bundles that are locally isomorphic to $U_i \times \mathbb{P}^1$, for a Zariski open cover $\{U_i\}$ of C. (See Section 14.5 for a bit more explanation of what this means.) We will see shortly that the invariants are the same as above, although they are not generally products. When $C = \mathbb{P}^1$, there are, up to isomorphism, countably many ruled surfaces. Here is a simple description. Let $C_n \subset \mathbb{P}^n$ be the closure of $\{(t, t^2, \ldots, t^n) \mid t \in \mathbb{C}\}$. Choose a point $p_0 \in \mathbb{P}^n - C_n$. Let F_n be the set of pairs $(q, p) \in \mathbb{P}^n \times C_n$ such that q lies on the line connecting p_0 to q.

Example 11.1.4. Let $X \subset \mathbb{P}^3$ be a smooth surface of degree d. Then $q = 0$. We will list the first few values of the remaining invariants:

d	p_g	h^{11}
2	0	2
3	0	7
4	1	20
5	4	45
6	10	86

These can be calculated using formulas given later (17.3.4).

A method of generating new examples from old is by blowing up. We start by describing the blowup of $\mathbb{C}^2$ at 0:

$$\mathrm{Bl}_0\mathbb{C}^2 = \{(x, \ell) \in \mathbb{C}^2 \times \mathbb{P}^1 \,|\, x \in \ell\}.$$

The projection $p_1 : \mathrm{Bl}_0\mathbb{C}^2 \to \mathbb{C}^2$ is one-to-one away from $0 \in \mathbb{C}^2$. This can be generalized to yield the blowup $\mathrm{Bl}_p X \to X$ of a surface X at the point p. Let $B \subset X$ be a coordinate ball centered at p. After identifying B with a ball in $\mathbb{C}^2$ centered at 0, we can let $\mathrm{Bl}_0 B$ be the preimage of B in $\mathrm{Bl}_0\mathbb{C}^2$. The boundary of $\mathrm{Bl}_0 B$ can be identified with the boundary of B. Thus we can glue $X - B \cup \mathrm{Bl}_0 B$ to form $\mathrm{Bl}_p X$. When X is algebraic, $\mathrm{Bl}_p X$ is again algebraic by Exercise 2.4.23.

Let us compute $H^*(\mathrm{Bl}_p X, \mathbb{Z})$. Set $Y = \mathrm{Bl}_p X$ and compare Mayer–Vietoris sequences:

$$\begin{array}{ccccc} H^i(X) & \longrightarrow & H^i(X-B')\oplus H^i(B) & \longrightarrow & H^i(X-B'\cap B) \\ \downarrow & & \downarrow & & \downarrow = \\ H^i(Y) & \longrightarrow & H^i(Y-\mathrm{Bl}_pB')\oplus H^i(\mathrm{Bl}_pB) & \longrightarrow & H^i(Y-\mathrm{Bl}_pB'\cap \mathrm{Bl}_pB) \end{array}$$

where $B' \subset B$ is a smaller ball. We thus have the following result:

Lemma 11.1.5. *$H^1(\mathrm{Bl}_pX) \cong H^1(X)$ and $H^2(\mathrm{Bl}_pX) = H^2(X) \oplus \mathbb{Z}$.*

Corollary 11.1.6. *q and p_g are invariant under blowing up. $h^{11}(\mathrm{Bl}_pX) = h^{11}(X)+1$.*

Proof. The lemma implies that $b_1 = 2q$ is invariant and $b_2(Y) = b_1(X)+1$. Since $b_2 = 2p_g + h^{11}$, the only possibilities are $h^{11}(Y) = h^{11}(X)+1$, $p_g(Y) = p_g(X)$, and $p_g(Y) < p_g(X)$. The last inequality means that there is a nonzero holomorphic 2-form on X that vanishes on $X-p$, but this is impossible. □

A *birational map* $\kappa : X \dashrightarrow Y$ is simply an isomorphism in the category of varieties and rational maps. In more explicit terms, it is given by an isomorphism of Zariski open sets $X \supset U \cong V \subset Y$. Blowups and their inverses ("blowdowns") are examples of birational maps. Two varieties are birationally equivalent if a birational map exists between them. For example, any two ruled surfaces over $\mathbb{P}^1$ are birationally equivalent to each other and to $\mathbb{P}^2$, because they all contain $\mathbb{A}^2$.

For surfaces, the structure of birational maps is explained by the following theorem:

Theorem 11.1.7 (Castelnuovo). *Any birational map between algebraic surfaces is given by a finite sequence of blowups and blowdowns.*

Proof. See [60, 9]. □

Corollary 11.1.8. *The numbers q and p_g depend only on the birational equivalence class of the surface.*

This implies that $p_g = g, q = 0$ for ruled surfaces over a genus-g curve, as claimed above.

Blowing up of singular points figures in the proof of the next important theorem.

Theorem 11.1.9 (Zariski). *Given a singular algebraic surface Y, there exist a nonsingular surface X and a morphism $\pi : X \to Y$, called a resolution of singularities, that is an isomorphism away from the singular points.*

Proof. See [9]. □

Corollary 11.1.10 (Zariski). *If $f : X \dashrightarrow V$ is a rational function from an algebraic surface to a variety V, then there is a finite sequence of blowups $Y \to X$ such that f extends to a holomorphic map $f' : Y \to V$.*

Proof. We can construct Y by resolving singularities of the closure of the graph of f. □

Analogues of Zariski's and Castelnuovo's theorems in higher dimensions have been established by Hironaka and Włodarczyk respectively. These are much harder.

An *elliptic surface* is a surface X that admits a surjective morphism $f : X \to C$ to a smooth projective curve such that all nonsingular fibers are elliptic curves.

Example 11.1.11. A simple example of an elliptic surface is given as follows: choose two distinct nonsingular cubics $E_0, E_1 \subset \mathbb{P}^2$ defined by $f_0(x,y,z)$ and $f_1(x,y,z)$. These generate a pencil of cubics $E_t = V(tf_1 + (1-t)f_0)$ with $t \in \mathbb{P}^1$. Define

$$X = \{(p,t) \in \mathbb{P}^2 \times \mathbb{P}^1 \mid p \in E_t\}.$$

Projection to $\mathbb{P}^1$ makes this an elliptic surface. X can be identified with the blowup of $\mathbb{P}^2$ at the nine points $E_0 \cap E_1$. So $q = p_g = 0$ and $h^{11} = 10$.

Example 11.1.12. Consider the family of elliptic curves in Legendre form

$$\mathscr{E} = \{([x,y,z],t) \in \mathbb{P}^2 \times \mathbb{C} - \{0,1\} \mid y^2z - x(x-z)(x-tz) = 0\} \to \mathbb{C}.$$

The above equation is meaningful if $t = 0, 1$, and it defines a rational curve with a single node. By introducing $s = t^{-1}$, we get an equation

$$sy^2z - x(x-z)(sx-z) = 0$$

that defines a union of lines when $s = 0$. In this way, we can extend $\mathscr{E}$ to a surface $\mathscr{E}' \to \mathbb{P}^1$. Unfortunately, $\mathscr{E}'$ is singular, and it is necessary to resolve singularities to get a nonsingular surface $\bar{\mathscr{E}}$ containing $\mathscr{E}$ (we can take the minimal desingularization, which for our purposes means that $b_2(\bar{\mathscr{E}})$ is chosen as small as possible).

Exercises

11.1.13. Finish the proof of Lemma 11.1.5.

11.1.14. Given a ruled surface X over a curve C, check that $e(X) = e(\mathbb{P}^1)e(C)$, and use this to verify that $h^{11}(X) = 2$.

11.1.15. Show that there is a nonsingular quartic $X \subset \mathbb{P}^3$ containing a line ℓ, which we can assume to be $x_2 = x_3 = 0$. Show that the map $\mathbb{P}^3 \dashrightarrow \mathbb{P}^1$ defined by $[x_0,\dots,x_3] \mapsto [x_0,x_1]$ determines a morphism $X \to \mathbb{P}^1$ that makes it an ellptic surface.

11.1.16. Given a smooth projective curve C, the symmetric product is given by $S^2C = C \times C/\sigma$, where σ is the involution interchanging factors. This has the structure of a smooth algebraic surface such that $H^i(S^2C,\mathbb{Q}) = H^i(C \times C,\mathbb{Q})^\sigma$. Compute the Betti and Hodge numbers for S^2C.

11.2 The Neron–Severi Group

Let X be an algebraic surface. The image of the first Chern class map

$$c_1 : \mathrm{Pic}(X) \to H^2(X,\mathbb{Z})$$

is the *Neron–Severi* group $\mathrm{NS}(X)$. The rank of this group is called the *Picard number* $\rho(X)$. By Lefschetz's Theorem 10.3.1, $\mathrm{NS}(X) = H^2(X,\mathbb{Z}) \cap H^{11}(X)$. Therefore $\rho \leq h^{11}$ with equality if $p_g = 0$.

A divisor on X is a finite integer linear combination $\sum n_i D_i$ of possibly singular irreducible curves $D_i \subset X$. We can define a line bundle $\mathscr{O}_X(D)$ as we did for Riemann surfaces in Section 6.3. If f_i are local equations of $D_i \cap U$ in some open set U, then

$$\mathscr{O}_X(D)(U) = \mathscr{O}_X(U)\frac{1}{f_1^{n_1} f_2^{n_2} \dots}$$

is a fractional ideal. In particular, when $n_i = 1$, $\mathscr{O}_X(-D)$ is the ideal sheaf of D, and an ideal sheaf of a subscheme supported on D when $n_i \geq 0$.

Lemma 11.2.1. *If D_i are smooth curves, then $c_1(\mathscr{O}_X(\sum n_i D_i)) = \sum n_i [D_i]$.*

Proof. This is an immediate consequence of Theorem 7.5.8. □

When D is singular, we simply define its fundamental class to be $c_1(\mathscr{O}_X(D))$.

The cup product pairing

$$H^2(X,\mathbb{Z}) \times H^2(X,\mathbb{Z}) \to H^4(X,\mathbb{Z}) \cong \mathbb{Z}$$

restricts to a pairing on $\mathrm{NS}(X)$ denoted by "$\cdot$". Note that the last isomorphism follows from a stronger form of Poincaré duality than what we proved earlier [61].

Lemma 11.2.2. *Given a pair of transverse smooth curves D and E,*

$$D \cdot E = \int_X c_1(\mathscr{O}(D)) \cup c_1(\mathscr{O}(E)) = \int_D c_1(\mathscr{O}_X(E))|_D = \#(D \cap E).$$

Proof. By Lemma 11.2.1 and Proposition 5.6.3, $D \cdot E$ is a sum of local intersection numbers $i_p(D,E)$. The numbers $i_p(D,E)$ are always $+1$ in this case by Exercise 5.6.6. □

If the intersection of the curves D and E is finite but not transverse, it is still possible to give a geometric meaning to the above product. Choose local coordinates centered at p, and local equations f and g for D and E respectively.

Definition 11.2.3. The local intersection number is given by $i_p(D,E) = \dim \mathscr{O}_p/(f,g)$. (This depends only on the ideals (f) and (g), so it is well defined.)

Proposition 11.2.4. *If D, E are curves such that $D \cap E$ is finite, then*

$$D \cdot E = \sum_{p \in X} i_p(D,E).$$

Proof. We assume for simplicity that D and E are smooth, although this argument can be made to work in general. As in the proof of Proposition 5.6.3, the number on the left is given by

$$\int_X \tau_D \wedge \tau_E$$

for appropriate representatives τ_D, τ_E for the Thom classes. In particular, we assume that the supports are small enough that it breaks up into a sum of integrals over disjoint coordinate neighborhoods of $p \in D \cap E$.

We need a convenient expression for the Thom classes. We first note that if $\rho : \mathbb{R}^+ \to [0,1]$ is a cutoff function, 1 in a neighborhood of 0 and 0 in a neighborhood of ∞, then $-\frac{1}{2\pi} d\rho(r) \wedge d\theta$ gives a local expression for the Thom class of $0 \in \mathbb{R}^2$. Thus after choosing local equations of D, E at $p \in D \cap E$ as above, we can assume that (locally) $\tau_D = \tau_f$ and $\tau_E = \tau_g$, where

$$\tau_f = -\frac{1}{2\pi\sqrt{-1}} d\rho(|f|) \wedge \frac{df}{f}.$$

Let $h(z_1, z_2) = (f(z_1, z_2), g(z_1, z_2))$. It maps a small ball $0 \in U \subset \mathbb{C}^2$ to another small ball $0 \in U'$. The *degree* of h is the number of points in the fiber $h^{-1}(y)$ for almost all y. This coincides with $i_p(E, D)$ by Lemma 1.3.3. Computing the integral by a change of variables gives

$$\int_U \tau_f \wedge \tau_g = \int_{U'} h^*(\tau_f \wedge \tau_g) = (\deg h) \int \tau_{z_1} \wedge \tau_{z_2} = \deg h = i_p(E, D). \qquad \square$$

Example 11.2.5. Recall that $H^2(\mathbb{P}^2, \mathbb{Z}) = \mathbb{Z}$, and the generator of $H^2(\mathbb{P}^2, \mathbb{Z})$ is the class of the line $[L]$. Since $[L] = [L']$ for any other line, we have $L^2 = L \cdot L' = 1$, where $L^2 = L \cdot L$.

Example 11.2.6. $H^2(\mathbb{P}^1 \times \mathbb{P}^1, \mathbb{Z}) = \mathbb{Z}^2$ with generators given by fundamental classes of the horizontal and vertical lines $H = \mathbb{P}^1 \times \{0\}$ and $V = \{0\} \times \mathbb{P}^1$. We see that $H^2 = V^2 = 0$ and $H \cdot V = 1$.

Given a curve $D \subset \mathbb{P}^2$ defined by a polynomial f, we let $\deg D = \deg f$.

Corollary 11.2.7 (Bézout). *If D, E are curves on $\mathbb{P}^2$ with a finite intersection, then*

$$\sum_{p \in X} i_p(D, E) = \deg(D) \deg(E).$$

Proof. We have $[D] = c_1(\mathscr{O}(\deg D)) = (\deg D)[L]$, and likewise $[E] = (\deg E)[L]$. Therefore $D \cdot E = \deg D \deg E (L^2) = \deg D \deg E$. $\square$

Corollary 11.2.8. *If D, E are distinct irreducible curves, then $D \cdot E \geq 0$, and equality holds only if they are disjoint.*

This nonnegativity can fail when $D = E$. For example, by Corollary 5.7.3, the diagonal in a product of curves $\Delta \subset C \times C$ has negative self-intersection as soon as the genus of C is greater than 1. From Lemma 11.2.1, we obtain the following:

Lemma 11.2.9. *If D is a smooth curve, then*

$$D^2 = \int_D c_1(\mathscr{O}_X(D)) = \deg(\mathscr{O}_D(D)).$$

Given a surjective morphism $f : X \to Y$ of algebraic surfaces, the pullback $f^* : H^2(Y,\mathbb{Z}) \to H^2(X,\mathbb{Z})$ preserves the Neron–Severi group and the intersection pairing. This can be interpreted directly in the language of divisors. Given an irreducible divisor D on Y, we can make the set-theoretic preimage $f^{-1}D$ into a divisor by pulling back the ideal, i.e.,

$$\mathscr{O}_X(-f^{-1}D) = \mathrm{im}[f^*\mathscr{O}_Y(-D) \to \mathscr{O}_X].$$

We extend this operation to all divisors by linearity. The operation satisfies $f^*\mathscr{O}_Y(D) = \mathscr{O}_X(f^{-1}D)$. Since c_1 is functorial, we get the following:

Lemma 11.2.10. *f^{-1} is compatible with f^* on* $\mathrm{NS}(X)$.

Exercises

11.2.11. Let $X = C \times C$ be the product of a curve with itself. Consider the divisors $H = C \times \{p\}, V = \{p\} \times C$ and the diagonal Δ. Compute their intersection numbers, and show that these are linearly independent in $\mathrm{NS}(X) \otimes \mathbb{Q}$. Thus the Picard number is at least 3. Show that this is at least 4 if C admits a nontrivial automorphism with the appropriate conditions.

11.2.12. Let $E = \mathbb{C}/(\mathbb{Z}+\mathbb{Z}\tau)$ be an elliptic curve, and let $X = E \times E$. Show that the Picard number is 3 for "most" τ, but that it is 4 for $\tau = \sqrt{-1}$.

11.2.13. The ruled surface F_1 can be described as the blowup $\pi : F_1 \to \mathbb{P}^2$ of $\mathbb{P}^2$ at some point p. Let L_1 be a line in $\mathbb{P}^2$ containing p, and L_2 another line not containing p. Show that $\pi^*L_1 = \pi^*L_2$ and $\pi^*L_1 = E + F$, where $E = \pi^{-1}(p)$ and F is the closure in Y of $L_1 - \{p\}$ (F is called the strict transform of L_1). Use all of this to show that $E^2 = -1$. Conclude that F_1 and $\mathbb{P}^1 \times \mathbb{P}^1$ are not isomorphic.

11.2.14. A divisor is called (very) ample if $\mathscr{O}_X(H)$ is. If H is ample, then prove that $H^2 > 0$ and that $H \cdot C > 0$ for any curve $C \subset X$. This pair of conditions characterizes ampelness (Nakai–Moishezon). Show that the first condition alone is not sufficient.

11.3 Adjunction and Riemann–Roch

In this section, we introduce two of the most basic tools of surface theory. The first result, called the adjunction formula, computes the genus of a curve on a surface.

To set things up, recall that the canonical divisor K of a smooth projective curve is a divisor such that $\mathcal{O}_C(K) \cong \Omega^1_C$. Since this determines K uniquely up to linear equivalence, we can talk about *the* canonical divisor class K_C. A canonical divisor K_X (class) on a surface X is divisor such that $\mathcal{O}_X(K_X) \cong \Omega^2_X$. The linear equivalence class is again well defined. For the present, we need only its image in $\mathrm{NS}(X)$, and this can be defined to be $c_1(\Omega^2_X)$.

Theorem 11.3.1. *If C is a smooth curve of genus g on an algebraic surface X, then*

$$2g-2 = (K_X + C)\cdot C.$$

Proof. Let $\Omega^2_X(\log C)$ be the $\mathcal{O}_X$-module generated by rational 2-forms of the form $\frac{dz_1\wedge dz_2}{f}$, where f is a local equation for C. This is the same as the tensor product $\Omega^2_X \otimes \mathcal{O}_X(C)$. Such expressions can be rewritten as $\alpha\wedge\frac{df}{f}$, with α holomorphic. We define the residue map $\Omega^2_X(\log C) \to \Omega^1_c$ by sending $\alpha\wedge\frac{df}{f}$ to $\alpha|_C$. The kernel consists of the holomorphic differentials Ω^2_X, leading to a sequence

$$0 \to \Omega^2_X \to \Omega^2_X(\log C) \to \Omega^1_C \to 0, \tag{11.3.1}$$

which is seen to be exact. The holomorphic forms Ω^2_X are spanned locally by $f\frac{dz_1\wedge dz_2}{f}$. Thus we can identify the inclusion in (11.3.1) with the tensor product of the map $\mathcal{O}_X(-C) \to \mathcal{O}_X$ with $\Omega^2_X \otimes \mathcal{O}_X(C)$. The cokernel of the map just described is $\mathcal{O}_C \otimes \Omega^2_X(C)$, that is, the restriction of $\Omega^2_X \otimes \mathcal{O}_X(C)$ to C. So in summary, Ω^1_C is isomorphic to the restriction of $\Omega^2_X \otimes \mathcal{O}_X(C)$ to C. Therefore

$$\int_C c_1(\Omega^1_C) = \int_C c_1(\Omega^2_X \otimes \mathcal{O}_X(C)) = \int_C (c_1(\Omega^2_X) + c_1(\mathcal{O}_X(C)) = (K_X + C)\cdot C.$$

The left side is the degree of the K_C, but this is $2g-2$ by Proposition 6.3.7. $\square$

We can use this to recover the formula for the genus of a degree-d curve $C \subset \mathbb{P}^2$. Since $\mathrm{NS}(\mathbb{P}^2) \subseteq H^2(\mathbb{P}^2) = \mathbb{Z}$, we can identify $K_{\mathbb{P}^2}$ with an integer k. Therefore

$$g = \frac{1}{2}(k+d)d + 1.$$

When $d = 1$, we know that $g = 0$, so $k = -3$. Thus

$$g = (d-1)(d-2)/2.$$

A fundamental, and rather difficult, problem in algebraic geometry is to estimate $\dim H^0(X, \mathcal{O}_X(D))$. As a first step, one can calculate the Euler characteristic

$$\chi(\mathcal{O}_X(D)) = \sum(-1)\dim H^i(X, \mathcal{O}_X(D))$$

by the Riemann–Roch formula given below. The higher cohomologies can then be controlled in some cases by other techniques. The advantage of the Euler characteristic is the additivity property:

Lemma 11.3.2. *If* $0 \to \mathscr{F}_1 \to \mathscr{F}_2 \to \mathscr{F}_3 \to 0$ *is an exact sequence of sheaves with* $\sum \dim H^i(X, \mathscr{F}_j) < \infty$, *then* $\chi(\mathscr{F}_2) = \chi(\mathscr{F}_1) + \chi(\mathscr{F}_3)$.

This is a consequence of following elementary lemma:

Lemma 11.3.3. *If*

$$\cdots \to A^i \to B^i \to C^i \to A^{i+1} \to \cdots$$

is a finite sequence of finite-dimensional vector spaces,

$$\sum(-1)^i \dim B^i = \sum(-1)^i \dim A^i + \sum(-1)^i \dim C^i.$$

Theorem 11.3.4 (Riemann–Roch). *If* D *is a divisor on a surface* X, *then*

$$\chi(\mathscr{O}_X(D)) = \frac{1}{2} D \cdot (D - K_X) + \chi(\mathscr{O}_X).$$

Proof. We prove this under the assumption that $D = \sum_i n_i D_i$ is a sum of smooth curves. (In fact, by a simple trick, it is always possible to reduce to this case. The basic idea can be found, for example, in the proof of [60, Chapter V, Theorem 1.1].) By induction, it suffices to prove Riemann–Roch for $D = D' \pm C$, where the formula holds for D' and C is smooth. The idea is to do induction on $\sum |n_i|$. We will use the Riemann–Roch theorem for C as given in Exercise 6.3.16. We treat the case of $D = D' + C$, leaving the remaining case for the exercises. Tensoring the sequence

$$0 \to \mathscr{O}_X(-C) \to \mathscr{O}_X \to \mathscr{O}_C \to 0$$

by $\mathscr{O}(D)$ yields

$$0 \to \mathscr{O}_X(D') \to \mathscr{O}_X(D) \to \mathscr{O}_C(D) \to 0.$$

Therefore, using this together with the adjunction formula and Riemann–Roch on C, we obtain

$$\begin{aligned} \chi(\mathscr{O}_X(D)) &= \chi(\mathscr{O}_X(D')) + \chi(\mathscr{O}_C(D)) \\ &= \frac{1}{2} D'(D' - K) + \chi(\mathscr{O}_X) + \deg(D|_C) + 1 - g(C) \\ &= \frac{1}{2} D'(D' - K) + C \cdot D - \frac{1}{2} C(C + K) + \chi(\mathscr{O}_X) \\ &= \frac{1}{2} D(D - K) + \chi(\mathscr{O}_X). \end{aligned}$$

□

Exercises

11.3.5. Find a formula for the genus of a curve in $\mathbb{P}^1 \times \mathbb{P}^1$ in terms of its bidegree.

11.3.6. Given a morphism $f : C \to D$, calculate the self-intersection of the graph Γ_f^2.

11.3.7. Do the case $D = D' - C$ in the proof of Riemann–Roch.

11.3.8. Find a formula for $D \cdot C$ in terms of $\chi(\mathscr{O}(D)), \chi(\mathscr{O}(C)), \chi(\mathscr{O}(C+D))$ and $\chi(\mathscr{O}_X)$.

11.3.9. Use the formula of the previous exercise to give another proof of Proposition 11.2.4.

11.4 The Hodge Index Theorem

The next result is Hodge-theoretic, so we work with a compact Kähler surface X. Let ω denote the Kähler form.

Lemma 11.4.1. *If α is a harmonic 2-form, then $\omega \wedge \alpha$ is again harmonic.*

Proof. By the Kähler identities, it is enough to prove that $\bar{\partial}(\omega \wedge \alpha) = 0$, which is trivially true, and $\bar{\partial}^*(\omega \wedge \alpha) = 0$. By Proposition 9.3.5 and some calculation,

$$\bar{\partial}^*(\omega \wedge \alpha) = C_1(\Lambda\partial - \partial\Lambda)(\omega \wedge \alpha) = C_1\partial\Lambda(\omega \wedge \alpha) = C_2\partial\alpha = 0$$

for appropriate constants C_1, C_2. □

Then the form ω is a closed real $(1,1)$-form. Therefore the Kähler class $[\omega]$ is an element of $H^{11}(X) \cap H^2(X, \mathbb{R})$.

Theorem 11.4.2 (Hodge index theorem). *Let X be a compact Kähler surface. Then the restriction of the cup product to $(H^{11}(X) \cap H^2(X, \mathbb{R})) \cap (\mathbb{R}[\omega])^\perp$ is negative definite.*

Proof. Around each point we have a neighborhood U such that $\mathscr{E}^{(1,0)}(U)$ is a free module. By Gram–Schmid, we can find an orthonormal basis $\{\phi_1, \phi_2\}$ for it. In this basis, over U, we have

$$\omega = \frac{\sqrt{-1}}{2}(\phi_1 \wedge \bar{\phi}_1 + \phi_2 \wedge \bar{\phi}_2)$$

and the volume form

$$d\text{vol} = \frac{\omega^2}{2} = -\frac{1}{4}\phi_1 \wedge \bar{\phi}_1 \wedge \phi_2 \wedge \bar{\phi}_2,$$

using Exercise 10.1.18. It follows from this and the previous lemma that $d$vol is the unique harmonic 4-form up to scalar multiples. Choose an element $\alpha \in (H^{11}(X) \cap H^2(X, \mathbb{R}))$ and represent it by a harmonic real $(1,1)$-form. Then over U,

$$\alpha = \sqrt{-1}\sum a_{ij}\phi_i \wedge \bar{\phi}_j$$

with

$$a_{ji} = \bar{a}_{ij}. \tag{11.4.1}$$

By the previous lemma,

$$\alpha \wedge \omega = 2(a_{11} + a_{22})d\text{vol}$$

is also a harmonic 4-form, and therefore a scalar multiple of $d$vol. If α is chosen in $(\mathbb{R}[\omega])^{\perp}$, then $\int \alpha \wedge \omega = 0$, so that

$$a_{11} + a_{22} = 0. \tag{11.4.2}$$

Combining (11.4.1) and (11.4.2) yields

$$\alpha \wedge \alpha = -8(|a_{11}|^2 + |a_{12}|^2)d\text{vol},$$

so globally, $\alpha \wedge \alpha$ is a negative multiple of $d$vol. Therefore

$$\int_X \alpha \wedge \alpha < 0.$$

□

Corollary 11.4.3. *If H is an ample divisor on an algebraic surface, the intersection pairing is negative definite on* $(\text{NS}(X) \otimes \mathbb{R}) \cap (\mathbb{R}[H])^{\perp}$.

Proof. By Corollary 10.1.10, $[H]$ is a Kähler class. □

Corollary 11.4.4. *If H, D are divisors on an algebraic surface such that $H^2 > 0$ and $D \cdot H = 0$, then $D^2 < 0$ unless $[D] = 0$.*

Proof. This is an exercise in linear algebra using the fact that the intersection form on $(\text{NS}(X) \otimes \mathbb{R})$ has signature $(+1, -1, \ldots, -1)$. □

Exercises

11.4.5. Prove that the restriction of the cup product to $(H^{20}(X) + H^{02}(X)) \cap H^2(X, \mathbb{R})$ is positive definite.

11.4.6. Conclude that (the matrix representing) the cup product pairing has $2p_g + 1$ positive eigenvalues. Therefore p_g is a topological invariant.

11.4.7. Let $f : X \to Y$ be a morphism from a smooth algebraic surface to a possibly singular projective surface. Consider the set $\{D_i\}$ of irreducible curves that map to points under f. Prove a theorem of Mumford that the matrix $(D_i \cdot D_j)$ is negative definite.

11.5 Fibered Surfaces*

Let us say that a surface X is *fibered* if it admits a nonconstant holomorphic map $f : X \to C$ to a nonsingular curve. For example, ruled surfaces and elliptic surfaces are fibered. Not all surfaces are fibered. However, any surface can be fibered after blowing up: Choose a nontrivial rational function $X \dashrightarrow \mathbb{P}^1$, then blow up X to get a morphism. The fiber over $p \in C$ is the closed set $f^{-1}(p)$. Let z be a local coordinate at p. Then $f^{-1}(p)$ is defined by the equation $f^*z = 0$. We can regard $f^{-1}(p)$ as a divisor $\sum n_i D_i$, where D_i are its irreducible components and n_i is the order of vanishing of f^*z along D_i. Call a divisor on X *vertical* if its irreducible components are contained in the fiberes.

Lemma 11.5.1. *Suppose that $f : X \to C$ is a fibered surface.*

(a) If $F = f^{-1}(p)$ and D is a vertical divisor, then $F \cdot D = 0$. In particular, $F^2 = 0$.
(b) If D is vertical, then $D^2 \leq 0$.

Proof. Since the fundamental class of $f^{-1}(p)$ is independent of p, we can assume that F and D are disjoint. This proves (a).

Suppose that $D^2 > 0$, when combined with (a), we would get a contradiction to the Hodge index theorem. □

Corollary 11.5.2. *A necessary condition for a surface to be fibered is that there be an effective divisor with $F^2 = 0$. In particular, $\mathbb{P}^2$ is not fibered.*

We can give a complete characterization of surfaces fibered over curves of genus greater than one.

Theorem 11.5.3 (Castelnuovo–de Franchis). *Suppose that X is an algebraic surface. A necessary and sufficient condition for X to admit a nonconstant holomorphic map to a smooth curve of genus $g \geq 2$ is that there exist two linear independent forms $\omega_i \in H^0(X, \Omega^1_X)$ such that $\omega_1 \wedge \omega_2 = 0$.*

Proof. The necessity is easy. If $f : X \to C$ is a holomorphic map onto a curve of genus at least 2, then it possesses at least two linearly independent holomorphic 1-forms ω_i'. Set $\omega_i = f^*\omega_i'$. By writing this in local coordinates, we see that these are nonzero. But $\omega_1 \wedge \omega_2 = f^*(\omega_1' \wedge \omega_2') = 0$.

We will sketch the converse. A complete proof can be found in [9, pp. 123–125]. Choosing local coordinates, we can write

$$\omega_i = f_i(z_1, z_2)dz_1 + g_i(z_1, z_2)dz_2.$$

The condition $\omega_1 \wedge \omega_2 = 0$ forces

$$(f_1 g_2 - f_2 g_1)dz_1 \wedge dz_2 = 0.$$

Therefore $f_2/f_1 = g_2/g_1$. Call the common value F. Thus $\omega_2 = F\omega_1$. Since the ω_i are globally defined, $F = \omega_2/\omega_1$ defines a global meromorphic function $X \dashrightarrow \mathbb{P}^1$.

By Corollary 11.1.10, there exists $Y \to X$ that is a composition of blowups such that F extends to a holomorphic function $F' : Y \to \mathbb{P}^1$. The fibers of F' need not be connected. Stein's factorization theorem [60] shows that the map can be factored as

$$Y \stackrel{\Phi}{\to} C \to \mathbb{P}^1,$$

where Φ has connected fibers and $C \to \mathbb{P}^1$ is a finite-to-one map of smooth projective curves. To avoid too much notation, let us denote the pullbacks of ω_i to Y by ω_i as well. We now have a relation $\omega_2 = \Phi\omega_1$.

We claim that the ω_i are pullbacks of holomorphic 1-forms on C. We will check this locally around a general point $p \in Y$. Since the ω_i are harmonic (by Exercise 10.2.9) and therefore closed,

$$d\omega_2 = d\Phi \wedge \omega_1 = 0. \tag{11.5.1}$$

Let t_1 be a local coordinate centered at $\Phi(p) \in C$. Let us also denote the pullback of this function to neighborhood of p by t_1. Then we can choose a function t_2 such that t_1, t_2 give local coordinates at p. Then (11.5.1) becomes $dt_1 \wedge \omega_1 = 0$. Consequently, $\omega_1 = f(t_1,t_2)dt_1$, for some function f. The relation $d\omega_1 = 0$ implies that f is a function of t_1 alone. Thus ω_1 is locally the pullback of a 1-form on C, as claimed. The same reasoning applies to ω_2. This implies that the genus of C is at least two.

The final step is to prove that blowing up was unnecessary. Let

$$Y = Y_1 \xrightarrow{\pi_1} Y_2 \to \cdots X$$

be a composition of blowups at points $p_i \in Y_i$. Then $\pi_1^{-1}(p_2) \cong \mathbb{P}^1$. Any map from $\mathbb{P}^1$ to C is constant, since it has positive genus. So we conclude that $Y \to C$ factors through Y_1, and then likewise through Y_2 and so on until we reach Y. □

An obvious corollary is the following:

Corollary 11.5.4. *If $q \geq 2$ and $p_g = 0$, then X admits a nonconstant map to a curve as above.*

Exercises

11.5.5. Given an elliptic surface X, show that K_X is vertical. Conclude that $K_X^2 \leq 0$.

11.5.6. Show that X maps onto a curve of genus ≥ 2 if $q > p_g + 1$. (This bound can be sharpened to $(p_g + 3)/2$, but the argument is more delicate; cf. [9, IV, 4.2].)

Chapter 12
Hodge Structures and Homological Methods

Our next goal is to make the Hodge decomposition functorial with respect to holomorphic maps. This is not immediate, since the pullback of a harmonic form along a holomorphic map is almost never harmonic. The trick is to state things in a way that depends only on the complex structure: a cohomology class is of type (p,q) if it can be represented by a form with p dz_i's and by a form with q $d\bar{z}_j$'s. Of course, just making a definition is not enough. There is something to be proved. The main ingredients are the previous Hodge decomposition for harmonic forms together with some homological algebra, which we develop here.

Although projective manifolds are Kähler, there are examples of algebraic manifolds that are not. One benefit of this homological approach is that it will allow us to extend the decomposition to these manifolds where harmonic theory alone would be insufficient.

The articles and books by Deligne [24], Griffiths and Schmid [50], Peters and Steenbrink [95], and Voisin [115, 116] cover this material in more detail.

12.1 Pure Hodge Structures

It is useful to isolate the purely linear algebraic features of the Hodge decomposition. We define a *pure real Hodge structure* of weight m to be a finite-dimensional complex vector space with a real structure $H_{\mathbb{R}}$, and a bigrading

$$H = \bigoplus_{p+q=m} H^{pq}$$

satisfying $\bar{H}^{pq} = H^{qp}$. We generally use the same symbol for Hodge structure and the underlying vector space. A (pure weight m) Hodge structure is a real Hodge structure H together with a choice of a finitely generated abelian group $H_{\mathbb{Z}}$ and an isomorphism $H_{\mathbb{Z}} \otimes \mathbb{R} \cong H_{\mathbb{R}}$. Even though the abelian group $H_{\mathbb{Z}}$ may have torsion, it is helpful to think of it as a "lattice" in $H_{\mathbb{R}}$. Rational Hodge structures are defined in a similar way.

Before continuing with the abstract development of Hodge structures, we need to ask the obvious question. Why is it useful to consider these things? More specifically, why is it useful for algebraic geometry? To answer, we observe that algebraic varieties tend to come in families. For example, we may simply allow the coefficients of the defining equations to vary. Thus varieties tend to come with natural "continuous" parameters. The cohomological invariants considered up to now are discrete. Hodge structures, however, also have continuous parameters that sometimes match those coming from geometry. The simplest example is very instructive. Start with an elliptic curve $X_\tau = \mathbb{C}/(\mathbb{Z}+\mathbb{Z}\tau)$ with τ in the upper half plane H. We can identify $\iota : H^1(X_\tau,\mathbb{C}) \cong \mathbb{C}^2$ by mapping a closed differential form α to $(\int_0^1 \alpha, \int_0^\tau \alpha)$, where the paths of integration are lines. Then

$$H^{10} = \mathbb{C}\iota(dz) = \mathbb{C}(1,\tau).$$

So in this case, we recover the basic parameter τ and therefore the curve itself from its Hodge structure.

Given a pure Hodge structure, define the Hodge filtration by

$$F^p H_{\mathbb{C}} = \bigoplus_{p' \geq p} H^{pq}.$$

In many situations, the Hodge filtration is the more natural object to work with. This determines the bigrading thanks to the following lemma:

Lemma 12.1.1. *If H is a pure Hodge structure of weight m, then*

$$H_{\mathbb{C}} = F^p \oplus \bar{F}^{m-p+1}$$

for all p. Conversely, if $F^\bullet$ is a descending filtration satisfying $F^a = H_{\mathbb{C}}$ and $F^b = 0$ for some $a,b \in \mathbb{Z}$ and satisfying the above identity, then

$$H^{pq} = F^p \cap \bar{F}^q$$

defines a pure Hodge structure of weight m.

The most natural examples of Hodge structures come from compact Kähler manifolds: if $H_{\mathbb{Z}} = H^i(X,\mathbb{Z})$ with the Hodge decomposition on $H^i(X,\mathbb{C}) \cong H_Z \otimes \mathbb{C}$, then we get a Hodge structure of weight i. It is easy to manufacture other examples. For every integer i, there is a rank-one Hodge structure $\mathbb{Z}(i)$ of weight $-2i$. Here the underlying space is $\mathbb{C}$, with $H^{(-i,-i)} = \mathbb{C}$ and lattice $H_{\mathbb{Z}} = (2\pi\sqrt{-1})^i\mathbb{Z}$ (these factors should be ignored on first reading). The collection of Hodge structures forms a category HS, where a morphism is a linear map f preserving the lattices and the bigradings. In particular, morphisms between Hodge structures with different weights must vanish. This category has the following operations: direct sums of Hodge structures of the same weight (we will eventually relax this), and (unrestricted) tensor products and duals. Explicitly, given Hodge structures H and G of weights n and m, their tensor product $H \otimes_{\mathbb{Z}} G$ is equipped with a weight-$(n+m)$ Hodge structure with bigrading

$$(H \otimes G)^{pq} = \bigoplus_{\substack{p'+p''=p \\ q'+q''=q}} H^{p'q'} \otimes G^{p''q''}.$$

If $m = n$, their direct sum $H \oplus G$ is equipped with the weight-m Hodge structure

$$(H \oplus G)^{pq} = \bigoplus_{p+q=m} H^{pq} \oplus G^{pq}.$$

The dual $H^* = \mathrm{Hom}(H, \mathbb{Z})$ is equipped with a weight-$(-n)$ Hodge structure with bigrading

$$(H^*)^{pq} = (H^{-p,-q})^*.$$

The operation $H \mapsto H(i) = H \otimes \mathbb{Z}(i)$ is called the Tate twist. It has the effect of leaving H unchanged and shifting the bigrading by $(-i, -i)$.

Exercises

12.1.2. Show that there are no free rank-one pure Hodge structures of odd weight, and up to isomorphism a unique rank-one Hodge structure for every even weight.

12.1.3. Show that $\mathrm{Hom}_{\mathrm{HS}}(\mathbb{Z}(0), H^* \otimes G) \cong \mathrm{Hom}_{\mathrm{HS}}(H, G)$.

12.1.4. Prove Lemma 12.1.1.

12.1.5. Given a g-dimensional complex torus T, use Exercise 10.3.9, that $T \cong \mathrm{Alb}(T)$, to conclude that T can be recovered from the Hodge structure $H = H^1(T)$.

12.2 Canonical Hodge Decomposition

The Hodge decomposition involved harmonic forms, so it is tied up with the Kähler metric. It is possible to reformulate it so as to make it independent of the choice of metric. Let us see how this works for a compact Riemann surface X. We have an exact sequence

$$0 \to \mathbb{C}_X \to \mathscr{O}_X \to \Omega^1_X \to 0,$$

and we saw in Lemma 6.2.8 that the induced map

$$H^0(X, \Omega^1_X) \to H^1(X, \mathbb{C})$$

is injective. If we define

$$F^0H^1(X,\mathbb{C}) = H^1(X,\mathbb{C}),$$
$$F^1H^1(X,\mathbb{C}) = \operatorname{im}[H^0(X,\Omega^1_X) \to H^1(X,\mathbb{C})],$$
$$F^2H^1(X,\mathbb{C}) = 0,$$

then this together with the isomorphism $H^1(X,\mathbb{C}) = H^1(X,\mathbb{Z}) \otimes \mathbb{C}$ determines a pure Hodge structure of weight 1. To see this, choose a metric, which is automatically Kähler because $\dim X = 1$. Then $H^1(X,\mathbb{C})$ is isomorphic to a direct sum of the space of harmonic $(1,0)$-forms, which maps to F^1, and the space of harmonic $(0,1)$-forms, which maps to $\bar{F}^1$.

Before proceeding with the higher-dimensional version, we need some facts from homological algebra. Let

$$C^\bullet = \to \cdots C^a \xrightarrow{d} C^{a+1} \to \cdots$$

be a complex of vectors spaces (or modules or ...). It is convenient to allow the indices to vary over $\mathbb{Z}$, but we will require that it be *bounded below*, which means that $C^a = 0$ for all $a \ll 0$. Let us suppose that each C^i is equipped with a filtration $F^pC^i \supseteq F^{p+1}C^i \supseteq \cdots$, that is preserved by d, i.e., $dF^pC^i \subseteq F^pC^{i+1}$. This implies that each $F^pC^\bullet$ is a subcomplex. We suppose further that $F^\bullet$ *biregular*, which means that for each i there exist a and b with $F^aC^i = C^i$ and $F^bC^i = 0$. We get a map on cohomology

$$\phi^p : \mathcal{H}^\bullet(F^pC^\bullet) \to \mathcal{H}^\bullet(C^\bullet),$$

and we let $F^p\mathcal{H}^\bullet(C^\bullet)$ be the image. Define $Gr^p\mathcal{H}^i(C^\bullet) = F^p\mathcal{H}^i(C^\bullet)/F^{p+1}\mathcal{H}^i(C^\bullet)$. When $C^\bullet$ is a complex of vector spaces, there are noncanonical isomorphisms

$$\mathcal{H}^i(C^\bullet) = \bigoplus_p Gr^p\mathcal{H}^i(C^\bullet).$$

The filtration is said to be *strictly compatible with differentials* of $C^\bullet$, or simply just *strict*, if all the ϕ^p's are injective. Let $Gr^p_FC^\bullet = Gr^pC^\bullet = F^pC^\bullet/F^{p+1}C^\bullet$. Then we have a short exact sequence of complexes

$$0 \to F^{p+1}C \to F^pC \to Gr^pC \to 0,$$

from which we get a connecting map $\delta : \mathcal{H}^i(Gr^pC^\bullet) \to \mathcal{H}^{i+1}(F^{p+1}C^\bullet)$. This can be described explicitly as follows. Given $\bar{x} \in \mathcal{H}^i(Gr^pC^\bullet)$, it can be lifted to an element $x \in F^pC^i$ such that $dx \in F^{p+1}C^{i+1}$. Then $\delta(\bar{x})$ is represented by dx.

Proposition 12.2.1. *The following are equivalent:*

(1) F is strict.
(2) $F^pC^{i+1} \cap dC^i = dF^pC^i$ for all i and p.
(3) The connecting maps $\delta : \mathcal{H}^i(Gr^pC^\bullet) \to \mathcal{H}^{i+1}(F^{p+1}C^\bullet)$ vanish for all i and p.

Proof. This proof is due to Su-Jeong Kang.

(1)$\Rightarrow$ (2). Suppose that $z \in F^pC^{i+1} \cap dC^i$. Then $z = dx \in F^pC^{i+1}$ for some $x \in C^i$. Thus we have $z \in \ker[d : F^pC^{i+1} \to F^pC^{i+2}]$. Let $\bar{z} \in \mathcal{H}^{i+1}(F^pC^\bullet)$ denote the cohomology class of z. Note that $\phi^p(\bar{z}) = 0$, since $z = dx$. Hence from the assumption that F is strict, $\bar{z} = 0$ in $\mathcal{H}^{i+1}(F^pC^\bullet)$, or equivalently $z = dy$ for some $y \in F^pC^i$. This shows that $F^pC^{i+1} \cap dC^i \subseteq dF^pC^i$. The reverse inclusion is clear.

(2) $\Rightarrow$ (3). Let $\bar{x} \in \mathcal{H}^i(Gr^pC^\bullet)$. This lifts to an element $x \in F^pC^i$ with $dx \in F^{p+1}C^{i+1}$ as above. Then from the assumption (2), we have

$$dx \in F^{p+1}C^{i+1} \cap dC^i = dF^{p+1}C^i.$$

Since $\delta(\bar{x})$ is represented by dx, we have $\delta(\bar{x}) = 0$ in $\mathcal{H}^{i+1}(F^{p+1}C^\bullet)$.

(3) $\Rightarrow$ (1). For each i, ϕ^p can be expressed as finite a composition

$$\mathcal{H}^i(F^pC^\bullet) \to \mathcal{H}^i(F^{p-1}C^\bullet) \to \mathcal{H}^i(F^{p-2}C^\bullet) \to \cdots.$$

These maps are all injective by assumption, since their kernels are the images of the connecting maps. □

Corollary 12.2.2. *$Gr^p\mathcal{H}^i(C^\bullet)$ is a subquotient of $\mathcal{H}^i(Gr^pC^\bullet)$, which means that there is a diagram*

$$\mathcal{H}^i(Gr^pC^\bullet) \supseteq I^{i,p} \to Gr^p\mathcal{H}^i(C^\bullet)$$

where the last map is onto. Isomorphisms $Gr^p\mathcal{H}^i(C^\bullet) \cong I^{i,p} \cong \mathcal{H}^i(Gr^pC^\bullet)$ hold for all i, p if and only if F is strict.

Proof. Let $I^{i,p} = \mathrm{im}[\mathcal{H}^i(F^pC^\bullet) \to \mathcal{H}^i(Gr^pC^\bullet)]$. Then the surjection $\mathcal{H}^i(F^pC^\bullet) \to Gr^p\mathcal{H}^i(C^\bullet)$ factors through I. The remaining statement follows from (3) and a diagram chase. □

Corollary 12.2.3. *Suppose that $C^\bullet$ is a complex of vector spaces over a field such that $\dim \mathcal{H}^i(Gr^pC^\bullet) < \infty$ for all i, p. Then*

$$\dim \mathcal{H}^i(C^\bullet) \leq \sum_p \dim \mathcal{H}^i(Gr^pC^\bullet),$$

and equality holds for all i if and only if F is strict, in which case we also have

$$\dim F^p\mathcal{H}^i(C^\bullet) = \sum_{p' \geq p} \dim \mathcal{H}^i(Gr^{p'}C^\bullet).$$

Proof. We have

$$\dim \mathcal{H}^i(C^\bullet) = \sum_p \dim Gr^p\mathcal{H}^i(C^\bullet) \leq \sum_p \dim I^{ip} \leq \sum_p \dim \mathcal{H}^i(Gr^pC^\bullet),$$

and equality is equivalent to strictness of F by the previous corollary. The last statement is left as an exercise. □

These results are usually formulated in terms of spectral sequences, which we have chosen to avoid. In this language, the last corollary says that F is strict if and only if the associated spectral sequence degenerates at E_1. This is partially explained in the exercises.

Let X be a complex manifold. Then the de Rham complex $\mathscr{E}^\bullet(X)$ has a filtration called the Hodge filtration:

$$F^p\mathscr{E}^\bullet(X) = \sum_{p'\geq p} \mathscr{E}^{p'q}(X).$$

Its conjugate equals

$$\bar{F}^q\mathscr{E}^\bullet(X) = \sum_{q'\geq q} \mathscr{E}^{pq'}(X).$$

Theorem 12.2.4. *If X is compact Kähler, the Hodge filtration is strict. The associated filtration $F^\bullet H^i(X,\mathbb{C})$, on cohomology, gives a Hodge structure*

$$H^i(X,\mathbb{Z})\otimes\mathbb{C}\cong H^i(X,\mathbb{C}) = \bigoplus_{p+q=i} H^{pq}(X)$$

of weight i, where

$$H^{pq}(X) = F^p\mathscr{H}^i(X,\mathbb{C})\cap\bar{F}^q H^i(X,\mathbb{C})\cong H^q(X,\Omega^p_X).$$

Proof. Dolbeault's theorem (Corollary 9.2.3) implies that $\mathscr{H}^q(Gr^p\mathscr{E}^\bullet(X)) = \mathscr{H}^q(\mathscr{E}_X^{(p,\bullet)}(X))$ is isomorphic to $H^q(X,\Omega^p_X)$. Therefore F is strict by Corollary 12.2.3 and the Hodge decomposition. By conjugation, we see that $\bar{F}$ is also strict. Furthermore, these facts together with Corollary 12.2.3 give

$$\dim F^p H^i(X,\mathbb{C}) = h^{p,i-p}(X) + h^{p+1,i-p-1}(X) + \cdots \tag{12.2.1}$$

and

$$\dim \bar{F}^{i-p+1} H^i(X,\mathbb{C}) = h^{p-1,i-p+1}(X) + h^{p-2,i-p+2}(X) + \cdots. \tag{12.2.2}$$

A cohomology class lies in $F^pH^i(X,\mathbb{C})$ (respectively $\bar{F}^{i-p+1}H^i(X,\mathbb{C})$) if and only if it can be represented by a form in $F^p\mathscr{E}^\bullet(X)$ (respectively $\bar{F}^{i-p+1}\mathscr{E}^\bullet(X)$). Thus $H^i(X,\mathbb{C})$ is the sum of these subspaces. Using (12.2.1) and (12.2.2), we see that it is a direct sum. Therefore the filtrations determine a pure Hodge structure of weight i on $H^i(X,\mathbb{C})$. □

Even though harmonic theory is needed to verify that this Hodge structure, it should be clear that it involves only the holomorphic structure and not the metric. Thus we have obtained a canonical Hodge decomposition. The word canonical is really synonymous with functorial:

Corollary 12.2.5. *If $f : X \to Y$ is a holomorphic map of compact Kähler manifolds, then the pullback map $f^* : H^i(Y,\mathbb{Z}) \to H^i(X,\mathbb{Z})$ is compatible with the Hodge structures.*

Corollary 12.2.6. *If X is compact Kähler, the maps*

$$H^q(X,\Omega_X^p) \to H^q(X,\Omega_X^{p+1})$$

induced by differentiation vanish. In particular, global holomorphic differential forms on X are closed.

Proof. This follows from strictness, as will be explained in the exercises. □

This corollary, and hence the theorem, can fail for compact complex non-Kähler manifolds. An explicit example is described in the exercises.

Theorem 12.2.7. *If X is a compact Kähler manifold, the cup product*

$$H^i(X)\otimes H^j(X) \to H^{i+j}(X)$$

is a morphism of Hodge structures.

The proof comes down to the observation that

$$F^p\mathscr{E}^\bullet \wedge F^q\mathscr{E}^\bullet \subseteq F^{p+q}\mathscr{E}^\bullet.$$

For the corollaries, we work with rational Hodge structures. We have compatibility with Poincaré duality:

Corollary 12.2.8. *If $\dim X = n$, then Poincaré duality gives an isomorphism of Hodge structures*

$$H^i(X) \cong [H^{2n-i}(X)^*](-n).$$

We have compatibility with the Künneth formula:

Corollary 12.2.9. *If X and Y are compact Kähler manifolds, then*

$$\bigoplus_{i+j=k} H^i(X)\otimes H^j(Y) \cong H^k(X\times Y)$$

is an isomorphism of Hodge structures.

We have compatibility with the Gysin map:

Corollary 12.2.10. *If $f: X\to Y$ is a holomorphic map of compact Kähler manifolds of dimension n and m respectively, the Gysin map is a morphism*

$$H^i(X) \to H^{i+2(m-n)}(Y)(n-m).$$

Exercises

12.2.11. Finish the proof of Corollary 12.2.3.

12.2.12. Let $C^\bullet$ be a bounded-below complex with biregular filtration $F^\bullet$. Define $E_1^{pq} = H^{p+q}(Gr^p C^\bullet)$ and $d_1 : E_1^{pq} \to E_1^{p+1,q}$ as the connecting map associated to

$$0 \to Gr^{p+1}C^\bullet \to F^pC^\bullet/F^{p+2}C^\bullet \to Gr^pC^\bullet \to 0.$$

Show that $d_1 = 0$ if $F^\bullet$ is strict. (The converse is not quite true, as we will see shortly.)

12.2.13. When $C^\bullet = \mathscr{E}^\bullet(X)$ with its Hodge filtration, show that $E_1^{pq} \cong H^q(X, \Omega_X^p)$ and that d_1 is induced by $\alpha \mapsto \partial\alpha$ on $\bar\partial$-cohomology. Conclude that these maps vanish when X is compact Kähler.

12.2.14. Continuing the notation from Exercise 12.2.12. Suppose that $d_1 = 0$ for all indices. Construct a map $d_2 : E_1^{pq} \to E_1^{p+2,q-1}$ that fits into the commutative diagram

$$\begin{array}{ccccc} & & H^{p+q}(Gr^p) & & \\ & & \uparrow & \searrow^{d_2} & \\ H^{p+q}(F^p/F^{p+3}) & \longrightarrow & H^{p+q}(F^p/F^{p+2}) & \longrightarrow & H^{p+q+1}(Gr^{p+2}) \end{array}$$

Show that $d_2 = 0$ if $F^\bullet$ is strict. Optional messy part: If $d_1 = d_2 = 0$, define $d_3 : E_1^{pq} \to E_1^{p+3,q-2}$ etc. in the same way, and check that strictness is equivalent to the vanishing of whole lot.

12.2.15. Given a commutative ring R, let $U_3(R)$ be the space of upper triangular 3×3 matrices

$$\begin{pmatrix} 1 & x & z \\ 0 & 1 & y \\ 0 & 0 & 1 \end{pmatrix}$$

with entries in R. The Iwasawa manifold is the quotient $U_3(\mathbb{C})/U_3(\mathbb{Z} + \mathbb{Z}\sqrt{-1})$. Verify that this is a compact complex manifold with a nonclosed holomorphic form $dz - xdy$.

12.3 Hodge Decomposition for Moishezon Manifolds

A compact complex manifold need not have any nonconstant meromorphic functions at all. At the other extreme, a compact manifold X is called *Moishezon* if its field of meromorphic functions is as large as possible, that is, if it has transcendence degree equal to $\dim X$ (this is the maximum possible by a theorem of Siegel [104]). This is a very natural class of manifolds, which includes smooth proper algebraic varieties. Moishezon manifolds need not be Kähler; explicit examples due to Hironaka can be found in [60, Appendix B]. Nevertheless, Theorem 12.2.4 holds for these manifolds. A somewhat more general result is true. Let us say that a holomorphic map between complex manifolds is bimeromorphic if it is a biholomorphism between dense open sets.

Theorem 12.3.1. *Suppose that X is a compact complex manifold for which there exist a compact Kähler manifold and a surjective holomorphic bimeromorphic map $f : \tilde{X} \to X$. Then X possesses a canonical Hodge decomposition on cohomology described exactly as in Theorem 12.2.4.*

Corollary 12.3.2. *The Hodge decomposition holds for Moishezon manifolds.*

Proof. Moishezon [88] proved that that there exists a bimeromorphic map $\tilde{X} \to X$, in fact a blowup, with $\tilde{X}$ smooth projective. □

We have already proved a special case of Serre duality for Kähler manifolds. In fact, the result holds for a general compact complex manifold X. There is a pairing

$$\langle , \rangle : H^q(X, \Omega_X^p) \otimes H^{n-q}(X, \Omega_X^{n-p}) \to \mathbb{C} \tag{12.3.1}$$

induced by

$$(\alpha, \beta) \to \int_X \alpha \wedge \beta.$$

Theorem 12.3.3. *Suppose X is a compact complex manifold. Then*

(a) (Cartan) $\dim H^q(X, \Omega_X^p) < \infty$.
(b) (Serre) The pairing (12.3.1) *is perfect.*

Proof. Both results can be deduced from the Hodge decomposition theorem for the $\bar{\partial}$-operator, which works regardless of the Kähler condition. See [49]. □

We outline the proof of Theorem 12.3.1. Further details can be found in [23, 28].

Proof. Let $n = \dim X$. There is a map

$$f^* : H^q(X, \Omega_X^p) \to H^q(\tilde{X}, \Omega_{\tilde{X}}^p)$$

that is induced by the map $\alpha \mapsto f^*\alpha$ of (p,q)-forms. We claim that the map f^* is injective. To see this, define a map

$$f_* : H^q(\tilde{X}, \Omega_{\tilde{X}}^p) \to: H^q(X, \Omega_X^p),$$

analogous to the Gysin map, as the adjoint $\langle f_*\alpha, \beta \rangle = \langle \alpha, f^*\beta \rangle$. We leave it as an exercise to check that $f_* f^*(\alpha) = \alpha$. This proves injectivity of f^* as claimed. By similar reasoning, $f^* H^i(X, \mathbb{C}) \to H^i(\tilde{X}, \mathbb{C})$ is also injective.

We claim that F is strict. As we saw in previous exercises (12.2.12,12.2.14), this is equivalent to the vanishing of the differentials $d_1, d_2 \ldots$. We check only the first case, but the same reasoning works in general. Consider the commutative diagram

$$\begin{array}{ccc} H^q(\Omega_X^p) & \xrightarrow{d_1} & H^q(\Omega_X^{p+1}) \\ {\scriptstyle f^*}\downarrow & & \downarrow{\scriptstyle f^*} \\ H^q(\Omega_{\tilde{X}}^p) & \xrightarrow{d_1} & H^q(\Omega_{\tilde{X}}^{p+1}) \end{array}$$

Since the bottom d_1 vanishes, the same goes for the top.

The filtration $\bar{F}$ can also be shown to be strict. We can now argue as in the proof of Theorem 12.2.4 that the filtrations give a Hodge structure on $H^i(X)$. □

Exercises

12.3.4. Check that $\langle f^*\alpha, f^*\beta\rangle = \langle\alpha,\beta\rangle$, and deduce the identity $f_*f^*\alpha = \alpha$ used above.

12.4 Hypercohomology*

At this point, it is convenient to give a generalization of the constructions from Chapter 4. Recall that a complex of sheaves is a possibly infinite sequence of sheaves

$$\cdots \to \mathscr{F}^i \xrightarrow{d^i} \mathscr{F}^{i+1} \xrightarrow{d^{i+1}} \cdots$$

satisfying $d^{i+1}d^i = 0$. We say that the complex is *bounded* (below) if finitely many of these sheaves are nonzero (or if $\mathscr{F}^i = 0$ for $i \ll 0$). Given any sheaf $\mathscr{F}$ and natural number n, we get a bounded complex $\mathscr{F}[n]$ consisting of $\mathscr{F}$ in the $-n$th position, and zeros elsewhere. The collection of bounded (respectively bounded below) complexes of sheaves on a space X form a category $C^b(X)$ (respectively $C^+(X)$), where a morphism of complexes $f : \mathscr{E}^\bullet \to \mathscr{F}^\bullet$ is defined to be a collection of sheaf maps $\mathscr{E}^i \to \mathscr{F}^i$ that commute with the differentials. This category is abelian. We define additive functors $\mathscr{H}^i : C^+(X) \to \mathrm{Ab}(X)$

$$\mathscr{H}^i(\mathscr{F}^\bullet) = \ker(d^i)/\operatorname{im}(d^{i-1}).$$

A morphism $f : \mathscr{E}^\bullet \to \mathscr{F}^\bullet$ in $C^+(X)$ is a called a *quasi-isomorphism* if it induces isomorphisms $\mathscr{H}^i(\mathscr{E}^\bullet) \cong \mathscr{H}^i(\mathscr{F}^\bullet)$ on all the sheaves.

Theorem 12.4.1. *Let X be a topological space. Then there are additive functors $\mathbb{H}^i : C^+(X) \to \mathrm{Ab}$, with $i \in \mathbb{N}$, such that*

(1) For any sheaf $\mathscr{F}$, $\mathbb{H}^i(X, \mathscr{F}[n]) = H^{i+n}(X, \mathscr{F})$.
(2) If $0 \to \mathscr{E}^\bullet \to \mathscr{F}^\bullet \to \mathscr{G}^\bullet \to 0$ is exact, then there is an exact sequence

$$0 \to \mathbb{H}^0(X, \mathscr{E}^\bullet) \to \mathbb{H}^0(X, \mathscr{F}^\bullet) \to \mathbb{H}^0(X, \mathscr{G}^\bullet) \to \mathbb{H}^1(X, \mathscr{E}^\bullet) \to \cdots.$$

(3) If $\mathscr{E}^\bullet \to \mathscr{F}^\bullet$ is a quasi-isomorphism, then the induced map $\mathbb{H}^i(X, \mathscr{E}^\bullet) \to \mathbb{H}^i(X, \mathscr{F}^\bullet)$ is an isomorphism.

$\mathbb{H}^i(X, \mathscr{E}^\bullet)$ is called the ith hypercohomology group of $\mathscr{E}^\bullet$.

Proof. We outline the proof. Further details can be found in [44], [68], or [118]. We start by redoing the construction of cohomology for a single sheaf $\mathscr{F}$. The functor $\mathbf{G}$ defined in Section 4.1, gives a flasque sheaf $\mathbf{G}(\mathscr{F})$ with monomorphism $\mathscr{F} \to \mathbf{G}(\mathscr{F})$. The sheaf $\mathbf{C}^1(\mathscr{F})$ is the cokernel of this map. Applying $\mathbf{G}$ again yields a sequence

$$\mathscr{F} \to \mathbf{G}(\mathscr{F}) \to \mathbf{G}(\mathbf{C}^1(\mathscr{F})).$$

By continuing as above, we get a resolution by flasque sheaves

$$\mathscr{F} \to \mathbf{G}^0(\mathscr{F}) \to \mathbf{G}^1(\mathscr{F}) \to \cdots.$$

Theorem 5.1.4 shows that $H^i(X,\mathscr{F})$ is the cohomology of the complex $\Gamma(X,\mathbf{G}^\bullet(\mathscr{F}))$, and this gives a clue how to generalize the construction. The complex $\mathbf{G}^\bullet$ is functorial. So given a complex

$$\cdots \to \mathscr{F}^i \xrightarrow{d} \mathscr{F}^{i+1} \to \cdots,$$

we get a commutative diagram

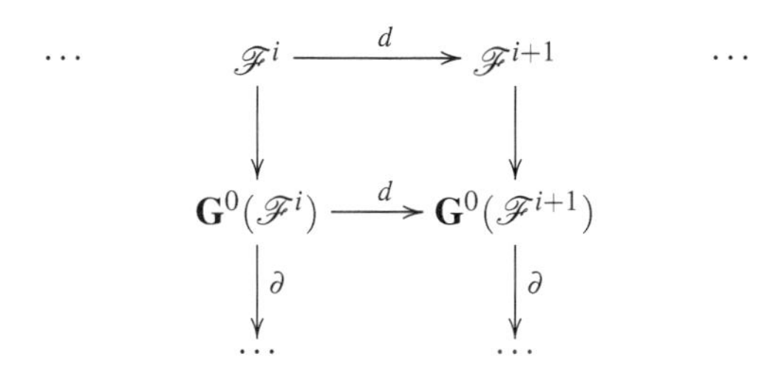

We define the total complex

$$\mathscr{T}^i(\mathscr{F}^\bullet) = \bigoplus_{p+q=i} \mathbf{G}^p(\mathscr{F}^q)$$

with a differential $\delta = d + (-1)^q\partial$. We can now define

$$\mathbb{H}^i(X,\mathscr{F}^\bullet) = H^i(\Gamma(X,\mathscr{T}^\bullet(\mathscr{G}))).$$

When applied to $\mathscr{F}[n]$, this yields $H^i(\Gamma(X,G^\bullet(\mathscr{F}))[n])$, which as we have seen is $H^i(X,\mathscr{F})$, and this proves (1).

(2) can be deduced from the exact sequence

$$0 \to \mathscr{T}^\bullet(\mathscr{F}^\bullet) \to \mathscr{T}^\bullet(\mathscr{G}^\bullet) \to \mathscr{T}^\bullet(\mathscr{F}^\bullet) \to 0$$

given in the exercises.

We now turn to the last statement, and prove it for bounded complexes. For any complex $\mathscr{E}^\bullet$ of sheaves (or elements of an abelian category), we can introduce the truncation operator given by the subcomplex

$$\tau_{\leq p}\mathscr{E}^i = \begin{cases} \mathscr{E}^p & \text{if } i < p, \\ \ker(\mathscr{E}^p \to \mathscr{E}^{p+1}) & \text{if } i = p, \\ 0 & \text{otherwise.} \end{cases}$$

Truncation yields an increasing filtration $\tau_{\leq p}$ or a decreasing filtration $\tau_{\leq -p}$. The key property is given in the following lemma:

Lemma 12.4.2. *There is an exact sequence of complexes*

$$0 \to \tau_{\leq q-1}\mathscr{E}^\bullet \to \tau_{\leq q}\mathscr{E}^\bullet \to \mathscr{H}^q(\mathscr{E}^\bullet)[-q] \to 0$$

for each q.

A quasi-isomorphism $\mathscr{E}^\bullet \to \mathscr{F}^\bullet$ induces a quasi-isomorphism $\tau_{\leq q}\mathscr{E}^\bullet \to \tau_{\leq q}\mathscr{F}^\bullet$ for each q. Thus the lemma can be applied to get a diagram with exact rows:

$$\begin{array}{ccccccccc} \cdots \longrightarrow & \mathbb{H}^i(\tau_{\leq q-1}\mathscr{E}^\bullet) & \longrightarrow & \mathbb{H}^i(\tau_{\leq q}\mathscr{E}^\bullet) & \longrightarrow & H^{i-q}(\mathscr{H}^q(\mathscr{E}^\bullet)) & \longrightarrow \cdots \\ & \downarrow & & \downarrow & & \downarrow \cong & \\ \cdots \longrightarrow & \mathbb{H}^i(\tau_{\leq q-1}\mathscr{F}^\bullet) & \longrightarrow & \mathbb{H}^i(\tau_{\leq q}\mathscr{F}^\bullet) & \longrightarrow & H^{i-q}(\mathscr{H}^q(\mathscr{E}^\bullet)) & \longrightarrow \cdots \end{array}$$

Thus (3) follows by induction on q. □

The precise relationship between the various (hyper) cohomology groups is usually expressed by the spectral sequence

$$E_1^{pq} = H^q(X, \mathscr{E}^p) \Rightarrow \mathbb{H}^{p+q}(\mathscr{E}^\bullet).$$

There are a number of standard consequences that we can prove directly. The first is a refinement of Theorem 5.1.4.

Corollary 12.4.3. *If $\mathscr{E}^\bullet$ is a bounded complex of acyclic sheaves, then $\mathbb{H}^i(X, \mathscr{E}^\bullet) = H^i(\Gamma(X, \mathscr{E}^\bullet))$.*

Proof. There is a map of complexes $\Gamma(X, \mathscr{E}^\bullet) \to \Gamma(X, \mathscr{T}^\bullet(\mathscr{F}^\bullet))$ inducing a map $H^i(\Gamma(X, \mathscr{E}^\bullet)) \to \mathbb{H}^i(X, \mathscr{E}^\bullet)$. We have to check that this is an isomorphism. We do this by induction on the length, or number of nonzero terms, of $\mathscr{E}^\bullet$. With the help of the "stupid" filtration,

$$\sigma^p\mathscr{E}^\bullet = \mathscr{E}^{\geq p} = \cdots \to 0 \to \mathscr{E}^p \to \mathscr{E}^{p+1} \to \cdots$$

is gotten by dropping the first $p-1$ terms of the complex. We have an exact sequence

$$0 \to \mathscr{E}^{\geq p+1} \to \mathscr{E}^{\geq p} \to \mathscr{E}^p[-p] \to 0 \tag{12.4.1}$$

leading to a commutative diagram

$$\begin{array}{ccccccc}
\cdots \longrightarrow H^i(\Gamma(X,\mathscr{E}^{\geq p+1})) & \longrightarrow & H^i(\Gamma(X,\mathscr{E}^{\geq p})) & \longrightarrow & H^i(\mathscr{E}^p[-p]) & \longrightarrow \cdots \\
\downarrow \cong & & \downarrow f & & \downarrow \cong & \\
\longrightarrow \mathbb{H}^i(X,\mathscr{E}^{\geq p+1}) & \longrightarrow & \mathbb{H}^i(X,\mathscr{E}^{\geq p}) & \longrightarrow & \mathbb{H}^i(X,\mathscr{E}^p[-p]) & \longrightarrow
\end{array}$$

with exact rows. The arrows marked by $\cong$ are isomorphisms by induction. Therefore f is an isomorphism by the 5-lemma. □

Corollary 12.4.4. *Suppose that $\mathscr{E}^\bullet$ is a bounded complex of sheaves of vector spaces. Then*

$$\dim \mathbb{H}^i(\mathscr{E}^\bullet) \leq \sum_{p+q=i} \dim H^q(X,\mathscr{E}^p).$$

Proof. The corollary follows by induction on the length (number of nonzero entries) of $\mathscr{E}^\bullet$ using the long exact sequences on hypercohomology coming from (12.4.1). □

Corollary 12.4.5. *Suppose that $\mathscr{E}^\bullet$ is a bounded complex with $H^q(X,\mathscr{E}^p) = 0$ for all $p+q=i$. Then $\mathbb{H}^i(\mathscr{E}^\bullet) = 0$.*

We can extract one more corollary, using Lemma 11.3.3.

Corollary 12.4.6. *If $\sum \dim H^q(X,\mathscr{E}^p) \leq \infty$, then*

$$\sum (-1)^i \dim \mathbb{H}^i(\mathscr{E}^\bullet) = \sum (-1)^{p+q} \dim H^q(X,\mathscr{E}^p).$$

In order to facilitate the computation of hypercohomology, we need a criterion for when two complexes are quasi-isomorphic. We will say that a filtration

$$\mathscr{E}^\bullet \supseteq F^p\mathscr{E}^\bullet \supseteq F^{p+1}\mathscr{E}^\bullet \supseteq \cdots$$

is finite (of length $\leq n$) if $\mathscr{E}^\bullet = F^a\mathscr{E}^\bullet$ and $F^{a+n}\mathscr{E}^\bullet = 0$ for some a.

Lemma 12.4.7. *Let $f : \mathscr{E}^\bullet \to \mathscr{F}^\bullet$ be a morphism of bounded complexes. Suppose that $F^p\mathscr{E}^\bullet$ and $G^p\mathscr{F}^\bullet$ are finite filtrations by subcomplexes such that $f(F^p\mathscr{E}^\bullet) \subseteq G^p\mathscr{F}^\bullet$. If the induced maps*

$$Gr_F^p(\mathscr{E}^\bullet) \to Gr_G^p(\mathscr{F}^\bullet)$$

are quasi-isomorphisms for all p, then f is a quasi-isomorphism.

Exercises

12.4.8. If $\mathscr{F}^\bullet$ is a bounded complex with zero differentials, show that $H^i(X,\mathscr{F}^\bullet) = \oplus_j H^{i-j}(X,\mathscr{F}^j)$.

12.4.9. Prove Lemma 12.4.7 by induction on the length.

12.5 Holomorphic de Rham Complex*

Let X be a C^∞ manifold. We can resolve $\mathbb{C}_X$ by the complex of C^∞ forms $\mathscr{E}_X^\bullet$. In other words, $\mathbb{C}_X$ and $\mathscr{E}_X^\bullet$ are quasi-isomorphic. Since $\mathscr{E}_X^\bullet$ is acyclic, it follows that

$$H^i(X,\mathbb{C}_X) = \mathbb{H}^i(X,\mathbb{C}_X[0]) \cong \mathbb{H}^i(X,\mathscr{E}_X^\bullet) \cong H^i(\Gamma(X,\mathscr{E}_X^\bullet)).$$

We have just re-proved de Rham's theorem.

Now suppose that X is a (not necessarily compact) complex manifold. Then we define a subcomplex

$$F^p\mathscr{E}_X^\bullet = \sum_{p'\geq p} \mathscr{E}_X^{p'q}.$$

The image of the map

$$\mathbb{H}^i(X,F^p\mathscr{E}_X^\bullet) \to \mathbb{H}^i(X,\mathscr{E}_X^\bullet)$$

is the filtration introduced just before Theorem 12.2.4. We want to reinterpret this purely in terms of holomorphic forms. We define the holomorphic de Rham complex by

$$\mathscr{O}_X \to \Omega_X^1 \to \Omega_X^2 \to \cdots.$$

We have a natural map $\Omega_X^\bullet \to \mathscr{E}_X^\bullet$ that takes σ^p to F^p, where $\sigma^p\Omega_X^\bullet = \Omega_X^{\geq p}$. Dolbeault's Theorem 9.2.3 implies that F^p/F^{p+1} is quasi-isomorphic to $\sigma^p/\sigma^{p+1} = \Omega_X^p[-p]$. Therefore, Lemma 12.4.7 implies that $\Omega_X^\bullet \to \mathscr{E}_X^\bullet$, and more generally $\sigma^p\Omega_X^\bullet \to F^p\mathscr{E}_X^\bullet$, are quasi-isomorphisms.

Lemma 12.5.1. *$H^i(X,\mathbb{C}) \cong \mathbb{H}^i(X,\Omega_X^\bullet)$ and $F^pH^i(X,\mathbb{C})$ is the image of $\mathbb{H}^i(X,\Omega_X^{\geq p})$.*

When X is compact Kähler, Theorem 12.2.4 implies that the map

$$\mathbb{H}^i(X,\Omega_X^{\geq p}) \to \mathbb{H}^i(X,\Omega_X^\bullet)$$

is injective.

From Corollaries 12.4.4, 12.4.5, 12.4.6 we obtain the following result.

Corollary 12.5.2. *If X is compact, the ith Betti number satisfies*

$$b_i(X) \leq \sum_{p+q=i} \dim H^q(X,\Omega_X^p),$$

and the Euler characteristic satisfies

$$e(X) = \sum(-1)^i b_i(X) = \sum(-1)^{p+q}\dim H^q(X,\Omega_X^p).$$

Corollary 12.5.3. *If $H^q(X,\Omega_X^p) = 0$ for all $p+q=i$, then $H^i(X,\mathbb{C}) = 0$.*

The next corollary uses the notion of Stein manifold that will be discussed later, in Section 16.1. For the time being, we note that Stein manifolds include smooth affine varieties. The above results give nontrivial topological information for this class of manifolds.

Corollary 12.5.4. *Let X be a Stein manifold, or in particular a smooth affine variety. Then $H^i(X,\mathbb{C}) = 0$ for $i > \dim X$.*

Proof. This follows from Theorem 16.3.3. □

Exercises

12.5.5. Suppose that $H^i(X,\mathscr{F}) = 0$ for $i > N$ and any locally free sheaf $\mathscr{F}$. Show that $b_i(X) = 0$ for $i > N + \dim X$.

12.5.6. Show that the inequality in Corollary 12.5.2 is strict for the Iwasawa manifold defined in Exercise 12.2.15).

12.6 The Deligne–Hodge Decomposition*

We fix the following: a smooth hypersurface (also called a smooth divisor) $X \subset Y$ in a projective smooth variety. Let $U = Y - X$. Our goal is to understand the cohomology and Hodge theory of U. This can be calculated using C^∞ differential forms $\mathscr{E}^\bullet_U$, but it will more useful to compute this with forms having controlled singularities. We define $\Omega^p_Y(*X)$ to be the sheaf of meromorphic p-forms that are holomorphic on U. This is not coherent, but it is a union of coherent subsheaves $\Omega^p_Y(mX)$ of mermorphic p-forms with at worst poles of order m along X. We also define $\Omega^p_Y(\log X) \subset \Omega^p_Y(1X)$ as the subsheaf of meromorphic forms α such that both α and $d\alpha$ have simple poles along X. If we choose local coordinates $z_1,\dots,z_n$ so that X is defined by $z_1 = 0$, then the sections of $\Omega^p_Y(\log X)$ are locally spanned as an $\mathscr{O}_X$ module by

$$\{dz_{i_1} \wedge \cdots \wedge dz_{i_p} \mid i_j > 1\} \cup \left\{ \frac{dz_1 \wedge dz_{i_2} \wedge \cdots \wedge dz_{i_p}}{z_1} \right\};$$

$\Omega^\bullet_X(\log D) \subset \Omega^\bullet_Y(*X)$ is a subcomplex.

Proposition 12.6.1. *There are isomorphisms*

$$H^i(U,\mathbb{C}) \cong \mathbb{H}^i(Y,\Omega^\bullet_Y(\log X)) \cong \mathbb{H}^i(Y,\Omega^\bullet_Y(*X)).$$

Proof. Details can be found in [49, pp. 449–454]. The key point is to show that the inclusions

$$\Omega^\bullet_Y(\log X) \subset \Omega^\bullet_Y(*X) \subset j_*\mathscr{E}^\bullet_U \tag{12.6.1}$$

are quasi-isomorphisms, where $j : U \to Y$ is the inclusion. This can be reduced to a calculation in which Y is replaced by a disk with coordinate z, and with X corresponding to the origin. Then (12.6.1) becomes

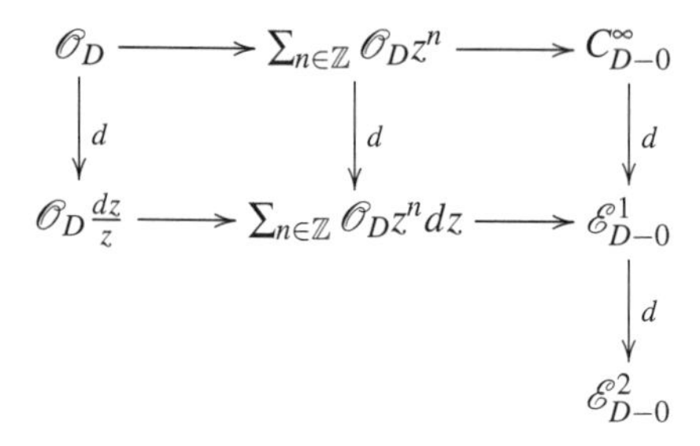

The cohomology in each column is $\mathbb{C}$ in degrees $0,1$, and horizontal maps induce isomorphisms between these. $\square$

The spaces in the proposition on the right carry natural filtrations. The Hodge filtration

$$F^pH^{i+1}(U)=\mathrm{im}[\mathbb{H}^{i+1}(0\to\Omega_Y^p(\log X)\to\Omega_Y^{p+1}(\log X)\to\cdots)\to\mathbb{H}^{i+1}(\Omega_Y^\bullet(\log X))]$$

and the pole filtration induced by

$$\mathrm{Pole}^pH^{i+1}(U)=\mathrm{im}[\mathbb{H}^{i+1}(\cdots\to 0\to\Omega_Y^p(X)\to\Omega_Y^{p+1}(2X)\to\cdots)\to\mathbb{H}^{i+1}(\Omega_Y^\bullet(*X))].$$

It follows more or less immediately that $F^p\subseteq \mathrm{Pole}^p$. Equality need not hold in general, but it does in an important case studied later, in Section 17.5.

In order to relate this to the cohomology of X, we use residues. We have a map

$$\mathrm{Res}:\Omega_Y^p(\log X)\to\Omega_X^{p-1}, \tag{12.6.2}$$

called the *Poincaré residue* map, given by

$$\mathrm{Res}\left(\alpha\wedge\frac{dz_1}{z_1}\right)=\alpha|_X.$$

Res commutes with d. Therefore it gives a map of complexes

$$\Omega_Y^\bullet(\log X)\to\Omega_X^\bullet[-1],$$

where $[-1]$ indicates a shift of indices by -1. This induces a map

$$H^i(U,\mathbb{C})=\mathbb{H}^i(\Omega_Y^\bullet(\log X))\to H^{i-1}(X,\mathbb{C}).$$

After normalizing this by a factor of $\frac{1}{2\pi\sqrt{-1}}$, it takes integer cohomology to integer cohomology (modulo torsion). This can be described topologically as a composition

$$H^i(U)\to H^i(\mathrm{Tube})\xrightarrow{\sim} H^{i-1}(X),$$

where Tube is a tubular neighborhood and the second map is the inverse of the Thom isomorphism, §5.5. The residue map is an epimorphism of sheaves, and the kernel is precisely the sheaf of holomorphic forms. So we have an exact sequence

$$0 \to \Omega_Y^\bullet \to \Omega_Y^\bullet(\log X) \to \Omega_X^\bullet[-1] \to 0, \tag{12.6.3}$$

which leads to a long exact sequence

$$\cdots \to H^q(\Omega_Y^p) \to H^q(\Omega_Y^p(\log X)) \to H^q(\Omega_X^{p-1}) \to H^{q+1}(\Omega_Y^p) \to \cdots \tag{12.6.4}$$

$$\cdots \to H^i(Y,\mathbb{C}) \to H^i(U,\mathbb{C}) \to H^{i-1}(X,\mathbb{C}) \xrightarrow{\gamma} H^{i+1}(Y,\mathbb{C}) \to \cdots. \tag{12.6.5}$$

The second is called the Gysin sequence. Indeed, γ is the Gysin map.

Theorem 12.6.2 (Deligne). *The Hodge filtration on $H^i(U,\mathbb{C})$ is strict, i.e., the maps*

$$\mathbb{H}^{i+1}(0 \to \Omega_Y^p(\log X) \to \Omega_Y^{p+1}(\log X) \to \cdots) \to \mathbb{H}^{i+1}(\Omega_Y^\bullet(\log X))$$

are injective. In particular, there is a (noncanonical) decomposition

$$H^i(U,\mathbb{C}) \cong \bigoplus_{p+q=i} H^q(Y,\Omega_Y^p(\log X)).$$

Proof. By Corollary 12.2.3, it is enough to prove that

$$\dim H^i(U) = \sum_{p+q=i} \dim H^q(Y,\Omega_Y^p(\log X)).$$

From (12.6.4) and (12.6.5), we get

$$\begin{aligned}\dim H^q(\Omega_Y(\log X)) = \dim\ker[H^q(\Omega_X^{p-1}) \to H^{q+1}(\Omega_Y^p)]\\ + \dim\operatorname{im}[H^{q-1}(\Omega_X^{p-1}) \to H^q(\Omega_Y^p)]\end{aligned}$$

and

$$\dim H^i(U) = \dim\ker[H^{i-1}(X) \to H^{i+1}(Y)] + \dim\operatorname{im}[H^{i-2}(X) \to H^i(Y)].$$

Combining the last equation with Corollaries 12.2.5 and 12.2.10 shows that

$$\begin{aligned}\dim H^i(U) &= \sum \dim\ker[H^q(\Omega_X^{p-1}) \to H^{q+1}(\Omega_Y^p)]\\ &\quad + \sum \dim\operatorname{im}[H^{q-1}(\Omega_X^{p-1}) \to H^q(\Omega_Y^p)]\\ &= \sum \dim H^q(Y,\Omega_Y^p(\log X)).\end{aligned}$$

□

Corollary 12.6.3 (Weak Kodaira vanishing). *If X is a smooth divisor such that $U = X - Y$ is affine, then $H^i(Y,\Omega_Y^n \otimes \mathcal{O}_Y(X)) = 0$ for $i > 0$.*

Proof. By Theorem 12.6.2 and Corollary 12.5.4,

$$\dim H^i(Y,\Omega_Y^n(\log X)) \le \dim H^{i+n}(U,\mathbb{C}) = 0$$

when $i > 0$. A direct calculation shows that

$$\Omega_Y^n(\log X) = \Omega_Y^n(X) \cong \Omega_Y^n \otimes \mathscr{O}_Y(X). \qquad \square$$

Here is a more useful form.

Corollary 12.6.4 (Kodaira vanishing). *If L is an ample line bundle, then*

$$H^i(Y, \Omega_Y^n \otimes L) = 0$$

for $i > 0$.

Proof. Here is the outline. By assumption, $L^{\otimes m} = \mathscr{O}_Y(1)$ for some $m > 0$. By Bertini's theorem, we can choose a hyperplane $H \subset \mathbb{P}^n$ such that $X = Y \cap H$ is smooth. Note that $\mathscr{O}(X) \cong L^{\otimes m}$ and $Y - X$ is affine. So when $m = 1$, we can apply the previous corollary. In general, one can construct a nonsingular cover $\pi : Y' \to Y$ branched over X, which locally is given by $y^m = f$, where $f = 0$ is the local equation for X. A precise construction can be found in [77, vol. I, Proposition 4.1.6], along with the proof of the following properties:

1. The set-theoretic preimage $X' = \pi^{-1}X$ is again smooth,
2. $L' = \pi^* L$ has a smooth section vanishing along X' without multiplicity, or to be more precise, $\mathscr{O}(X') \cong L'$.
3. The cohomology $H^i(Y, \Omega_Y^n \otimes L)$ injects into $H^i(Y', \Omega_{Y'}^n \otimes L')$.

The last property follows from [77, Lemma 4.1.14] plus Serre duality [60, Chapter III, Corollary 7.7]. Thus the result follows from the previous corollary applied to (Y', X'). $\square$

Remark 12.6.5. Kodaira proved a slightly different statement, where ampleness was replaced by positivity in a differential-geometric sense (cf. [49, Chapter 1§2]). This form was used in the proof of the Kodaira embedding theorem, Theorem 10.1.11. The embedding theorem then implies that positivity and ampleness are, a posteriori, equivalent conditions for line bundles.

Deligne [24] proved (a refinement of) Theorem 12.6.2 en route to constructing a *mixed Hodge structure* on cohomology. This is roughly something given by gluing pure Hodge structures of different weights together. More formally, a mixed Hodge structure is given by a lattice H with two filtrations W and F defined over $\mathbb{Q}$ and $\mathbb{C}$ respectively so that F induces a pure rational Hodge structure of weight k on W_k/W_{k-1} for each k. In the case of $H^i(U)$, where U is the complement of a smooth divisor in a smooth projective variety X, we have

$$W_k H^i(U, \mathbb{Q}) = \begin{cases} 0 & \text{if } k < i, \\ \operatorname{im} H^i(Y, \mathbb{Q}) & \text{if } k = i, \\ H^i(U, \mathbb{Q}) & \text{otherwise.} \end{cases}$$

A fairly detailed introduction to mixed Hodge theory can be found in the book by Peters and Steenbrink [95]. Since we can barely scratch the surface, we will be content to give a simple example to indicate the power of these ideas.

Example 12.6.6. Given an elliptic curve E, we have seen that $H^1(E)$ with its Hodge structure determines E. Now remove the origin 0 and a nonzero point p and consider the mixed Hodge structure on $H^1(E - \{0,p\})$. This determines the complement $E - \{0,p\}$ by the work of Carslon [16].

Exercises

12.6.7. Work out (12.6.4) and (12.6.5) explicitly when $Y = \mathbb{P}^2$ and X is a smooth curve of degree d.

12.6.8. Show that forms in $H^0(Y, \Omega_Y^p(\log X))$ are closed. Is this necessarily true for forms in $H^0(Y, \Omega_Y^p(kX))$?

12.6.9. Verify the isomorphism $\Omega_Y^n(\log X) \cong \Omega_Y^n \otimes \mathcal{O}_Y(X)$ used in the proof of Corollary 12.6.3.

Chapter 13
Topology of Families

In this chapter, we make a brief detour and study some aspects of the topology of families of algebraic varieties that will be used later on. As we will see, a family of smooth projective varieties is locally trivial topologically; in particular, all the fibers are diffeomorphic. However, the global topology may be nontrivial. An important measure of the nontriviality is the monodromy, which roughly tells us what happens to cycles when they are transported around a loop. This can also be understood using sheaf theory.

13.1 Topology of Families of Elliptic Curves

We briefly encountered elliptic surfaces earlier. Let us analyze the topology of a couple of examples in some detail. We start with a local example over a disk. Recall from Section 1.4 that the curve $E_\tau = \mathbb{C}/\mathbb{Z}+\mathbb{Z}\tau$ is given by the Weierstrass equation

$$y^2 = 4x^3 - g_2(\tau)x - g_3(\tau), \tag{13.1.1}$$

where the coefficients $g_k(\tau)$ are constant multiples of the Eisenstein series

$$\sum_{(m,n)\neq(0,0)} \frac{1}{(m+n\tau)^{2k}} = 2\sum_1^\infty \frac{1}{m^{2k}} + 2\sum_{n=1}^{\infty}\sum_{m=-\infty}^{\infty} \frac{1}{(m+n\tau)^{2k}}.$$

Since these are invariant under $\tau \mapsto \tau+1$, they can be expanded in a Fourier series, or equivalently in a Laurent series in $q = \exp(-2\pi i\tau)$. The explicit formulas, which can be found in [106, Chapter 1, §7], show that the $g_k(\tau)$ are holomorphic functions $\gamma_k(q)$.

Example 13.1.1. Therefore, we have a surface given by

$$\mathscr{E} = \{([x,y,z],q) \in \mathbb{P}^2 \times D \mid zy^2 = 4x^3 - z^2\gamma_2(q)x - z^3\gamma_3(q)\},$$

which maps to the unit disk D via projection, denoted by π.

The fiber over $q \neq 0$ is just the elliptic curve E_τ for any (normalized) logarithm τ of q. The fiber over $q = 0$ is a nodal cubic. Topologically, this is obtained by gluing together two points on a sphere.

In order to get a better feeling for this space, let us give a more direct construction of $\mathscr{E}^* = \mathscr{E} - \pi^{-1}(0)$. Let Γ be the semidirect product $\mathbb{Z}^2 \rtimes \mathbb{Z}$, where $k \in \mathbb{Z}$ acts by the matrix $\left(\begin{smallmatrix} 1 & k \\ 0 & 1 \end{smallmatrix}\right)$. More explicitly, the elements are triples $(m,n,k) \in \mathbb{Z}^3$ with multiplication

$$(m,n,k) \cdot (m',n',k') = (m + m' + kn', n + n', k + k').$$

Let Γ act on the product $\mathbb{C} \times H$ of the complex and upper half-planes by

$$(m,n,k) : (z,t) \mapsto (z + m + n\tau, \tau + k).$$

The quotient gives back the same family

$$\mathscr{E}^* \to H/\mathbb{Z} \overset{\tau \to q}{\cong} D^*$$

over the punctured disk. As Riemann surfaces, the fibers $\mathscr{E}^* \to D^*$ are all nonisomorphic. But the C^∞ picture is much simpler. All the fibers are diffeomorphic to the same torus, which we will call T. In fact, $\mathscr{E}^* \to D^*$ is a locally trivial fiber bundle with T as fiber. In other words, it is locally diffeomorphic with a product of T with D^*.

Let us analyze what happens in the limit as $\tau \to \infty$ or equivalently as $q \to 0$. The fiber E_τ is isomorphic to $\mathbb{C}/(\tau^{-1}\mathbb{Z} + \mathbb{Z})$. Let $a(\tau)$ be the image in E_τ of the line segment joining 0 to 1, and let $b(\tau)$ be the image of the segment joining 0 to τ^{-1}. If we orient these curves so that $a \cdot b = 1$, they form a basis of $H_1(E_\tau, \mathbb{Z})$. The curve $b(t)$ is called a *vanishing cycle*, since it shrinks to the node as $t \to 0$; see Figure 13.1.

In the C^∞ category, the bundle $\mathscr{E}^*$ is locally trivial, but it is not globally trivial. Its restriction to the circle $S = \{t \mid |t| = \varepsilon\}$ can be constructed directly by taking $T \times [0,1]$ and gluing the ends $T \times \{0\} \cong T \times \{1\}$ using a so-called Dehn twist about the vanishing cycle $b = b(t)$. This is a diffeomorphism that is the identity outside a neighborhood U of b and twists "once around" along b (see Figures 13.2 and 13.3; U is the shaded region).

The twist induces an automorphism $\mu : H^1(T) \to H^1(T)$ called monodromy. It is given explicitly by the Picard–Lefschetz formula:

$$\mu(a) = a + b, \ \mu(b) = b.$$

Let us study a global example considered earlier.

Example 13.1.2. The family of elliptic curves in Legendre form is

$$\mathscr{E} = \{([x,y,z],t) \in \mathbb{P}^2 \times \mathbb{C} - \{0,1\} \mid y^2 z - x(x-z)(x-tz) = 0\} \to \mathbb{C}.$$

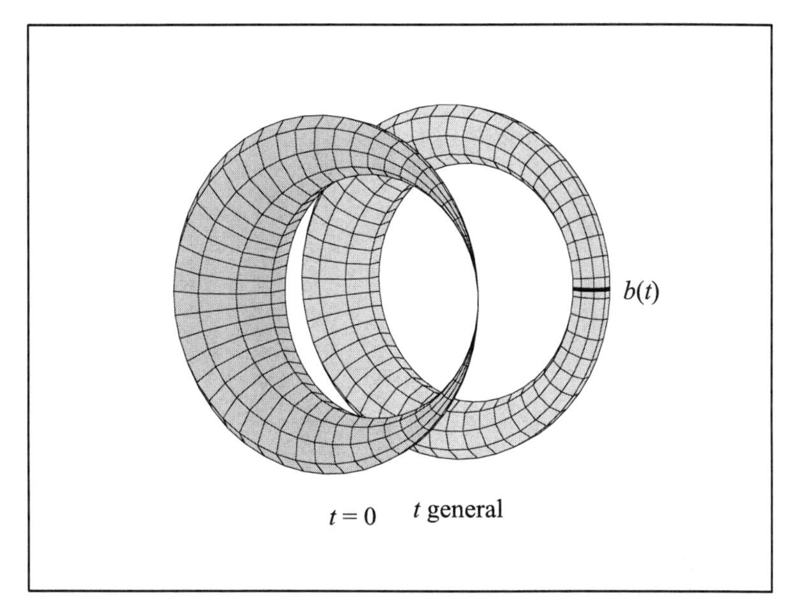

Fig. 13.1 Vanishing cycle.

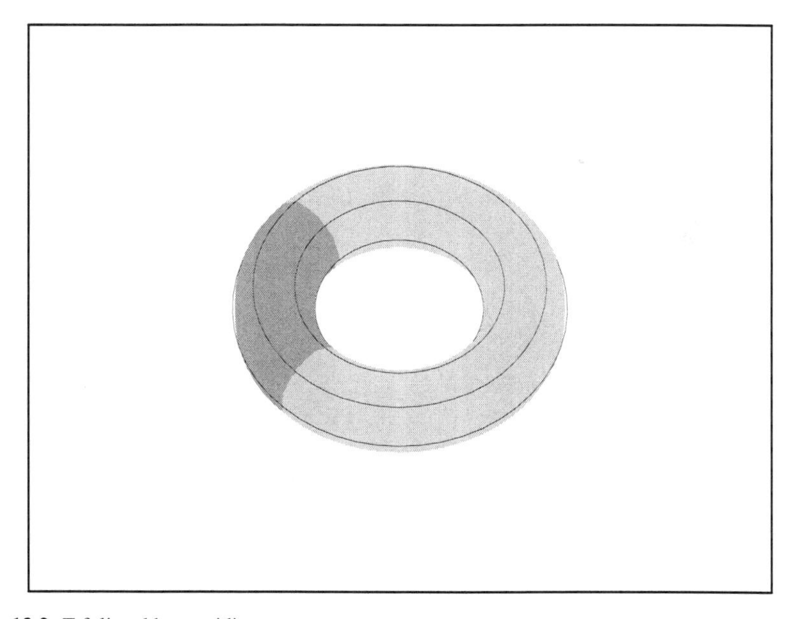

Fig. 13.2 T foliated by meridians.

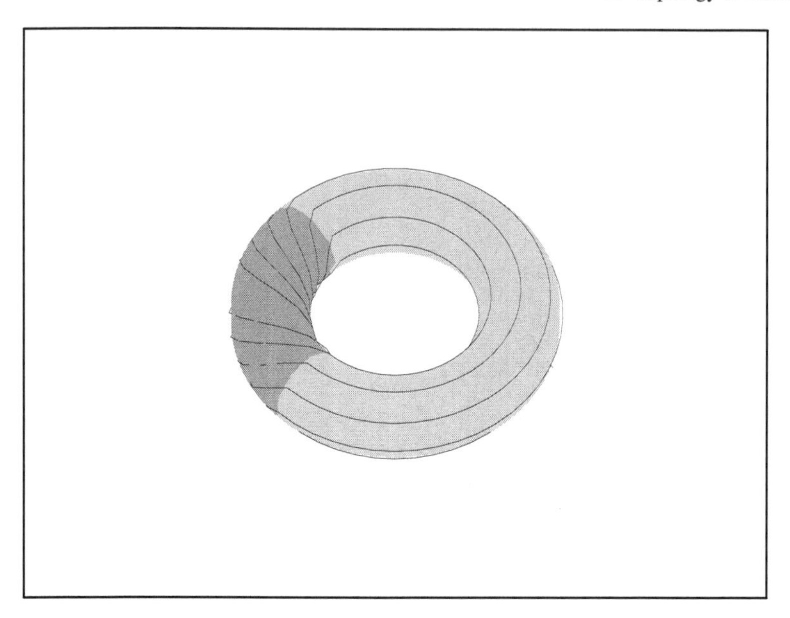

Fig. 13.3 T foliated by images of meridians under a Dehn twist.

We can complete it to a projective surface, and then resolve singularities to obtain a nonsingular surface $\bar{\mathscr{E}} \to \mathbb{P}^1$. Although we will mostly be interested in the original family $\mathscr{E}$. We claim that this is a fiber bundle, topologically. This is easy enough to check directly, but at this point we may as well state a much more general result.

First, let us start with some precise definitions. A C^∞ map $f : X \to Y$ of manifolds is called a *fiber bundle* if it is locally a product of Y with another manifold F (called the fiber). This means that there exist an open cover $\{U_i\}$ and diffeomorphisms $f^{-1}U_i \cong U_i \times F$ compatible with the projections. If $X \cong Y \times F$, the bundle is called trivial. Bundles over S^1 can be constructed as follows. Let F be a manifold with a diffeomorphism $\phi : F \to F$. Then glue $F \times \{0\}$ in $F \times [0,1]$ to $F \times \{1\}$ by identifying $(x,0)$ to $(\phi(x),1)$. This includes the familiar example of the Möbius strip, where $F = \mathbb{R}$ and ϕ is multiplication by -1. If the induced map $\phi^* : H^*(F) \to H^*(F)$, called the *monodromy transformation*, is nontrivial, then the fiber bundle is nontrivial. For more general bundles, $f : X \to Y$ monodromy can be defined by restricting to embedded circles $S^1 \subset Y$. We will give a more careful treatment in the next section. A C^∞ map $f : X \to Y$ is called a *submersion* if the map on tangent spaces is surjective. The fibers of such a map are submanifolds. A continuous map of topological spaces is called proper if the preimage of any compact set is compact.

Theorem 13.1.3 (Ehresmann). *Let $f : X \to Y$ be a proper submersion of C^∞ connected manifolds. Then f is a C^∞ fiber bundle; in particular, the fibers are diffeomorphic.*

Proof. We give a proof when $0 \in Y \subset \mathbb{R}$. This suffices to show that the fibers are diffe0omorphic in the general case, because any two points can be connected by a smooth path. A complete proof, in general, can be found in [89, 4.1].

The strategy is simple. We choose a nowhere-zero vector field v "normal" to the fibers, and the flow along it from $f^{-1}(0)$ to the neighboring fibers. To obtain the vector field, choose a Riemannian metric on X. The gradient $v = \nabla f$ is defined as the vector field dual to df under the inner product associated to the metric. By assumption, df, and therefore v, is nowhere zero. The existence and uniqueness theorem for ordinary differential equations [110, Chapter 5, Theorem 5] allows us to define a neighborhood U of 0, $\varepsilon > 0$, and a family of diffeomorphisms $\phi_t : f^{-1}(U) \to f^{-1}(U)$ parameterized by $|t| < \varepsilon$. This is given by flowing along v. In other words, $\phi_t(p) = \gamma(t)$, where $\gamma : (-\varepsilon, \varepsilon) \to X$ is a C^∞ path passing through p at time 0 with velocity $v(\gamma(s))$ for all s. Then the map $(-\varepsilon, \varepsilon) \times f^{-1}(0) \to X$, given by $(t, p) \mapsto \phi_t(p)$, gives the desired local trivialization.

There is one subtlety to the story that we skipped over, and that is the role of properness. The existence and uniqueness theorem quoted above is only a local statement. So in general, ϕ_t would be defined locally near a given $p \in f^{-1}(0)$ with constants ε_p. But by properness, we can assume that the covering by domains of the local ϕ_t's is finite and then choose $\varepsilon = \min(\varepsilon_p)$. The uniqueness guarantees that these local flows patch. □

Example 13.1.4. The projection $\mathbb{R}^2 - \{0\} \to \mathbb{R}$ shows that the result is false for nonproper submersions,

Let us return to Example 13.1.2 and see how to calculate the monodromy of going around 0 and 1. The fibers over these points are no longer nodal curves, so the Picard–Lefschetz formula will not apply. However, it is possible to calculate this from a different point of view. Recall that $H/\Gamma(2) = \mathbb{P}^1 - \{0, 1, \infty\}$. $\mathscr{E}$ can also be realized as a quotient of $\mathbb{C} \times H$ by an action of the semidirect product $\mathbb{Z}^2 \rtimes \Gamma(2)$ as above. The $\Gamma(2)$ action extends to $H^* = H \cup \mathbb{Q} \cup \{\infty\}$. We can choose a fundamental domain for $\Gamma(2)$ in H^* as depicted in Figure 13.4; the three cusps are $0, 1, \infty$.

The subgroup of $\Gamma(2)$ that fixes ∞ is generated by

$$\begin{pmatrix} 1 & 2 \\ 0 & 1 \end{pmatrix},$$

and it follows easily that this is the monodromy matrix for it. We will call $\bar{\mathscr{E}}$ the elliptic modular surface of level two.

Exercises

13.1.5. Calculate the monodromy matrices for the remaining cusps for $\Gamma(2)$ in Example 13.1.2.

13.1.6. Construct a family of elliptic curves over D^* with $-I$ as its monodromy.

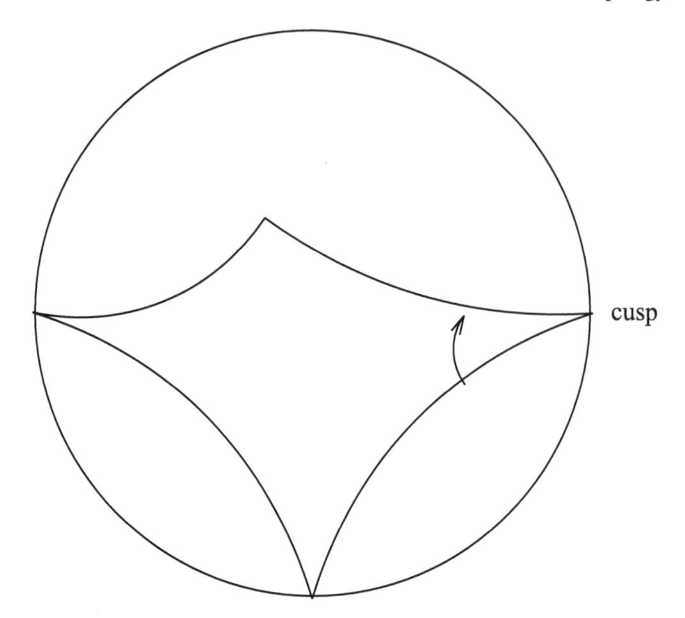

Fig. 13.4 Fundamental domain of $\Gamma(2)$.

13.2 Local Systems

In this section, we give a more formal treatment of monodromy. This notion is implicit in the usual proof that $\sqrt{z}$ is multivalued: analytically continuing this around a loop results in $-\sqrt{z}$. We want to extract the essence of this idea and extend it.

Definition 13.2.1. A sheaf $\mathscr{F}$ on a topological space X is locally constant if there exists an open cover $\{U_i\}$ such that $\mathscr{F}|_{U_i}$ is constant.

Here is a classical example of where this comes up:

Example 13.2.2. Let A be an $n \times n$ matrix of holomorphic functions defined on an open set $X \subset \mathbb{C}$. Then the sheaf of solutions of the differential equation

$$\mathscr{S}(U) = \{f \in \mathscr{O}(U)^n \mid f' = Af\}$$

is locally constant, because standard existence and uniqueness arguments show that $\mathscr{F}|_D \cong \mathbb{C}^n_D$ over a disk.

To see that this is generally not constant, take $X = \mathbb{C}^*$, $n = 1$, and $A = \frac{1}{2z}$. Any global solution is zero, since otherwise it would be a multiple of the multivalued function $\sqrt{z}$. As a first step we want to extend the language of "analytic continuation" to locally constant sheaves. Let X be a topological space. A path from $x \in X$ to $y \in X$ is a continuous map $\gamma\colon [0,1] \to X$ such that $\gamma(0) = x$ and $\gamma(1) = y$. Two paths γ, η are homotopic if there is a continuous map $\Gamma : [0,1] \times [0,1] \to X$ such that

$\gamma(t) = \Gamma(t,0)$, $\eta(t) = \Gamma(t,1)$, $\Gamma(0,s) = x$, and $\Gamma(1,s) = y$. This is an equivalence relation. We can compose paths: If γ is a path from x to y and η is a path from y to z, then $\gamma \cdot \eta$ is the path given by following one by the other at twice the speed. More formally,

$$\gamma \cdot \eta(t) = \begin{cases} \gamma(2t) & \text{if } t \leq 1/2, \\ \eta(2t-1) & \text{if } t > 1/2. \end{cases}$$

This operation is compatible with homotopy in the obvious sense, and the induced operation on homotopy classes is associative. We can define a category $\Pi(X)$ whose objects are points of X and whose morphisms are homotopy classes of paths. This makes $\Pi(X)$ into a *groupoid*, which means that every morphism is an isomorphism. In other words, every (homotopy class of a) path has an inverse. This is not a group, because it is not possible to compose any two paths. To get around this, we can consider loops, i.e., paths that start and end at the same place. Let $\pi_1(X,x)$ be the set of homotopy classes of loops based (starting and ending) at x. This is just $\mathrm{Hom}_{\Pi(X)}(x,x)$, and as such it inherits a composition law that makes it a group, called the fundamental group of (X,x). We summarize the standard properties that can be found in almost any algebraic topology textbook, e.g., [61, 108]:

1. π_1 is a functor on the category of path-connected spaces with base point and base-point-preserving continuous maps.
2. If X is path connected, then $\pi_1(X,x) \cong \pi_1(X,y)$ (consequently, we usually suppress the base point).
3. Two homotopy equivalent path-connected spaces have isomorphic fundamental groups.
4. Van Kampen's theorem: If X is a path-connected simplicial complex that is the union of two subcomplexes $X_1 \cup X_2$, then $\pi_1(X)$ is the free product of $\pi_1(X_1)$ and $\pi_1(X_2)$ amalgamated along the image of $\pi_1(X_1 \cup X_2)$.
5. If X is a connected locally path-connected space, then it has a universal cover $\pi : \tilde{X} \to X$. $\pi_1(X)$ is isomorphic to the group of deck transformations, i.e., self-homeomorphisms of $\tilde{X}$ that commute with π.

This already suffices to calculate the examples of interest to us. From item 5 above, we see that the fundamental group of the circle $\mathbb{R}/\mathbb{Z}$ is $\mathbb{Z}$. The complement in $\mathbb{C}$ of a set S of k points is homotopic to a wedge of k circles. Therefore $\pi_1(\mathbb{C}-S)$ is a free group on k generators.

Let X be a topological space. A *local system* of abelian groups is a functor $F : \Pi(X) \to \mathrm{Ab}$. A local system F gives rise to a $\pi_1(X,x)$-module, i.e., abelian group $F(x)$ with a $\pi_1(X,x)$ action. We define a sheaf $\mathscr{F}$ as follows:

$$\mathscr{F}(U) = \left\{ f : U \to \bigcup_x F(x) \mid f(x) \in F(x) \text{ and } \forall \gamma \in \Pi(U),\, f(\gamma(1)) = F(\gamma)(f(\gamma(0))) \right\}.$$

If every point $x \in X$ has a neighborhood U_x with trivial fundamental group (i.e., X is locally simply connected), then the restrictions $\mathscr{F}|_{U_x}$ are constant. Therefore $\mathscr{F}$ is locally constant.

Theorem 13.2.3. *Let X be a path-connected locally simply connected topological space. There is an equivalence of categories between*

(1) The category of $\pi_1(X)$-modules.
(2) The category of local systems and natural transformations.
(3) The full subcategory $\mathrm{Ab}(X)$ of locally constant sheaves on X.

Proof. See [68, Chapter IV, §9], [108, p. 360], or the exercises. □

In view of this theorem, we will treat local systems and locally constant sheaves as the same.

Let $f : X \to Y$ be a fiber bundle. We are going to construct a local system that takes y to $H^i(X_y, \mathbb{Z})$. Given a path $\gamma : [0,1] \to Y$ joining y_0 and y_1, the pullback $\gamma^*X = \{(x,t) \mid f(x) = \gamma(t)\}$ will be a trivial bundle over $[0,1]$. Therefore γ^*X will deformation retract onto both X_{y_0} and X_{y_1}, and so we have isomorphisms

$$H^i(X_{y_0}) \xleftarrow{\sim} H^i(\gamma^*X) \xrightarrow{\sim} H^i(X_{y_1}).$$

The map $H^i(X_{y_0}) \to H^i(X_{y_1})$ can be seen to depend only on the homotopy class of the path. Thus we have a local system. When $y_0 = y_1$, this is the monodromy transformation considered earlier. The corresponding locally constant sheaf will be described in the next section.

Exercises

13.2.4. For a path-connected space, show that $\Pi(X)$ and $\pi_1(X,x)$ (viewed as a groupoid) are equivalent as categories. Deduce the equivalence of (1) and (2) in Theorem 13.2.3.

13.2.5. Given a locally constant sheaf $\mathscr{F}$, show that $\mathscr{F}$ is isomorphic to the sheaf associated to a local system, and that this local system is determined up to isomorphism.

13.2.6. Let V be a holomorphic vector bundle on a Riemann surface X. Define a connection to be a $\mathbb{C}$-linear morphism $\nabla : V \to \Omega^1_X \otimes V$ such that $\nabla(fv) = f\nabla(v) + df \otimes V$. Show that $\ker(\nabla)$ is a locally constant sheaf. (Hint: check that locally it is the same as the sheaf in Example 13.2.2.)

13.2.7. Let $\mathscr{F}$ be a locally constant sheaf of finite-dimensional $\mathbb{C}$-vector spaces on a Riemann surface X. Construct a connection ∇ on $V = \mathscr{O}_X \otimes_{\mathbb{C}} \mathscr{F}$ such that $\ker(\nabla) = \mathscr{F}$.

13.3 Higher Direct Images*

Given a product $F \times Y$, which we can view as a trivial bundle over Y, the cohomology is the tensor product of the cohomology of F and Y by the Künneth formula

(Theorem 5.3.6). For general fiber bundles, the story is more complicated. Let us start with the simplest case of a fiber bundle over the circle. Here we see that monodromy enters in a fundamental way.

Theorem 13.3.1 (Wang sequence). *Let $X \to S^1$ be a fiber bundle with fiber F and monodromy μ. Then there is an exact sequence*

$$\cdots \to H^i(X,\mathbb{Q}) \to H^i(F,\mathbb{Q}) \xrightarrow{1-\mu} H^i(F,\mathbb{Q}) \to H^{i+1}(X,\mathbb{Q}) \to \cdots.$$

Proof. The sequence is constructed in [108, Chapter 8,§5]. But in order to see that the middle map is as claimed, it is better to do this explicitly. To make the job easier, we will make free use of standard properties of relative homology [108]. Let $\tilde{X} = F \times [0,1]$. Then we can identify $X = \tilde{X}/\sim$, where $\sim$ is the equivalence relation gluing the fibers $\tilde{X}_0 = F \times \{0\}$ and $\tilde{X}_1 = F \times \{1\}$ by the twist that determines monodromy. Although F and the fibers $\tilde{X}_t$ and X_t of $X \to S^1$ are all homeomorphic, it is useful to distinguish between them until the end. At that point, we can use the projections $F \leftarrow \tilde{X} \to X$ to identify $H_*(F) = H_*(\tilde{X}_t) = H_*(X_0)$ and $\mu : H_*(X_1) \cong H_*(F)$. Also $\tilde{X}$ is homotopy equivalent to F, so it has the same homology as F. By excision, we have an isomorphism of relative homology

$$H_*(X,X_0) \cong H_*(\tilde{X},\tilde{X}_0 \cup \tilde{X}_1). \tag{13.3.1}$$

The relative groups $H_*(\tilde{X},\tilde{X}_0 \cup \tilde{X}_1)$ are built using simplicial (or singular) chains with boundary in $\tilde{X}_0 \cup \tilde{X}_1$. These fit into a long exact sequence [108, Chapter 4, §5]

$$\cdots \xrightarrow{\mathrm{id}\oplus\mathrm{id}} H_{i+1}(\tilde{X}) \to H_{i+1}(\tilde{X},\tilde{X}_0 \cup \tilde{X}_1) \xrightarrow{\partial} H_i(\tilde{X}_0) \oplus H_i(\tilde{X}_1) \xrightarrow{\mathrm{id}\oplus\mathrm{id}} H_i(\tilde{X}) \to \cdots. \tag{13.3.2}$$

Since the maps labeled $\mathrm{id} \oplus \mathrm{id}$ are surjective, we can truncate the sequence. The map ∂ is the boundary operator. Since the boundary components F_0, F_1 have opposite orientations, the signs of $\partial\gamma$ on these components are opposite. Thus (13.3.2) reduces to

$$H_{i+1}(\tilde{X},\tilde{X}_0 \cup \tilde{X}_1) = \{(\alpha,-\alpha) \mid \alpha \in H_i(F)\} \cong H_i(F). \tag{13.3.3}$$

Now consider the exact sequence

$$\cdots \to H_{i+1}(X) \to H_{i+1}(X,X_0) \xrightarrow{\partial} H_i(X_0) \to H_i(X) \to \cdots. \tag{13.3.4}$$

The isomorphisms (13.3.1) and (13.3.3) allows us to identify ∂ with

$$1-\mu : H_i(X_0) \to H_i(X_0).$$

So we have obtained the Wang sequence in homology

$$\cdots \to H_{i+1}(X) \to H_i(F) \xrightarrow{1-\mu} H_i(F) \to H_i(X) \to \cdots.$$

Dualizing yields the sequence claimed in the theorem. □

Corollary 13.3.2. *The cohomology of X is isomorphic to the cohomology of the product $F \times S^1$ if the monodromy is trivial.*

Corollary 13.3.3. *If $X \to Y$ is a fiber bundle over a connected space, the image of the restriction of cohomology lies in the monodromy invariant part,*

$$\mathrm{im}[H^i(X,\mathbb{Q}) \to H^i(X_y,\mathbb{Q})] \subset H^i(X_y,\mathbb{Q})^{\pi_1(Y,y)}.$$

Proof. By restricting to loops in Y, this can be reduced to the case of $Y = S^1$, where it follows from the Wang sequence. □

The last corollary implies that the cohomology of a fiber bundle cannot look like that of a product unless the monodromy is trivial. Over more general bases, monodromy is not the only complication.

Example 13.3.4. Let $S^3 \subset \mathbb{C}^2$ be the unit sphere. This maps to $\mathbb{P}^1_{\mathbb{C}} = S^2$ by sending v to $[v]$. The map $S^3 \to S^2$ is a fiber bundle, with fiber S^1, called the Hopf fibration. There is no monodromy because S^2 is simply connected. We have $H^2(S^3) = 0$, but $H^2(S^1 \times S^2) \neq 0$, so they are not the same.

This example shows that there may be "higher twists" that affect cohomology. To explain this properly, we return to general sheaf theory and develop the necessary tools. Let $f : X \to Y$ be a continuous map and $\mathscr{F} \in \mathrm{Ab}(X)$ a sheaf. We can define the *higher direct images* by imitating the definition of $H^i(X,\mathscr{F})$ in Section 4.2:

$$\begin{aligned} R^0 f_*\mathscr{F} &= f_*\mathscr{F}, \\ R^1 f_*\mathscr{F} &= \mathrm{coker}[f_*\mathbf{G}(\mathscr{F}) \to f_*\mathbf{C}^1(\mathscr{F})], \\ R^{n+1} f_*\mathscr{F} &= R^1 f_*\mathbf{C}^n(\mathscr{F}). \end{aligned}$$

Note that when Y is a point, $Rf^i_*\mathscr{F}$ is just $H^i(X,\mathscr{F})$ viewed as a sheaf on it. We have an analogue of Theorem 4.2.3.

Theorem 13.3.5. *Given an exact sequence of sheaves*

$$0 \to \mathscr{A} \to \mathscr{B} \to \mathscr{C} \to 0,$$

there is a long exact sequence of sheaves

$$0 \to R^0 f_*\mathscr{A} \to R^0 f_*\mathscr{B} \to R^0 f_*\mathscr{C} \to R^1 f_*\mathscr{A} \to \cdots.$$

There is an alternative description that is a bit more convenient.

Lemma 13.3.6. *If $f : X \to Y$ is a continuous map, and $\mathscr{F} \in \mathrm{Ab}(X)$, then $R^i f_*\mathscr{F}(U)$ is the sheafification of the presheaf $U \mapsto H^i(U,\mathscr{F})$.*

Proof. Let $\mathscr{R}^i$ denote the presheaf $U \mapsto H^i(U,\mathscr{F})$. For $i = 0$, we have $\mathscr{R}^0 = f_*\mathscr{F}$ by definition of f_*. By our original construction of $H^1(U,\mathscr{F})$, we have an exact sequence

$$f_*\mathbf{G}(\mathscr{F})(U) \to f_*\mathbf{C}^1(\mathscr{F})(U) \to \mathscr{R}^1(U) \to 0.$$

This shows that $\mathscr{R}^1$ is the cokernel $f_*\mathbf{G}(\mathscr{F}) \to f_*\mathbf{C}^1(\mathscr{F})$ in the category of presheaves. Hence $(\mathscr{R}^1)^+ = R^1 f_*\mathscr{F}$. The rest follows by induction. □

Each element of $H^i(X,\mathscr{F})$ determines a global section of the presheaf $\mathscr{R}^i$ and hence of the sheaf $R^i f_*\mathscr{F}$. This map $H^i(X,\mathscr{F}) \to H^0(X,R^i f_*\mathscr{F})$ is often called an *edge homomorphism.*

Let us now assume that $f : X \to Y$ is a fiber bundle of triangulable spaces. Then choosing a contractible neighborhood U of y, we see that $H^i(U,\mathbb{Z}) \cong H^i(X_y,\mathbb{Z})$. Since such neighborhoods are cofinal, it follows that $R^i f_*\mathbb{Z}$ is locally constant. This coincides with the sheaf associated to the local system $H^i(X_y,\mathbb{Z})$ constructed in the previous section. The global sections of $H^0(Y,R^i f_*\mathbb{Z})$ can be identified with the space $H^i(X_y,\mathbb{Z})^{\pi_1(Y,y)}$ of cohomology classes of the fiber invariant under monodromy. We can construct elements of this space using the edge homomorphism, which corresponds to restriction

$$H^i(X) \to H^i(X_y)^{\pi_1(Y,y)} \subseteq H^i(X_y).$$

The importance of the higher direct images is that they provide a mechanism for computing the cohomology of X in terms of data on Y. More precisely, construct $\mathbf{G}^\bullet(\mathscr{F})$ as in Section 12.4. Then as we saw that $\Gamma(\mathbf{G}^\bullet(\mathscr{F}))$ is a complex whose cohomology groups are exactly $H^*(X,\mathscr{F})$. We can factor this construction through Y, by considering the complex of sheaves $f_*\mathbf{G}^\bullet(\mathscr{F})$ on Y. We will denote this by $\mathbb{R}f_*\mathscr{F}$ even though this is not technically quite correct. ($\mathbb{R}f_*\mathscr{F}$ is really the corresponding object in the *derived category*, whose definition can be found in [44, 118].) The interesting features of this complex can be summarized by the following:

Proposition 13.3.7.

(1) The ith cohomology sheaf $\mathscr{H}^i(\mathbb{R}f_*\mathscr{F})$ *is isomorphic to* $R^i f_*\mathscr{F}$.
(2) $\mathbb{H}^i(\mathbb{R}f_*\mathscr{F}) \cong H^i(X,\mathscr{F})$.

To make the relationship between $\mathbb{R}f_*\mathscr{F}$ and $R^i f_*\mathscr{F}$ clearer, let us use the truncation operators introduced in §12.4. From Lemma 12.4.2, we obtain the following result:

Lemma 13.3.8. *The sequence*

$$0 \to \tau_{\le q-1}\mathbb{R}f_*\mathscr{F} \to \tau_{\le q}\mathbb{R}f_*\mathscr{F} \to R^q f_*\mathscr{F}[-q] \to 0 \qquad (13.3.5)$$

is exact for each q.

From here it is a straightforward matter to construct the Leray spectral sequence

$$E_2^{pq} = H^p(Y,R^q f_*\mathscr{F}) \Rightarrow H^{p+q}(X,\mathscr{F}).$$

We would rather not get into this, but the exercises give some further hints. The spectral sequence has standard consequences that can deduced directly.

Proposition 13.3.9. *If $\mathscr{F}$ is a sheaf of vector spaces over a field such that $\sum \dim H^p(Y, R^q f_* \mathscr{F}) < \infty$, then*

$$\dim H^i(X, \mathscr{F}) \leq \sum_{p+q=i} \dim H^p(Y, R^q f_* \mathscr{F})$$

and

$$\sum_i \dim(-1)^i H^i(X, \mathscr{F}) = \sum_{p,q} (-1)^{p+q} \dim H^p(Y, R^q f_* \mathscr{F}).$$

Proof. This follows from Lemma 13.3.8 an induction. □

Corollary 13.3.10. *If $X \to Y$ is a fiber bundle induced from a map of finite simplicial complexes, then*

$$\dim H^i(X, \mathbb{Q}) \leq \sum_{p+q=i} \dim H^p(Y, R^q f_* \mathbb{Q}). \tag{13.3.6}$$

If, moreover, the monodromy acts trivially on the cohomology of the fibers, then

$$\dim H^i(X, \mathbb{Q}) \leq \sum_{p+q=i} \dim H^p(Y, \mathbb{Q}) \otimes H^q(X_y, \mathbb{Q}). \tag{13.3.7}$$

Example 13.3.4 shows that these inequalities may be strict, even when there is no monodromy.

Exercises

13.3.11. Prove Theorem 13.3.5.

13.3.12. Complete the proof of Proposition 13.3.9. (See Corollary 12.4.4 for some hints.)

13.3.13. Show that equality holds in Proposition 13.3.9 if and only if $\tau_{\leq \bullet}$ is strict.

13.3.14. Let $R^q = R^q f_* \mathscr{F}$ and $R = \mathbb{R} f_* \mathscr{F}$. Define the map $d_2 : H^p(Y, R^q) \to H^{p+2}(Y, R^{q-1})$ as the composition of the connecting map associated to (13.3.5), and the map $H^*(\tau_{\leq q-1} R) \to H^*(R^{q-1}[-q+1])$. This is the first step in the construction of the Leray spectral sequence.

(a) Show that d_2 vanishes if $\tau_{\leq \bullet}$ is strict. Therefore this gives an obstruction to equality in Proposition 13.3.9 by the previous exercise.
(b) Calculate d_2 in Example 13.3.4 and show that it is nonzero.

In fact, strictness is equivalent to the vanishing of a countable number of obstructions $d_2, d_3, \ldots$ (compare Exercise 12.2.14).

13.3.15.

(a) A locally constant sheaf L of vector spaces on S^1 is determined by the vector space $V = L_0$ and the monodromy operator $\mu : V \to V$. Using a Čech complex, show that

$$H^i(S^1, L) \cong \begin{cases} \ker[V \xrightarrow{1-\mu} V] & \text{if } i = 0, \\ \operatorname{coker}[V \xrightarrow{1-\mu} V] & \text{if } i = 1, \\ 0 & \text{otherwise.} \end{cases}$$

(b) If $f : X \to S^1$ is a fiber bundle, show that there are exact sequences

$$0 \to H^1(R^{q-1} f_* \mathbb{Q}) \to H^q(X, \mathbb{Q}) \to H^0(R^q f_* \mathbb{Q}) \to 0.$$

(c) Use this to construct the Wang sequence.

13.4 First Betti Number of a Fibered Variety*

We will return to geometry, and use the previous ideas to compute the first Betti number of an elliptic surface. Suppose that $f : X \to C$ is a morphism of a smooth projective variety onto a smooth projective curve. Assume that f has connected fibers. Let $S \subset C$ be the set of points for which the fibers are singular, and let U be its complement. The map $f^{-1}U \to U$ is a submersion and hence a fiber bundle. We have a monodromy representation of $\pi_1(U)$ on the cohomology of the fiber.

Theorem 13.4.1. $b_1(C) \leq b_1(X) \leq b_1(C) + \dim H^1(X_y, \mathbb{Q})^{\pi_1(U)}$.

The proof will be carried out in a series of steps. A (reduced) divisor in a complex manifold is a subset that is locally the zero set of a holomorphic function.

Lemma 13.4.2. *Let $S \subset W$ be a divisor in a complex manifold. Then $H^1(W, \mathbb{Q}) \to H^1(W - S, \mathbb{Q})$ is injective.*

Proof. We prove this under the extra assumption that W has finite topological type. Then using Mayer–Vietoris and induction, we can reduce to the case that W is a ball, where the lemma is trivially true, since $H^1(W) = 0$. □

Let $j : U \to C$ denote the inclusion, and let $\mathscr{F} = R^1 f_* \mathbb{Q}$. The restriction $j^* \mathscr{F}$ is locally constant, but $\mathscr{F}$ need not be. Nevertheless, we can still control it. There is a canonical morphism $\mathscr{F} \to j_* j^* \mathscr{F}$ induced by the restriction maps $\mathscr{F}(V) \to \mathscr{F}(V \cap U)$ as V ranges over open subsets of C.

Corollary 13.4.3. *$\mathscr{F} \to j_* j^* \mathscr{F}$ is a monomorphism.*

Proof. This follows from the injectivity of the restriction maps $H^1(V, \mathbb{Q}) \to H^1(V \cap U, \mathbb{Q})$. □

We can now prove the theorem.

Proof. Proposition 13.3.9 gives

$$b_1(X) \leq \dim H^1(C, f_*\mathbb{Q}) + \dim H^0(C, R^1 f_*\mathbb{Q}).$$

Since the fibers of f are connected, $f_*\mathbb{Q}$ is easily seen to be $\mathbb{Q}$. So the dimension of its first cohomology is just $b_1(C)$. The previous corollary implies that

$$H^0(C, R^1 f_*\mathbb{Q}) \to H^0(C, j_*j^*\mathscr{F}) = H^0(U, R^1 f_*\mathbb{Q}) = H^1(X_y, \mathbb{Q})^{\pi_1(U)}$$

is injective. This proves the upper inequality

$$b_1(X) \leq b_1(C) + \dim H^1(X_y, \mathbb{Q})^{\pi_1(U)}.$$

For the lower inequality, note that a nonzero holomorphic 1-form on C will pull back to a nonzero form on X. Thus $h^{10}(C) \leq h^{10}(X)$. □

Corollary 13.4.4. *If $\mathscr{E} \to C$ is an elliptic surface such that its monodromy is nontrivial, then $b_1(\mathscr{E}) = b_1(C)$.*

Proof. By assumption, $\dim H^1(\mathscr{E}_y)^{\pi_1(U)} \leq 1$. Therefore $b_1(C) \leq b_1(X) \leq b_1(C)+1$. Since $b_1(X)$ is even, this forces $b_1(\mathscr{E}) = b_1(C)$. □

Exercises

13.4.5. Let G be a finite group of automorphisms of a curve $\tilde{C}$. Suppose that G acts on a smooth projective variety F and therefore on $\tilde{C} \times F$ by the diagonal action. Let $f : X = (\tilde{C} \times F)/G \to C = \tilde{C}/G$ be the projection. Check that $b_1(X) = b_1(C) + \dim H^1(X_y, \mathbb{Q})^{\pi_1(C)}$.

Chapter 14
The Hard Lefschetz Theorem

The pioneering study of the topology of algebraic varieties was carried out by Lefschetz in his monograph [79] published in 1924. He suggested working inductively, by comparing the homology of a smooth projective variety with its intersection with a single hyperplane and eventually a suitable family of them. There were two basic results, nowadays called the weak and hard Lefschetz theorems. In this chapter, we discuss both of these but focus on the latter. In one of its incarnations, it gives the structure of cohomology under cup product with a hyperplane class. The first correct proof of this was due to Hodge using harmonic forms, and we present a version of this. The hard Lefschetz theorem has a number of important consequences for the topology of projective, and more generally Kähler, manifolds, and we will discuss a number of these. We also want to interpret this using Lefschetz's original more geometric point of view, which, remarkably, played a role in Deligne's proof of the Weil conjectures [25] and in his subsequent arithmetic proof of hard Lefschetz [27].

14.1 Hard Lefschetz

Let X be an n-dimensional compact Kähler manifold. Recall that L was defined by wedging (or more correctly cupping) with the Kähler class $[\omega]$. The space

$$P^i(X) = \ker[L^{n-i+1} : H^i(X,\mathbb{C}) \to H^{2n-i+2}(X,\mathbb{C})]$$

is called the *primitive* cohomology.

Theorem 14.1.1 (Hard Lefschetz). *For every i,*

$$L^i : H^{n-i}(X,\mathbb{C}) \to H^{n+i}(X,\mathbb{C})$$

is an isomorphism. For every i,

$$H^i(X,\mathbb{C}) = \bigoplus_{j=0}^{[i/2]} L^j P^{i-2j}(X).$$

We indicate the proof in the next section. As a simple corollary, we find that the Betti numbers satisfy $b_{n-i} = b_{n+i}$. Of course, this is nothing new, since this also follows from Poincaré duality. However, it is easy to extract some less trivial "numerology."

Corollary 14.1.2. *The Betti numbers satisfy* $b_{i-2} \leq b_i$ *for* $i \leq n/2$.

Proof. The theorem implies that the map $L : H^{i-2}(X) \to H^i(X)$ is injective. □

Suppose that $X = \mathbb{P}^n$ with the Fubini–Study metric of Example 10.1.5. The Kähler class ω is equal to $c_1(\mathcal{O}(1))$. The class $c_1(\mathcal{O}(1))^i \neq 0$ is the fundamental class of a codimension-i linear space (see Sections 7.2 and 7.5), so it is nonzero. Since all the cohomology groups of $\mathbb{P}^n$ are either 0- or 1-dimensional, this implies the hard Lefschetz theorem for $\mathbb{P}^n$. Things get much more interesting when $X \subset \mathbb{P}^n$ is a nonsingular subvariety with induced metric. By Poincaré duality and the previous remarks, we get a statement closer to what Lefschetz would have stated, namely, that any element of $H_{n-i}(X,\mathbb{Q})$ is homologous to the intersection of a class in $H_{n+i}(X,\mathbb{Q})$ with a codimension-i subspace.

The Hodge index theorem for surfaces generalizes to a set of inequalities, called the Hodge–Riemann bilinear relations, on an n-dimensional compact Kähler manifold X. Consider the pairing

$$H^i(X,\mathbb{C}) \times H^i(X,\mathbb{C}) \to \mathbb{C}$$

defined by

$$Q(\alpha,\beta) = (-1)^{i(i-1)/2} \int_X \alpha \wedge \beta \wedge \omega^{n-i}.$$

Theorem 14.1.3. $H^i(X) = \bigoplus H^{pq}(X)$ *is an orthogonal decomposition with respect to* Q. *If* $\alpha \in P^{p+q}(X) \cap H^{pq}(X)$ *is nonzero, then*

$$\sqrt{-1}^{p-q} Q(\alpha,\bar{\alpha}) > 0.$$

Proof. See [49, p. 123]. □

We define the *Weil operator* $C : H^i(X) \to H^i(X)$, which acts on $H^{pq}(X)$ by multiplication by $\sqrt{-1}^{p-q}$.

Corollary 14.1.4. *The form* $\tilde{Q}(\alpha,\beta) = Q(\alpha, C\bar{\beta})$ *on* $P^i(X)$ *is positive definite Hermitian.*

Exercises

14.1.5. When X is compact Kähler, show that Q gives a nondegenerate skew-symmetric pairing on $P^i(X)$ with i odd. Use this to give another proof that $b_i(X)$ is even.

14.1.6. Determine which products of spheres $S^n \times S^m$ can admit Kähler structures.

14.2 Proof of Hard Lefschetz

Let X be as in the previous section. We defined the operators L, Λ acting on forms $\mathscr{E}^\bullet(X)$ in Section 10.1. We define a new operator H that acts by multiplication by $n-i$ on $\mathscr{E}^i(X)$. Then we have the following additional Kähler identities:

Proposition 14.2.1. *The following hold:*

(1) $[\Lambda, L] = H$.
(2) $[H, L] = -2L$.
(3) $[H, \Lambda] = 2\Lambda$.

Furthermore, these operators commute with Δ.

Proof. See [49, pp. 115, 121]. □

This proposition plus the following theorem of linear algebra will prove the hard Lefschetz theorem.

Theorem 14.2.2. *Let V be a complex vector space with endomorphisms L, Λ, H satisfying the above identities. Then:*

(a) H is diagonalizable with integer eigenvalues.
(b) For each $a \in \mathbb{Z}$, let V_a be the space of eigenvectors of H with eigenvalue a. Then L^i induces an isomorphism between V_i and V_{-i}.
(c) If $P = \ker(\Lambda)$, then

$$V = P \oplus LP \oplus L^2P \oplus \cdots.$$

(d) If $\alpha \in P \cap V_i$ then $L^{i+1}\alpha = 0$.

We give the main ideas of the proof. Consider the Lie algebra $\mathrm{sl}_2(\mathbb{C})$ of traceless 2×2 complex matrices. This is a Lie algebra with a basis given by

$$\lambda = \begin{pmatrix} 0 & 1 \\ 0 & 0 \end{pmatrix}, \quad \ell = \begin{pmatrix} 0 & 0 \\ 1 & 0 \end{pmatrix}, \quad h = \begin{pmatrix} 1 & 0 \\ 0 & -1 \end{pmatrix}.$$

These matrices satisfy

$$[\lambda, \ell] = h, \quad [h, \lambda] = 2\lambda, \quad [h, \ell] = -2\ell.$$

So the hypothesis of the theorem is simply that the linear map $\mathrm{sl}_2(\mathbb{C}) \to \mathrm{End}(V)$ determined by $\lambda \mapsto \Lambda$, $\ell \mapsto L$, $h \mapsto H$, preserves the bracket, or equivalently, that V is a representation of $\mathrm{sl}_2(\mathbb{C})$. The theorem can then be deduced from the following two facts from representation theory of $\mathrm{sl}_2(\mathbb{C})$ (which can be found in almost any book on Lie theory, e.g., [43]):

1. Every representation of $\mathrm{sl}_2(\mathbb{C})$ is a direct sum of irreducible representations. Irreducibility means that the representation contains no proper nonzero subrepresentations, i.e., subspaces stable under the action of $\mathrm{sl}_2(\mathbb{C})$.
2. There is a unique irreducible representation of dimension $N+1$ for each $N \geq 0$, namely $S^N(\mathbb{C}^2)$, where $\mathbb{C}^2$ is the standard representation. Let $e_1 = (1,0)^T$ and $e_2 = (0,1)^T$ be the standard basis of $\mathbb{C}^2$. Then $e_1^N, e_1^{N-1}e_2, \ldots, e_2^N$ gives a basis of $S^N(\mathbb{C}^2)$, and the operators act by

$$\lambda(e_1^i e_2^j) = j e_1^{i+1} e_2^{j-1},$$
$$\ell(e_1^i e_2^j) = i e_1^{i-1} e_2^{j+1},$$
$$h(e_1^i e_2^j) = (i-j) e_1^i e_2^j.$$

It suffices to check the theorem when $V = S^N(\mathbb{C}^2)$. In this case, we see that the span of $e_1^i e_2^{N-i}$ is precisely the eigenspace V_{2i-N} of h, and all eigenspaces are of this form. The operators ℓ, λ shift these eigenspaces as pictured:

$$0 \xleftarrow{\lambda} V_N \cong \mathbb{C} \underset{\lambda\cong}{\overset{\ell\cong}{\rightleftarrows}} V_{N-2} \cong \mathbb{C} \underset{\lambda\cong}{\overset{\ell\cong}{\rightleftarrows}} \cdots \underset{\lambda\cong}{\overset{\ell\cong}{\rightleftarrows}} V_{-N} \cong \mathbb{C} \xrightarrow{\ell} 0$$

This implies (b). For the remaining properties, we have that $P = \ker(\lambda)$ is the span of e_1^N. Thus

$$V = P \oplus \ell P \oplus \ell^2 P \oplus \cdots,$$

as expected.

Exercises

14.2.3. Using the relations, check that $\lambda(V_a) \subseteq V_{a+2}$ and $\ell(V_a) \subseteq V_{a-2}$ for any representation V.

14.2.4. If V is a representation and $v \in \ker(\lambda)$, show that $\{v, \ell v, \ell^2, \ldots\}$ spans a subrepresentation of V. In particular, this spans V if it is irreducible.

14.3 Weak Lefschetz and Barth's Theorem

Recall that a hypersurface $Y \subset X$ in a projective variety is called ample if there exist a projective embedding and a hyperplane H such that $Y = X \cap H$. The version of the weak Lefschetz theorem, or Lefschetz hyperplane theorem, that we give here compares their rational cohomologies, although, in fact, a much stronger statement can be proved using methods from differential topology, namely Morse theory [86].

Theorem 14.3.1 (Weak Lefschetz). *Let X be a smooth projective variety with a smooth ample divisor $Y \subset X$.*

(a) Then the restriction map

$$H^i(X,\mathbb{Q}) \to H^i(Y,\mathbb{Q})$$

is an isomorphism when $i < \dim X - 1$ and an injection when $i = \dim X - 1$.

(b) The Gysin map

$$H^i(Y,\mathbb{Q}) \to H^{i+2}(X,\mathbb{Q})$$

is an isomorphism for $i > \dim X - 1$ and a surjection when $i = \dim X - 1$.

Proof. Observe that (a) and (b) are equivalent by Poincaré duality. So we prove (b). Also, by the universal coefficient theorem, we can switch to complex coefficients. Let $U = X - Y$. Then using the Gysin sequence (12.6.5)

$$\cdots \to H^{i+1}(U,\mathbb{C}) \to H^i(Y,\mathbb{C}) \to H^{i+2}(X,\mathbb{C}) \to \cdots,$$

we see that we have to prove $H^{i+1}(U,\mathbb{C}) = 0$ for $i+1 > \dim X$. Since U is a closed subset of a projective space minus a hyperplane, it is affine. So the required vanishing follows from Corollary 12.5.4. □

Lemma 14.3.2. *Let $\iota : Y \hookrightarrow X$ be an oriented submanifold of a compact oriented C^∞ manifold. Then the following identities hold:*

(a) $\iota^*(\beta \cup \gamma) = \iota^*\beta \cup \iota^*\gamma$.
(b) $\iota_!\iota^*\beta = [Y] \cup \beta$.
(c) $\iota^*\iota_!\alpha = \iota^*[Y] \cup \alpha$.

Proof. The first identity is simply a restatement of the functoriality of the cup product. The second is usually called the projection formula, and it can be checked by applying (5.5.2) and (5.6.1) to obtain

$$\int_X \iota_!\iota^*\beta \cup \gamma = \int_Y \iota^*\beta \cup \iota^*\gamma = \int_X [Y] \cup \beta \cup \gamma.$$

The same argument shows that (b) holds more generally for Y a compact submanifold of a noncompact manifold. In this case, $i_!$ and $[Y]$ would take values in $H_c^*(X)$.

We prove (c) with the help of Lemma 5.5.3, which shows that $\iota_!\alpha = \tau_Y \cup \pi^*\alpha$ on a tubular neighborhood $Y \overset{\pi}{\leftarrow} T \to X$, where τ_Y is the Thom class of T. Since τ_Y represents $[Y]$, we get $\iota^*\iota_!\alpha = \iota^*[Y] \cup \alpha$. □

Barth proved a variation on the weak Lefschetz theorem that the cohomologies of $\mathbb{P}^n$ and a nonsingular subvariety coincides in low degree. Hartshorne [59] gave a short elegant proof, which we reproduce here.

Theorem 14.3.3 (Barth). *If $Y \subset \mathbb{P}^n$ is a nonsingular complex projective variety, then $H^i(\mathbb{P}^n,\mathbb{Q}) \to H^i(Y,\mathbb{Q})$ is an isomorphism for $i \leq 2\dim Y - n$.*

Remark 14.3.4. This theorem is of course vacuous unless $\dim Y \geq n/2$. When $\dim Y = \dim X - 1$, Y is ample, so this is exactly what we get from weak Lefschetz.

Proof. Let $P = \mathbb{P}^n$, $m = \dim Y$, and let $\iota : Y \to P$ denote the inclusion. Let L be the Lefschetz operators associated to a hyperplane H and to $H|_Y$ (it will be clear from context which is which). Then $H^{2(n-m)}(P)$ is generated by H^{n-m}. Therefore $[Y] = d[H]^{n-m}$ with $d \neq 0$. Consider the diagram

$$\begin{array}{ccc} H^i(P) & \xrightarrow{L^{n-m}} & H^{i+2(n-m)}(P) \\ \downarrow \iota^* & \nearrow (1/d)\iota_! & \downarrow \iota^* \\ H^i(Y) & \xrightarrow{L^{n-m}} & H^{i+2(n-m)}(Y) \end{array} \tag{14.3.1}$$

The diagram commutes thanks to the previous lemma.

The map $L^{n-m} : H^i(P) \to H^{i+2(n-m)}(P)$ is an isomorphism because $P = \mathbb{P}^n$. So it follows that the restriction $\iota^* : H^i(X) \to H^i(Y)$ is injective. Therefore it is enough to prove that they have the same dimension. Hard Lefschetz for Y implies that $L^{n-m} : H^i(Y) \to H^{i+2(n-m)}(Y)$ is injective. Therefore the same is true of $\iota_!$ by (14.3.1). Thus

$$b_i(X) \leq b_i(Y) \leq b_{i+2(n-m)}(X) = b_i(X).$$ □

Exercises

14.3.5. Check that the bound in the weak Lefschetz theorem is sharp for $\dim X \leq 3$.

14.3.6. Show that a product of three curves of positive genus cannot embed into $\mathbb{P}^4$ or $\mathbb{P}^5$.

14.4 Lefschetz Pencils*

In this section, we explain Lefschetz's original approach to the hard Lefschetz theorem. Modern references for this material are [75], [30], and [116]. Given an n-dimensional smooth projective variety $X \subset \mathbb{P}^N$, Bertini's theorem [60] shows that for "most" hyperplanes H, $Y = X \cap H$ is smooth. The weak Lefschetz theorem yields an isomorphism $H^i(X,\mathbb{Q}) \cong H^i(Y,\mathbb{Q})$ for $i < n - 1$. To go beyond this, Lefschetz suggested letting H move in a sufficiently nice family H_t, and letting the cycles move as well. We then have two geometrically defined subspaces of $H^{n-1}(Y)$ (which can be identified with homology). There is the subspace of cycles $\mathrm{Van} \subset H^{n-1}(Y)$ that vanish as $X \cap H_t$ acquires a singularity, and the subspace $\mathrm{Inv} \subset H^{n-1}(Y)$ of cycles that stay invariant as H_t travels in a loop around such singularities. The original formulation of the hard Lefschetz theorem amounted to the following statement:

Theorem 14.4.1. $H^{n-1}(Y) = \mathrm{Inv} \oplus \mathrm{Van}$.

Our goal is to explain more precisely what this means, and then outline the proof. Let $\check{\mathbb{P}}^N$ be the dual projective space whose points correspond to hyperplanes of $\mathbb{P}^N$. Choose a general element $H \in \check{\mathbb{P}}^N$ and let $Y = X \cap H$. We define

$$I = \mathrm{im}[\iota^* : H^{n-1}(X, \mathbb{Q}) \to H^{n-1}(Y, \mathbb{Q})]$$

and

$$V = \ker[\iota_! : H^{n-1}(Y, \mathbb{Q}) \to H^{n+1}(X, \mathbb{Q})],$$

where $\iota : Y \to X$ is the inclusion. As we will see below, these will coincide with the spaces Inv and Van. So as a first step, we establish the following result:

Proposition 14.4.2. $H^{n-1}(Y) = I \oplus V$.

Proof. The composition

$$H^{n-1}(X) \to H^{n-1}(Y) \to H^{n+1}(X)$$

can be identified with the Lefschetz operator L, which is an isomorphism. Therefore given $\iota^*\beta \in I \cap V$, we get $\beta = L^{-1}\iota_!(\beta) = 0$. Furthermore, given $\alpha \in H^{n-1}(Y)$,

$$\alpha = \iota^* L^{-1} \iota_!(\alpha) + (\alpha - \iota^* L^{-1} \iota_!(\alpha))$$

decomposes it into an element of $I + V$. □

Let us try to make sense of the constructions indicated above. We can and will assume that $X \subset \mathbb{P}^N$ is nondegenerate, which means that X does not lie on a hyperplane. The dual variety

$$\check{X} = \{H \in \check{\mathbb{P}}^N \mid H \text{ contains a tangent space of } X\}$$

parameterizes hyperplanes such that $H \cap X$ is singular.

Proposition 14.4.3. *If $H \in \check{X}$ is a smooth point, then $H \cap X$ has exactly one singular point, which is a node (i.e., the completed local ring of the singularity is isomorphic to $\mathbb{C}[[x_1, \ldots, x_n]]/(x_1^2 + \cdots + x_n^2)$).*

Proof. See [30, Chapter XVII]. □

A line $\{H_t\}_{t \in \mathbb{P}^1} \subset \check{\mathbb{P}}^N$ is called a pencil of hyperplanes. Any two hyperplanes of the pencil will intersect in a common linear subspace called the base locus. A pencil $\{H_t\}$ is called a *Lefschetz pencil* if $H_t \cap X$ has at worst a single node for all $t \in \mathbb{P}^1$ and if the base locus $H_0 \cap H_\infty$ is transverse to X.

Corollary 14.4.4. *The set of Lefschetz pencils forms a nonempty Zariski open set of the Grassmannian of lines in $\check{\mathbb{P}}^N$.*

Proof. A general pencil will automatically satisfy the transversality condition. Furthermore, a general pencil will be disjoint from the singular set $\check{X}_{sing}$, since it has codimension at least two in $\check{\mathbb{P}}^N$. □

Given a pencil, we form an incidence variety $\tilde{X} = \{(x,t) \in X \times \mathbb{P}^1 \mid x \in H_t\}$. The second projection p gives a map onto $\mathbb{P}^1$ whose fibers are intersections $H_t \cap X$. There is a finite set S of $t \in \mathbb{P}^1$ with $\tilde{X}_t = p^{-1}t$ singular. Let $U = \mathbb{P}^1 - S$ and fix $t_0 \in U$. By Ehresmann's Theorem 13.1.3, $\tilde{X}$ is a fiber bundle over U. Thus we have a monodromy representation of $\pi_1(U,t_0)$ on the cohomology of the fiber $H^*(\tilde{X}_{t_0})$. We want to make this explicit. It will be convenient to switch back and forth between homology and cohomology using the Poincaré duality isomorphism $H_k(\tilde{X}_{t_0},\mathbb{Q}) \cong H^{2n-2-k}(\tilde{X}_{t_0},\mathbb{Q})$.

Consider the diagram

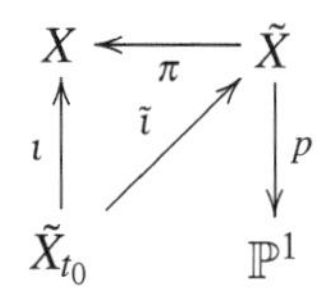

Choose small disks Δ_i around each $t_i \in S$, and connect these by paths γ_i to the base point t_0 (Figure 14.1).

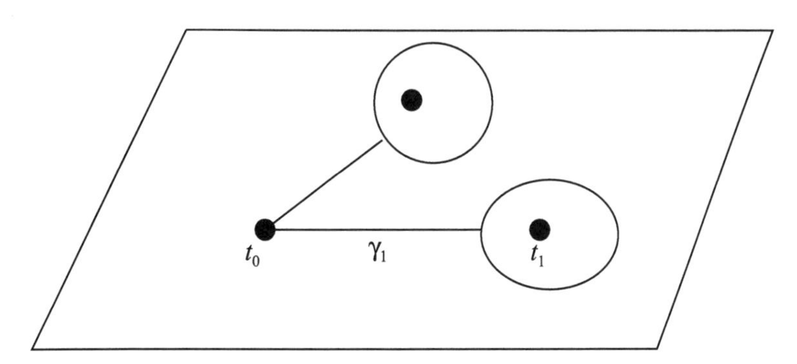

Fig. 14.1 Loops.

The space $p^{-1}(\gamma_i \cup \Delta_i)$ is homotopic to the singular fiber $\tilde{X}_{t_i} = p^{-1}(t_i)$.

Theorem 14.4.5. *Let $n = \dim X$. Then*

$$H_k(\tilde{X}_{t_0},\mathbb{Q}) \to H_k(p^{-1}(\gamma_i \cup \Delta_i),\mathbb{Q})$$

is an isomorphism if $k \neq n-1$, and it is surjective with a one-dimensional kernel if $k = n-1$.

Proof. Since $p^{-1}(\gamma_i \cup \Delta_i)$ is homotopic to $p^{-1}(\Delta_i)$, we can assume that t_0 is a point on the boundary of Δ_i. Let $y_i \in \tilde{X}_{t_i}$ be the singular point. We can choose coordinates about y_i such that p is given by $z_1^2 + z_2^2 + \cdots + z_n^2$. Pick $0 < \rho < \varepsilon \ll 1$, and let

$$B = \{(z_1, \ldots, z_n) \mid |z_1|^2 + |z_2|^2 + \cdots + |z_n|^2 \le \varepsilon,\ |z_1^2 + z_2^2 + \cdots + z_n^2| \le \rho\}.$$

We assume, after shrinking Δ_i if necessary, that Δ_i is the disk of radius ρ and choose $t_0 = \rho$. If $x = \mathrm{Re}(z)$, $y = \mathrm{Im}(z)$, then we can identify

$$\tilde{X}_{t_0} \cap B = \{(x,y) \in \mathbb{R}^n \times \mathbb{R}^n \mid ||x||^2 + ||y||^2 \le \varepsilon,\ ||x||^2 - ||y||^2 = \rho,\ \langle x, y\rangle = 0\}.$$

These inequalities imply that $||x|| \neq 0$ and $||y||^2 \le \frac{\varepsilon - \rho}{2}$. Therefore $(x,y) \mapsto (\frac{x}{||x||}, \frac{2y}{\varepsilon - \rho})$ gives a homeomorphism

$$\tilde{X}_{t_0} \cap B \cong \{(x,y) \in \mathbb{R}^n \times \mathbb{R}^n \mid ||x||^2 = 1,\ ||y||^2 \le 1,\ \langle x, y\rangle = 0\}.$$

The latter space deformation retracts onto the sphere $S^{n-1} = \{(x,0) \mid ||x|| = 1\}$. It follows that $H_k(\tilde{X}_{t_0} \cap B)$ is generated by the fundamental class of S^{n-1} when $k = n-1$ and is zero for all other $k > 0$. By Poincaré duality it follows that $H_c^{n-1}(\tilde{X}_{t_0} \cap B) \cong H^{n-1}(\tilde{X}_{t_0} \cap B) \neq 0$. Thus we can find a form α with compact support in $\tilde{X}_{t_0} \cap B$ such that

$$\int_{S^{n-1}} \alpha \neq 0.$$

Therefore the fundamental class δ_i of S^{n-1} in $\tilde{X}_{t_0}$ is nonzero.

Let B^o denote the interior of B. To conclude, we need to appeal to a refinement of Ehresmann's fibration theorem [75], which will imply that $p^{-1}\Delta_i - B^0 \to \Delta_i$ is trivial as a bundle of manifolds with boundary. This implies, by excision [61, 108], that there is an isomorphism of cohomologies of the pairs

$$H_k(\tilde{X}_{t_0}, \tilde{X}_{t_0} \cap B) \cong H_k(p^{-1}\Delta_i, B).$$

Thus by the exact sequence for a pair we have a commutative diagram with exact rows:

$$\begin{array}{ccccccc}
H_{k+1}(\tilde{X}_{t_0}, \tilde{X}_{t_0} \cap B) & \to & H_k(\tilde{X}_{t_0} \cap B) & \to & H_k(\tilde{X}_{t_0}) & \to & H_k(\tilde{X}_{t_0}, \tilde{X}_{t_0} \cap B) \\
\| & & \downarrow & & \downarrow & & \| \\
H_{k+1}(p^{-1}\Delta_i, B) & \to & H_k(B) = 0 & \to & H_k(p^{-1}\Delta_i) & \to & H_k(p^{-1}\Delta_i, B)
\end{array}$$

A straightforward diagram chase shows that $H_k(\tilde{X}_{t_0}) \to H_k(p^{-1}\Delta_i)$ is surjective with kernel spanned by δ_i when $k = n-1$ and an isomorphism otherwise. □

Let $\delta_i \in H^{n-1}(\tilde{X}_{t_0})$ be the fundamental class of S^{n-1} constructed above. It is called the *vanishing cycle* about t_i. We can now define $\mathrm{Van} \subset H^{n-1}(\tilde{X}_{t_0}, \mathbb{Q})$ to be the subspace spanned by all of the vanishing cycles δ_i. The subspace $\mathrm{Inv} = H^{n-1}(\tilde{X}_{t_0})^{\pi_1(U)}$ will be the space of classes invariant under $\pi_1(U)$. We will need to describe the monodromy explicitly. This is given by the Picard–Lefschetz formula (which refines the formula encountered in Section 13.1).

Theorem 14.4.6 (Picard–Lefschetz). *Let $\mu_i : H^{n-1}(\tilde{X}_{t_0},\mathbb{Q})$ denote the action of the loop going once around t_i. Then $\mu_i(\alpha) = \alpha + (-1)^{n(n+1)/2}\langle\alpha,\delta_i\rangle\delta_i$, where $\langle,\rangle$ denotes the cup product pairing on $H^{n-1}(\tilde{X}_{t_0})$.*

Proof. See [75] or [6, Section 2.4]. □

Corollary 14.4.7. *The orthogonal complement* Van$^\perp$ *coincides with* Inv.

Proof. $\mu_i(\alpha) = \alpha$ if and only if $\langle a,\delta_i\rangle = 0$. □

We now identify $Y = \tilde{X}_{t_0}$, and compare these spaces with the spaces V and I introduced earlier.

Proposition 14.4.8. $V =$ Van.

Proof. A proof can be found in [75]. We sketch the proof of one inclusion. Utilizing the Poincaré duality isomorphism $H_{n-1}(Y) \cong H^{n-1}(Y)$, as we have been doing, we note that the Gysin map $\iota_!$ corresponds to the natural (pushforward) map in homology $H_{n-1}(Y) \to H_{n-1}(X)$. Therefore its kernel can be identified with V. By Theorem 14.4.5, δ_i lies in $\ker[H_{n-1}(Y) \to H_{n-1}(p^{-1}(\gamma\cup\Delta_i))]$ and therefore in V. Therefore Van $\subset V$. □

Proposition 14.4.9. $I =$ Inv.

Proof. We prove this when n is odd, which implies that $\langle,\rangle$ is a nondegenerate symmetric pairing on $H^{n-1}(Y)$. Then by Corollary 14.4.7,

$$\dim \mathrm{Inv} = \dim H^{n-1}(Y) - \dim \mathrm{Van}.$$

The image of $H^{n-1}(X,\mathbb{Q})$ lies in $\mathrm{Inv} = H^{n-1}(\tilde{X}_{t_0})^{\pi_1(U)}$, since it factors through $H^{n-1}(\tilde{X},\mathbb{Q})$, and $H^{n-1}(\tilde{X},\mathbb{Q}) \to H^{n-1}(\tilde{X}_{t_0})$ is the edge homomorphism; see Section 13.3. Thus it suffices to prove that I and Inv have the same dimension. But this is a consequence of Propositions 14.4.2 and 14.4.8 and the previous equation:

$$\dim I = \dim H^{n-1}(Y) - \dim V = \dim \mathrm{Inv}.$$ □

With the above identifications, Theorem 14.4.1 will follow from Proposition 14.4.2. We finally state the following for later use.

Proposition 14.4.10. *V is an irreducible $\pi_1(U)$-module, i.e., it has no $\pi_1(U)$-submodules other than 0 or V.*

Proof. We give a proof modulo the basic fact that any two vanishing cycles are conjugate up to sign under the action of $\pi_1(U)$ [75, 7.3.5]. Suppose that $W \subset V$ is a $\pi_1(U)$ submodule, and let $w \in W$ be a nonzero element. We claim that $\langle w,\delta_i\rangle \neq 0$ for some i. If not, we would have $w \in V^\perp = I$, which would force it to be zero. Therefore

$$\mu_i(w) - w = \pm\langle w,\delta_i\rangle\delta_i \in W$$

implies that W contains δ_i, and thus all the vanishing cycles by the above fact. Therefore $W = V$. □

14.5 Cohomology of Smooth Projective Maps*

In this final section, we give some subtle applications of hard Lefschetz. The first, due to Deligne [23], analyzes the cohomology of a family of smooth projective varieties. A morphism $f : X \to Y$ of algebraic varieties is called *projective* if it factors through a projection $Y \times \mathbb{P}^N \to Y$. We also have an analogue of submersion in algebraic geometry. When X, Y are nonsingular, the morphism f is *smooth* if the induced maps on Zariski tangent spaces are surjective. This definition is really provisional; a better definition is given later, in Section 18.1. Ehresmann's theorem implies that smooth projective maps are C^∞ fiber bundles. In general, the cohomology of a fiber bundle can be very complicated. However, Deligne shows that in the absence of monodromy the cohomology of a smooth projective map is as simple as it can be.

Theorem 14.5.1 (Deligne). *Let $f : X \to Y$ be a smooth projective map of smooth complex algebraic varieties. Then the inequalities in* (13.3.6) *are equalities, i.e.,*

$$\dim H^i(X, \mathbb{Q}) = \sum_{p+q=i} \dim H^p(Y, R^q f_* \mathbb{Q}).$$

Here is a somewhat more concrete consequence.

Corollary 14.5.2. *If the monodromy action of $\pi_1(Y, y)$ on the cohomology of the fiber X_y is trivial (e.g., if Y is simply connected), then the Betti numbers are the same as for a product, i.e.,*

$$b_i(X) = \sum_{p+q=i} b_p(Y) b_q(X_y).$$

We sketch the proof of the theorem. A somewhat more detailed treatment can be found in [49, pp. 462–468].

Proof. In Exercise 13.3.14, we constructed a map $d_2 : H^p(Y, R^q) \to H^{p+2}(Y, R^{q-1})$, where $R^q = R^q f_* \mathbb{Q}$. The vanishing of d_2 and the higher differentials is equivalent to the conclusion of the theorem. We check this for d_2 only.

Let n be the dimension of the fibers. By assumption, there is an inclusion $X \hookrightarrow \mathbb{P}^N \times Y$ that gives Lefschetz operators on the fibers. For each y, we have a corresponding Lefschetz decomposition

$$H^i(X_y, \mathbb{Q}) = \bigoplus_{j=0}^{[i/2]} L^j P^{i-2j}(X_y).$$

In fact, we get a decomposition of sheaves

$$L^i : R^{n-i} \cong R^{n+i}, \quad R^i = \bigoplus_{j=0}^{[i/2]} L^j P^{i-2j},$$

where

$$P^i = \ker[L^{n-i+1} : R^i \to R^{2n-i+2}].$$

This allows us to decompose

$$H^p(Y,R^q) \cong \bigoplus H^p(Y,P^{q-2j}).$$

Thus it suffices to check the vanishing of the restrictions of d_2 to these factors. Consider the diagram

$$\begin{array}{ccc} H^p(Y,P^{n-k}) & \xrightarrow{d_2} & H^{p+2}(Y,R^{n-k-1}) \\ {\scriptstyle 0}\downarrow{\scriptstyle L^{k+1}} & & {\scriptstyle \cong}\downarrow{\scriptstyle L^{k+1}} \\ H^p(Y,P^{n+k+2}) & \xrightarrow{d_2} & H^{p+2}(Y,R^{n+k+1}) \end{array}$$

The first vertical arrow is zero by the definition of P, and the second vertical arrow is an isomorphism by hard Lefschetz. Therefore the top d_2 vanishes. □

A projective morphism $p : P \to X$ of varieties is said to be a *Brauer–Severi* morphism if it is smooth and all the fibers are isomorphic to projective space. The cohomology is easy to analyze using the previous results.

Lemma 14.5.3. *If $\pi : P \to X$ is Brauer–Severi with n-dimensional fibers, then*

$$H^i(P,\mathbb{Q}) \cong \bigoplus_{j=0}^{n} H^{i-2j}(X,\mathbb{Q}).$$

Proof. Choose an embedding $\iota : P \hookrightarrow \mathbb{P}^N \times X$ that commutes with the projection. Then $H^2(\mathbb{P}^N,\mathbb{Q}) = \mathbb{Q}$ maps nontrivially and therefore isomorphically to $H^2(P_x,\mathbb{Q}) = \mathbb{Q}$. Since $H^2(P_x)$ generates $H^*(P_x)$, the monodromy action is trivial. Therefore the lemma follows from Corollary 14.5.2. □

Note that the proof gives the somewhat stronger result that $H^*(P,\mathbb{Q})$ is generated as an algebra by $H^*(P,\mathbb{Q})$ and $H^2(\mathbb{P}^N,\mathbb{Q})$. If h denotes the positive generator $c_1(\iota^*\mathscr{O}_{\mathbb{P}^N}(1))$ of the latter space, then in fact

$$H^i(P,\mathbb{Q}) = \bigoplus_{j=0}^{n} H^{i-2j}(X,\mathbb{Q}) \cup h^j. \tag{14.5.1}$$

The Brauer–Severi morphisms of interest to us arise as follows. Let E be a rank-$(n+1)$ algebraic vector bundle on X in the original geometric sense (as opposed to a locally free sheaf). Define

$$\mathbb{P}(E) = \{\ell \mid \ell \text{ a line in some } E_x\} = \bigcup \mathbb{P}(E_x).$$

This comes with a projection $\pi : \mathbb{P}(E) \to X$, which sends ℓ to x. The fiber is precisely $\mathbb{P}(E_x)$. The map π is projective, and therefore carries a class h as above. This can be constructed directly as follows. Let g_{ij} be a 1-cocycle for E with respect to a cover $\{U_i\}$. Then we construct $\mathbb{P}(E)$ by gluing $(x,[v]) \in U_i \times \mathbb{P}^n$ to $(x,[g_{ij}v]) \in U_j \times \mathbb{P}^n$. Consider the line bundle

$$L = \{(v,\ell) \mid v \in \ell,\ \ell \text{ a line in some } E_x\}$$

on $\mathbb{P}(E)$. We define $\mathscr{O}_{\mathbb{P}(E)}(-1)$ as the sheaf of its sections, and $\mathscr{O}_{\mathbb{P}(E)}(1) = \mathscr{O}_{\mathbb{P}(E)}(-1)^*$. This need not be $\iota^*\mathscr{O}_{\mathbb{P}^N}(1)$ as above; however, a power will be. Therefore their first Chern classes will be proportional. Therefore we have the following corollary:

Corollary 14.5.4. *The decomposition* (14.5.1) *is valid for* $X = \mathbb{P}(E)$ *and* $h = c_1(\mathscr{O}_{\mathbb{P}(E)}(1))$.

Remark 14.5.5. It is possible to give a much more elementary proof using the Leray–Hirsch theorem [61, Theorem 4D.1]. This has the advantage of working over the integers and for complex C^∞ vector bundles.

Although the additive structure is now determined, the multiplicative structure is more subtle. If $E = \mathscr{O}_Y^{n+1}$ is trivial, then $\mathbb{P}(E) = \mathbb{P}^n \times X$. Therefore $h^{n+1} = 0$ by Künneth's formula. In general, it can be nontrivial. We can express h^{n+1} as a linear combination of $1, h, \ldots, h^n$ with coefficients in $H^*(X)$. These coefficients are, by definition, the Chern classes of E. More precisely, we define the ith Chern class $c_i(E) \in H^{2i}(X,\mathbb{Q})$ by

$$h^{n+1} - c_1(E) \cup h^n + \cdots + (-1)^{n+1} c_{n+1}(E) = 0. \qquad (14.5.2)$$

By the above remark, these classes can be defined in $H^{2i}(Y,\mathbb{Z})$. This method of defining *Chern classes*, which is due to Grothendieck [52], is one of many. Regardless of the method, the key point is the following characterization.

Theorem 14.5.6. *Chern classes satisfy the following properties:*

(a) *For a line bundle L, $c_1(L)$ is given by the connecting map* $\mathrm{Pic}(X) \to H^2(X,\mathbb{Z})$ *and $c_i(L) = 0$ for $i > 1$.*
(b) *The classes are functorial, i.e., $c_i(f^*E) = f^*c_i(E)$ for any map $f : Y \to X$.*
(c) *If $0 \to E_1 \to E \to E_2 \to 0$ is an exact sequence of vector bundles, then*

$$1 + c_1(E) + c_2(E) + \cdots = (1 + c_1(E_1) + \cdots) \cup (1 + c_1(E_2) + \cdots).$$

Moreover, any assignment of cohomology classes to vector bundles satisfying these properties coincides with the theory of Chern classes.

Proof. We give a sketch. For the first property, note that when L is a line bundle, $\mathbb{P}(L) = X$ and $\mathscr{O}_{\mathbb{P}(E)}(1) = L$. The second is more or less immediate from the construction. A special case of (c) will be given in the exercises.

The uniqueness is based on the splitting principle [42, §3.2], which says that given X, E, there exists a map $f : Y \to X$ that induces an injection in cohomology such that f^*E is a successive extension of line bundles L_i. Then (c) will imply that the pullback of the Chern classes are determined by $c_1(L_i)$. □

We want to end this chapter with one last application of hard Lefschetz. A vector bundle E is called *negative* if $\mathscr{O}_{\mathbb{P}(E)}(1)$ is ample, and E is *ample* if E^* is negative. We should point out that our definition of $\mathbb{P}(E)$ is dual to the one given in [55, 60], so our definitions appear to be the opposites of the usual ones. Also, our signs for the Chern classes have been adjusted accordingly. The following is a very special case of a result of Bloch and Gieseker [13].

Theorem 14.5.7 (Bloch–Gieseker). *If E is a negative vector bundle of rank $n+1$ on a smooth projective variety X with $d = \dim X \leq n+1$, then $c_d(E) \neq 0$.*

Proof. Let $h = c_1(\mathscr{O}_{\mathbb{P}(E)}(1))$, and let

$$\eta = h^{d-1} - c_1(E)h^{d-1} + \cdots \pm c_{d-1}(E) \in H^{2d-2}(\mathbb{P}(E)).$$

Since $\mathscr{O}_{\mathbb{P}(E)}(1)$ is ample, the hard Lefschetz theorem guarantees that

$$h^{n+2-d} \cup : H^{2d-2}(\mathbb{P}(E)) \to H^{2n+2}(\mathbb{P}(E))$$

is injective. We have

$$\eta \cup h^{n+2-d} = \pm c_d(E) \cup h^{n+1-d}$$

by (14.5.2) and the fact that $c_i(E) = 0$ for $i > d$. Therefore $c_d(E)$ cannot vanish. □

Exercises

14.5.8. Let $\pi : \mathbb{C}^2 - \{0\} \to \mathbb{P}^1_{\mathbb{C}}$ be the usual projection. This is smooth, and in fact a Zariski locally trivial $\mathbb{C}^*$-bundle. The restriction $S^3 \to \mathbb{P}^1_{\mathbb{C}}$ to the unit sphere is called the Hopf fibration. Show that the conclusion of Corollary 14.5.2 fails for π and the Hopf fibration.

14.5.9. Show that $b_1(X) = b_1(C) + \dim H^1(X_y, \mathbb{Q})^{\pi_1(C)}$ when $f : X \to C$ is smooth and projective.

14.5.10. Show that there is an epimorphism $\pi^*E \to \mathscr{O}_{\mathbb{P}(E)}(1)$.

14.5.11. Let $E = L_1 \oplus L_2$ be a sum of line bundles.

1. Show that the divisors $Z_i \subset \mathbb{P}(E)$ defined by the sections $\sigma_i \in H^0(\mathscr{O}_{\mathbb{P}(E)}(1) \otimes L_i^{-1})$ corresponding to the maps $\pi^*L_i \to \pi^*E \to \mathscr{O}_{\mathbb{P}(E)}(1)$ are disjoint.
2. Deduce that the product $[Z_1] \cup [Z_2]$ is zero.
3. Conclude that (3) holds for $E = L_1 \oplus L_2$, i.e., that $c_1(E) = c_1(L_1) + c_1(L_2)$ and $c_2(E) = c_1(L_1) \cup c_1(L_2)$.

14.5.12. Let $G = \mathbb{G}(2,n)$ and $F = \{(x,\ell) \in \mathbb{P}^{n-1} \times G \mid x \in \ell\}$. Show that both projections $\mathbb{P}^{n-1} \leftarrow F \to G$ are Brauer–Severi morphisms. Use this to calculate the Betti numbers of G.

14.5.13. F above can be identitified with $\mathbb{P}(V)$, where V is the vector constructed in Exercise 2.6.15. With the help of the previous exercise, show that the cohomology of $H^2(\mathbb{G}(2,4))$ is spanned by $c_1(V)^2$ and $c_2(V)$.

14.5.14. The obstruction for a Brauer–Severi morphism to be given by $\mathbb{P}(E)$, for some vector bundle E, lies in the Brauer group $H^2(Y,\mathcal{O}_Y^*)$ (this is the obstruction for lifting a class from $\check{H}^1(Y,\mathrm{PGL}_n(\mathcal{O}_Y))$ to $\check{H}^1(Y,\mathrm{GL}_{n+1}(\mathcal{O}_Y)))$. Show that the Brauer group vanishes when Y is a Riemann surface. However, the obstruction can be nontrivial in general.

Part IV
Coherent Cohomology

Chapter 15
Coherent Sheaves

Chow showed that every complex submanifold of $\mathbb{P}^n_{\mathbb{C}}$ is an algebraic variety. The eventual goal of this chapter and the next is to outline the proof of a refined version of this due to Serre [101], usually referred to as "GAGA," which is an acronym derived from the title of his paper. The first part of the theorem gives a correspondence between certain objects on $\mathbb{P}^n_{\mathbb{C}}$ viewed as an algebraic variety and objects on $\mathbb{P}^n_{\mathbb{C}}$ viewed as a complex manifold. These objects are coherent sheaves that are $\mathscr{O}$-modules that are locally finitely presented in a suitable sense. Some of the formal properties of coherent sheaves are given here. Over affine and projective spaces there is a complete description of coherent sheaves in elementary algebraic terms, which makes this class particularly attractive. Chow's theorem is recovered by applying GAGA to ideal sheaves, which are coherent.

As general references, we mention Hartshorne [60] and Serre [100] for the algebraic side of the story, and Grauert and Remmert [47] for the analytic.

15.1 Coherence on Ringed Spaces

In order to deal simultaneously with the analytic and algebraic cases, we return to the general setting of a ringed space $(X,\mathscr{R})$. We will be interested in $\mathscr{R}$-modules that have good finiteness properties. We ought to say what "good" should actually mean here. We would like the property to satisfy the *two out of three principle*: in any short exact sequence where two sheaves satisfy the property, the third does as well. So in particular, the class of sheaves satisfying the property should be stable under direct sums. The weakest finiteness condition is the following:

Definition 15.1.1. An $\mathscr{R}$-module $\mathscr{E}$ has finite type if $\mathscr{E}$ is locally finitely generated, i.e., each point has a neighborhood U such that there is an epimorphism $\mathscr{R}|_U^n \to \mathscr{E}_U$ for some $n < \infty$.

The class of sheaves of finite type is certainly stable under direct sums. However, the two out of three principle will fail in general.

Example 15.1.2. Let R be a non-Noetherian commutative ring regarded as a ringed space over a point. A nonfinitely generated ideal I gives a counterexample, because it fits into the exact sequence $0 \to I \to R \to R/I \to 0$.

The solution is to impose finiteness on kernels whenever they arise.

Definition 15.1.3. An $\mathscr{R}$-module $\mathscr{E}$ is coherent if and only if:

(1) $\mathscr{E}$ is of finite type,
(2) If $\mathscr{R}|_U^n \to \mathscr{E}|_U$ is any morphism defined over an open set U, the kernel has finite type.

Proposition 15.1.4.

(a) A submodule of finite type of a coherent module is coherent.
(b) The two out of three principle holds for coherence.

Proof. The first statement is automatic. We do one case of (b), leaving the rest for the exercises. Suppose that

$$0 \to \mathscr{E}_1 \to \mathscr{E}_2 \to \mathscr{E}_3 \to 0$$

is exact with $\mathscr{E}_1$ and $\mathscr{E}_2$ coherent. Then clearly $\mathscr{E}_3$ has finite type. Now choose a morphism $\phi : \mathscr{R}^n|_U \to \mathscr{E}_3|_U$. We have to show that $\ker\phi$ has finite type. Since this is a local condition, we are free to shrink U to a neighborhood of any given point $x \in U$. After doing this, we can assume that ϕ lifts to a morphism $\psi : \mathscr{R}^n|_U \to \mathscr{E}_2|_U$. After shrinking U again, we can choose an epimorphism $\mathscr{R}_U^m \to \mathscr{E}_3|_U$. By the snake lemma, there is an epimorphism of $\mathscr{E}_3|_U$, and hence of $\mathscr{R}^m|_U$, onto the cokernel of $\ker\psi \to \ker\phi$ over U. We can lift this to a morphism $\mathscr{R}^m|_U \to \ker\phi$ such that images of $\ker\psi$ and $\mathscr{R}^m|_U$ generate $\ker\phi$. This implies that $\ker\phi$ has finite type because $\ker\psi$ does. □

Corollary 15.1.5. *Given a morphism of coherent $\mathscr{R}$-modules, the kernel, image, and cokernel are coherent. The sum of two coherent modules is coherent. So in particular, the collection of coherent $\mathscr{R}$-modules and morphisms between them forms an abelian category.*

Proof. Suppose that $\phi : \mathscr{E} \to \mathscr{F}$ is a morphism of coherent modules. Coherence of the image follows from (a). Coherence of the kernel and cokernel follow by applying (b) to

$$0 \to \ker\phi \to \mathscr{E} \to \operatorname{im}\phi \to 0,$$
$$0 \to \operatorname{im}\phi \to \mathscr{F} \to \operatorname{coker}\phi \to 0.$$ □

While the structure sheaf $\mathscr{R}$ of a ringed space obviously has finite type, it need not be coherent in general. However, it is true for a scheme in our sense by straightforward arguments (Exercise 15.2.8). In the analytic case, this is somewhat deeper.

Theorem 15.1.6 (Oka). *If $(X, \mathscr{O}_X)$ is a complex manifold, then $\mathscr{O}_X$ is coherent.*

Proof. See [47, p. 59] □

Corollary 15.1.7. *Let X be a complex manifold. Any locally free $\mathscr{O}_X$-module is coherent. The ideal sheaf of a submanifold $Y \subset X$ is coherent.*

Exercises

15.1.8. Finish the proof of Proposition 15.1.4.

15.1.9. Prove that if $\mathscr{M}_i$ are coherent, then so are $\mathscr{M}_1 \otimes \mathscr{M}_2$ and $\mathscr{H}om(\mathscr{M}_1, \mathscr{M}_2)$.

15.1.10. Suppose that $\mathscr{R}$ is coherent and let $\mathscr{M}$ be an $\mathscr{R}$-module. Prove that $\mathscr{M}$ is coherent if and only if it is locally finitely presented, i.e., there is a covering $\{U_i\}$ with exact sequences

$$\mathscr{R}^{m_i}|_{U_i} \to \mathscr{R}^{n_i}|_{U_i} \to \mathscr{M}|_{U_i} \to 0.$$

15.1.11. Let $f : \mathscr{F} \to \mathscr{G}$ be a morphism of coherent sheaves over $(X, \mathscr{R})$. Suppose that f_{x_0} is an injection (respectively surjection) for some $x_0 \in X$. Then prove that f_{x_0} is an injection (respectively surjection) for all x in a neighborhood of x_0. Is this property true for morphisms of arbitrary $\mathscr{R}$-modules?

15.1.12. A ring R is called coherent if the kernel of any map $R^n \to R$ is finitely generated. Give an example of a noncoherent ring. Show that the structure sheaf of the associated ringed space given in Example 15.1.2 will not be coherent.

15.2 Coherent Sheaves on Affine Schemes

Coherent sheaves over affine schemes have a particularly nice description. They correspond to finitely generated modules over the coordinate ring. Let R be an affine algebra over an algebraically closed field. Let us first recall that in Definition 3.5.11, we constructed a sheaf of modules $\tilde{M}$ on $\mathrm{Spec}_m R$ associated to an R-module M. On basic open sets, we have

$$\tilde{M}(D(f)) = R[1/f] \otimes_R M \cong M[1/f],$$

where the localization $M[1/f]$ of M is constructed as for rings [8]. Important examples of such sheaves include ideal sheaves associated to subschemes of $\mathrm{Spec}_m R$.

Lemma 15.2.1. *If M is a finitely generated R-module, $\tilde{M}$ is coherent.*

Proof. This is Exercise 15.2.8. □

It convenient to extend this class of sheaves $\tilde{M}$ to more general schemes.

Definition 15.2.2. Let $(X, \mathcal{O}_X)$ be a scheme in our sense over an algebraically closed field k. A sheaf of modules $\mathcal{E}$ over X is called quasicoherent if there is a covering by affine open sets $\{U_i \cong \mathrm{Spec}_m R_i\}$ such that $\mathcal{E} \cong \tilde{M}_i$ for R_i-modules M_i.

We shall see shortly that coherent sheaves are quasicoherent, as the terminology suggests. Clearly sheaves of the form $\tilde{M}$ on $\mathrm{Spec}_m R$ are quasicoherent. The converse statement is also true, although it is not entirely trivial.

Proposition 15.2.3. *If R is an affine k-algebra, any quasicoherent sheaf $\mathcal{E}$ on $X = \mathrm{Spec}_m R$ is isomorphic to $\tilde{M}$ with $M = \Gamma(X, \mathcal{E})$.*

Proof. The elements of $M[1/f]$ can be represented by fractions m/f^n, with $m \in M$. Each such element determines a section $\frac{1}{f^n} m|_{D(f)} \in \mathcal{E}(D(f))$. In this way, we get a homomorphism $r_f : \tilde{M}(D(f)) \to \mathcal{E}(D(f))$, which gives rise to a morphism of sheaves $r : \tilde{M} \to \mathcal{E}$. We have to prove that this is an isomorphism. In fact, we will prove that each r_f is an isomorphism, and this is sufficient.

By assumption, there is an affine open covering $\{U_i\}$ of X such that $\mathcal{E}|_{U_i} = \tilde{M}_i$. By quasicompactness, we can assume that this is a finite covering by basic open sets $U_i = D(g_i)$. Suppose that $m \in M$ is a section mapping to 0 in $\mathcal{E}(D(f))$. Then the images $m_i \in M_i$. vanish in $M_i[1/f_i]$. This implies that $f^{n_i} m_i = 0$, for some n_i, because the kernel of $M_i \to M[1/f]$ consists of elements annihilated by f. Therefore taking $n = \max n_i$, we see that $f^n m = 0$ or that the image of m vanishes in $M[1/f]$. The upshot is that $r_f : M[1/f] \to \mathcal{E}(D(f))$ is injective.

Suppose that $e \in \mathcal{E}(D(f))$. The images of e in $M_i[1/f]$ can be written as m_i/f^n, where $m_i \in M_i$ and the exponent n can be chosen independently of i. Now m_i and m_j agree with $f^n e$, and therefore each other, on the intersection $D(fg_ig_j) = D(f) \cap D(g_i) \cap D(g_j)$. Therefore $f^N(m_i - m_j) = 0$ for some N independent of i, j. Thus we can patch the $f^N m_i$ to obtain a section $m \in M$. It follows that $m/f^{n+N} \in M[1/f]$ maps to e. That is, r_f is surjective. □

The operation $M \to \tilde{M}$ clearly extends to a functor. The above results show that $\mathcal{E} \to \Gamma(\mathcal{E})$ is an inverse. Therefore we may state the following theorem:

Theorem 15.2.4. *If R is an affine k-algebra and $X = \mathrm{Spec}_m R$, then $M \mapsto \tilde{M}$ gives an equivalence between the category of R-modules and quasicoherent modules on X.*

Corollary 15.2.5. *The functors $M \mapsto \tilde{M}$ and $\mathcal{E} \mapsto \Gamma(X, \mathcal{E})$ on the above categories are exact, i.e., if*

$$0 \to M_1 \to M_2 \to M_3 \to 0$$

is exact, then so is

$$0 \to \tilde{M}_1 \to \tilde{M}_2 \to \tilde{M}_3 \to 0,$$

and likewise for the second functor.

Lemma 15.2.6. *Coherent sheaves on a scheme are quasicoherent.*

Proof. Any coherent sheaf is locally of the form

$$\mathrm{coker}[\mathscr{O}_U^n \to \mathscr{O}_U^m] = \mathrm{coker}[R^n \to R^m]^{\sim}$$

for an affine open $U = \mathrm{Spec}_m R$. □

Theorem 15.2.7. *With $X = \mathrm{Spec}_m R$ as above, the functor $M \mapsto \tilde{M}$ gives an equivalence between the category of finitely generated R-modules and coherent modules on X.*

Proof. This is a consequence of the previous results and Exercise 15.2.9. □

Exercises

15.2.8. Let $X = \mathrm{Spec}_m R$, with R affine. Prove that $\mathscr{O}_X$ is coherent, and therefore conclude that $\tilde{M}$ is coherent if M is finitely generated.

15.2.9. If $\mathscr{E}$ is coherent on $\mathrm{Spec}_m R$, then show $M = \Gamma(\mathscr{E})$ is finitely generated.

15.2.10. Prove that quasicoherence on any scheme satisfies the "two out of three" property.

15.2.11. Let $i : Y \hookrightarrow X$ be a closed subscheme of a scheme. If $\mathscr{M}$ is a (quasi)coherent sheaf on Y, show that $i_*\mathscr{M}$ is (quasi)coherent on X.

15.2.12. Let $j : U \hookrightarrow X$ be a proper open subset of an affine variety. Show that the sheaf

$$j_!\mathscr{O}_U = \begin{cases} \mathscr{O}(V) & \text{if } V \subseteq U, \\ 0 & \text{otherwise,} \end{cases}$$

is an $\mathscr{O}_X$-module that is not isomorphic to $\tilde{M}$ for any M. In particular, it is not coherent.

15.3 Coherent Sheaves on $\mathbb{P}^n$

Our goal in this section is to describe all coherent sheaves on $\mathbb{P}^n = \mathbb{P}^n_k$. Let $S = k[x_0,\dots,x_n]$ be its homogeneous coordinate ring, with its standard grading $S = \bigoplus S_i$. Then the category of coherent sheaves on $\mathbb{P}^n$ will turn out to be almost equivalent to the category of finitely generated graded S-modules. We will see this by making explicit constructions, but let us first explain why we might expect this on a more conceptual level. Projective space is the quotient $\mathbb{P}^n = (\mathbb{A}^{n+1} - \{0\})/k^*$, so a coherent sheaf on it ought to be the same as a coherent sheaf on $\mathbb{A}^{n+1} - \{0\}$

on which k^* acts in the appropriate sense. Then taking global sections would produce an S-module M with a good k^*-action. The eigenspaces for this action yield a grading and conversely.

A $\mathbb{Z}$-graded S-module is an S-module M with a decomposition $M = \bigoplus_{i\in\mathbb{Z}} M_i$ into subspaces such that $S_iM_j \subset M_{i+j}$. If M is also finitely generated as an S-module, then each of the components M_i is finite-dimensional, and $M_i = 0$ for $i \ll 0$. Given a graded module M, we define a new grading on the same underlying module by

$$M(d)_i = M_{i+d}.$$

The collection of graded modules becomes a category, where morphisms $f: M \to N$ are S-module maps such that $f(M_i) \subseteq N_i$.

Starting with a graded S-module M, we can construct a sheaf of modules on $\mathbb{P}^n$ by doing a graded version of the construction given in Definition 3.5.11. Given an open set $U \subseteq \mathbb{P}^n$, $\mathscr{O}_{\mathbb{P}^n}(U) \subset k(x_0,\dots,x_n)$ carries the grading given by $\deg(f/g) = \deg f - \deg g$. The tensor product $M \otimes_S \mathscr{O}_{\mathbb{P}^n}(U)$ carries an induced grading $\deg(m \otimes (f/g)) = \deg m + \deg(f/g)$. Let

$$\widetilde{M}(U) = (M \otimes_S \mathscr{O}_{\mathbb{P}^n}(U))_0 \tag{15.3.1}$$

denote the subgroup of graded 0 elements. This is obviously a presheaf of modules, and in fact it can be seen to be a sheaf. To see that this is coherent, recall that a basis of the topology of $\mathbb{P}^n$ is given by $D(f) = \{a \in \mathbb{P}^n \mid f(a) \neq 0\}$ as f varies over the set of homogeneous polynomials. We can identify $D(f) = \mathrm{Spec}_m S[1/f]_0$. Then by comparing formulas, we can see that the following lemma most hold:

Lemma 15.3.1. *Suppose that $f \in S$ is homogeneous. Let $\tilde{N}$ denote the sheaf defined on $D(f)$ by $N = M[1/f]_0$. Then $\widetilde{M}|_{D(f)} = \tilde{N}$.*

Corollary 15.3.2. *The sheaf $\widetilde{M}$ is coherent if M is finitely generated, and quasicoherent in general.*

It is clear that $M \mapsto \widetilde{M}$ yields a functor from the category of graded (respectively finitely generated graded) modules to the category of quasicoherent (respectively coherent) sheaves on $\mathbb{P}^n$.

Lemma 15.3.3. *The functor $M \mapsto \widetilde{M}$ is exact.*

Proof. This follows from the previous lemma and Corollary 15.2.5. □

We have already seen examples of this construction previously.

Lemma 15.3.4. $\widetilde{S(d)} \cong \mathscr{O}_{\mathbb{P}^n}(d)$.

Proof. This follows from (3.6.2) and (15.3.1). □

Example 15.3.5. Let $I \subset S$ be a homogenous ideal. Then $\widetilde{I}$ coincides with the sheaf defined in Example 3.5.4. This defines a closed subscheme of $\mathbb{P}^n$ that is a subvariety if I is prime.

Definition 15.3.6. A collection of global sections $f_i \in H^0(X, \mathscr{F})$ generates the sheaf $\mathscr{F}$ if their germs span the stalk $\mathscr{F}_x$ for every $x \in X$. A coherent sheaf $\mathscr{F}$ on X is generated by global sections or globally generated if there exist sections with this property.

Theorem 15.3.7 (Serre).

(a) If $\mathscr{F}$ is a coherent sheaf on $\mathbb{P}^n_k$, then there exists d_0 such that for any $d \geq d_0$, $\mathscr{F}(d) = \mathscr{F} \otimes \mathscr{O}_{\mathbb{P}^n}(d)$ is globally generated.
(b) Every coherent sheaf on $\mathbb{P}^n_k$ is isomorphic to a sheaf of the form $\widetilde{M}$ for some finitely generated graded module M.

We make a few preliminary remarks before starting the proof. The sheaf $\mathscr{O}(-1)$ is the ideal sheaf of the hyperplane $x_i = 0$. Thus global sections of $\mathscr{O}(d) = (\mathscr{O}(-1)^*)^{\otimes d}$, when $d > 0$, can be interpreted as regular functions on U_i with poles of order d along x_i. Therefore $H^0(U_i, \mathscr{F})$ can be identified with the directed union $\bigcup_d H^0(\mathbb{P}^n, \mathscr{F}(d))$.

Proof. Let $\mathscr{F}$ be coherent. Then there are finitely generated modules M_i such that the restrictions $\mathscr{F}|_{U_i}$ are isomorphic to $\tilde{M}_i$. Choose a finite set of generators $\{m_{ji}\}$ for each M_i. From the above remarks, we can view these as global sections of $\mathscr{F}(d)$ for some fixed $d \gg 0$. These sections generate $\mathscr{F}(d)$.

Let $M' = \bigoplus_{e=0}^{\infty} H^0(\mathscr{F}(e))$. The argument of the previous paragraph shows that for some $d > 0$, $\mathscr{F}(d)$ is generated by finitely many sections $m_i \in H^0(\mathscr{F}(d)) = M'_d$. Let $M \subseteq M'$ be the submodule generated by these sections. Consider the subsheaf $\mathscr{G} = \widetilde{M} \subseteq \mathscr{F}$. Since the sections $m_i \in H^0(\mathscr{G}(d))$ generate $\mathscr{F}(d)$, we see that the stalks $\mathscr{G}(d)_x$ equal $\mathscr{F}(d)_x$. Therefore $\mathscr{G}(d) = \mathscr{F}(d)$. Tensoring by $\mathscr{O}(-d)$ implies that $\mathscr{G} = \mathscr{F}$. □

As we will see in the exercises, the map from isomorphism classes of graded modules to isomorphism classes of coherent sheaves is many-to-one. Fortunately, the ambiguity is easy to describe. Given two graded modules M, N, let us say that they are *stably isomorphic*, or $M \sim N$, if for some i_0,

$$\bigoplus_{i \geq i_0} M_i \cong \bigoplus_{i \geq i_0} N_i$$

as graded S-modules.

Now suppose that $\mathscr{M}$ is a coherent $\mathscr{O}_{\mathbb{P}}$-module. Define

$$\Gamma_*(\mathscr{M}) = \bigoplus_{d=-\infty}^{\infty} \Gamma(\mathbb{P}^n, \mathscr{M}(d)).$$

We state without proof some refinements of the last theorem.

Theorem 15.3.8.

1. $\Gamma_*(\widetilde{M}) \sim M$.
2. $\widetilde{\Gamma_*(\mathscr{M})} \cong \mathscr{M}$.

Proof. See [100, p. 258, Proposition 4] for the proof of the first statement, and [60, Chapter II, Proposition 5.15] for the second. □

Theorem 15.3.9. *Any coherent sheaf $\mathscr{E}$ on $\mathbb{P}^n$ fits into an exact sequence*

$$0 \to \mathscr{E}_r \to \mathscr{E}_{r-1} \to \cdots \to \mathscr{E}_0 \to \mathscr{E} \to 0,$$

where $r \leq n$ and each $\mathscr{E}_i$ is a sum of a finite number of line bundles $\mathscr{O}_{\mathbb{P}^n}(j)$.

Proof. The Hilbert syzygy theorem [33, 1.13] says that any finitely generated graded S-module has a finite free graded resolution of length at most n. The theorem is a consequence of this and Lemma 15.3.3. □

For many standard examples, it is possible to construct explicit resolutions of this type.

Example 15.3.10. The tangent sheaf fits into an exact sequence

$$0 \to \mathscr{O}_{\mathbb{P}^n} \xrightarrow{\begin{pmatrix} x_0 \\ x_1 \\ \dots \end{pmatrix}} \mathscr{O}_{\mathbb{P}^n}(1)^{n+1} \to \mathscr{T}_{\mathbb{P}^n} \to 0.$$

This will be justified later.

Exercises

15.3.11. Check that if $M_i = 0$ for $i \gg 0$, i.e., that if $M \sim 0$, then $\widetilde{M} = 0$. Use this to show that $M \sim N$ implies $\widetilde{M} \cong \widetilde{N}$.

15.3.12. Given a pair of graded S-modules M and N, define a grading on their tensor product $M \otimes N = M \otimes_S N$ by

$$(M \otimes N)_i = \sum_{a+b=i} M_a \otimes N_b.$$

Prove that $\widetilde{M \otimes N} \cong \widetilde{M} \otimes_{\mathscr{O}_{\mathbb{P}^n}} \widetilde{N}$.

15.3.13. Given a graded S-module M, show that

$$\pi_*(\widetilde{M}|_{\mathbb{A}^{n+1}-\{0\}}) = \bigoplus_{d \in \mathbb{Z}} \widetilde{M}(d),$$

where $\pi : \mathbb{A}^{n+1} - \{0\} \to \mathbb{P}^n$ is the canonical projection.

15.4 GAGA, Part I

We now want to compare the algebraic and analytic points of view. A nonsingular subvariety $X \subset \mathbb{A}^n_{\mathbb{C}}$, determines a complex submanifold $X_{an} \subset \mathbb{C}^n$ defined by the same equations. We refer to $(X_{an}, \mathscr{O}_{X_{an}})$ as the associated complex manifold; this does not depend on the embedding. We can extend this to any nonsingular algebraic variety X over $\mathbb{C}$, choosing an affine covering $\{U_i\}$ and then gluing $U_{i,an}$ to $U_{j,an}$ by the original transition maps, which are holomorphic. When applied to $(\mathbb{P}^n_{\mathbb{C}}, \mathscr{O}_{\mathbb{P}^n})$, we obtain the complex projective $(\mathbb{P}^n_{an}, \mathscr{O}_{\mathbb{P}^n_{an}})$ viewed as a complex manifold.

For any nonsingular variety X, the analytic topology is finer than the Zariski topology, and the regular functions are holomorphic. Therefore we have a morphism of ringed spaces

$$\iota : (X_{an}, \mathscr{O}_{X_{an}}) \to (X, \mathscr{O}_X)$$

given by the inclusion

$$\mathscr{O}_X(U) \subset \mathscr{O}_{X_{an}}(U).$$

Given an $\mathscr{O}_X$-module $\mathscr{E}$, define the $\mathscr{O}_{X_{an}}$-module $\mathscr{E}_{an} = \iota^*\mathscr{E}$. Then $\mathscr{E}_{an}$ is coherent if $\mathscr{E}$ is. This operation gives a functor between the categories of coherent $\mathscr{O}_X$- and $\mathscr{O}_{X_{an}}$-modules. We will sometimes refer to a coherent $\mathscr{O}_X$-module (respectively $\mathscr{O}_{X_{an}}$-module) as a coherent algebraic (respectively analytic) sheaf.

For locally free sheaves, we can also describe this operation as follows. A locally free sheaf $\mathscr{E}$ is the sheaf of sections of an algebraic vector bundle $V \to X$. Then $\mathscr{E}_{an}$ is the sheaf of sections of the associated complex analytic vector bundle $V_{an} \to \mathbb{P}_{an}$. This applies, in particular, to the line bundles $\mathscr{O}_{\mathbb{P}^n}(d)$ and the bundle of p-forms $(\Omega^p_X)_{an} = \Omega^p_{X_{an}}$.

Lemma 15.4.1. *On a nonsingular algebraic variety X, the functor $\mathscr{E} \mapsto \mathscr{E}_{an}$ is exact and conservative (the last condition means that $\mathscr{E} \neq 0 \Rightarrow \mathscr{E}_{an} \neq 0$). It takes locally free sheaves to locally free sheaves of the same rank.*

Proof. The second statement follows from what we said above. It suffices to check the first statement at the stalk level, and this can then be reduced to some standard commutative algebra [8]. Given $p \in X$, let $R = \mathscr{O}_{X,p}$ denote the local ring. This is the localization of the ring of regular functions in a neighborhood of p at the maximal ideal m corresponding to this point. Choose a minimal set of generators $x_1, \ldots, x_n$ for m. Note that $n = \dim X$ because R is regular. The local ring $R_{an} = \mathscr{O}_{X_{an},p}$ is the ring of convergent power series $\mathbb{C}\{x_1, \ldots, x_n\}$. We can identify the m-adic completions of both rings with the ring of formal power series $\hat{R} = \mathbb{C}[[x_1, \ldots, x_n]]$. On stalks, the operation $\mathscr{E} \mapsto \mathscr{E}_{an}$ is given by extension of scalars $E \mapsto R_{an} \otimes_R E$. This operation is exact and conservative precisely when the ring extension $R \subset R_{an}$ is faithfully flat by definition. This can be reduced to the faithful flatness of the completions $R \subset \hat{R}$ and $R_{an} \subset \hat{R}$ [8, Chapter 10, Exercise 7] or [55, Chapter 0, 7.3.5]. □

Corollary 15.4.2. *Given an ideal sheaf $\mathscr{I} \subseteq \mathscr{O}_X$, the support of $\mathscr{O}_{X_{an}}/\mathscr{I}_{an}$ coincides with $V(\mathscr{I})$.*

We can now state the following remarkable GAGA theorem, of Serre, which compares the categories of coherent algebraic and analytic sheaves on projective space.

Theorem 15.4.3 (Serre). *The functor $\mathscr{E} \mapsto \mathscr{E}_{an}$ induces an equivalence between the categories of coherent $\mathscr{O}_{\mathbb{P}^n}$ and $\mathscr{O}_{\mathbb{P}^n_{an}}$-modules. In particular, any coherent $\mathscr{O}_{\mathbb{P}^n_{an}}$-module arises from an $\mathscr{O}_{\mathbb{P}^n}$-module, which is unique up to isomorphism.*

We will defer discussion of the proof until Section 16.4, but we note some important corollaries.

Corollary 15.4.4 (Chow's theorem). *Every complex submanifold of $\mathbb{P}^n$ is a nonsingular projective algebraic subvariety.*

Proof. The analytic ideal sheaf $\mathscr{J}$ of a submanifold X of $\mathbb{P}^n$ is a coherent $\mathscr{O}_{\mathbb{P}^n_{an}}$-module. Thus $\mathscr{J} = \mathscr{I}_{an}$, for some coherent submodule $\mathscr{I} \subset \mathscr{O}_{\mathbb{P}^n}$. The subvariety $V(\mathscr{I})$ defined by $\mathscr{I}$ is easily seen to coincide with X as a set. It is nonsingular by 2.5.15. □

A more elementary noncohomological proof of Chow's theorem can be found in [104, Chapter VIII §3]. Note that it implies the assertion we made some time ago:

Corollary 15.4.5. *Any compact Riemann surface X is a projective algebraic curve.*

Proof. This is a consequence of Chow's theorem, because X embeds into projective space by Corollary 6.3.9. □

Corollary 15.4.6. *A holomorphic map between nonsingular projective algebraic varieties is a morphism of varieties.*

Proof. Apply Chow's theorem to the graph of the map. □

Exercises

15.4.7. Show that a meromorphic function on a smooth projective curve is the same thing as a rational function. This is also true in higher dimensions, but the proof would be harder.

Chapter 16
Cohomology of Coherent Sheaves

In this chapter, we continue the study of coherent sheaves, by studying their cohomology. The first key result is that the higher cohomology groups for coherent sheaves vanish for affine schemes. Using this we can compute cohomology for projective spaces using the Čech complex for the standard open affine cover, and establish finite-dimensionality and other basic results. We also consider analogous statements for complex manifolds. With these results in hand, we complete our discussion of the GAGA theorems. The second basic result is that if $\mathscr{E}$ is a coherent algebraic sheaf on $\mathbb{P}^n_{\mathbb{C}}$, its cohomology is isomorphic to the cohomology of $\mathscr{E}_{an}$. Thus the calculation of the latter reduces to a purely algebraic problem. This cohomological result is also needed for the proof of the first GAGA theorem stated in the previous chapter.

16.1 Cohomology of Affine Schemes

Corollary 15.2.5 tells us that the global section functor for quasicoherent sheaves on affine schemes is exact. Since higher sheaf cohomology measures the obstruction to exactness of this functor, we might guess that there is no higher cohomology. This is correct, as will see shortly, but the result is not quite automatic. The key piece of algebra is the following.

Definition 16.1.1. Given a commutative ring R, an R-module I is injective if for any for injective map $N \to M$ of R-modules, the induced map $\mathrm{Hom}_R(M,I) \to \mathrm{Hom}_R(N,I)$ is surjective.

Example 16.1.2. If $R = \mathbb{Z}$, then I is injective, provided that it is divisible, i.e., if $a = nb$ has a solution for every $a \in I$ and $n \in \mathbb{Z} - 0$.

A standard result of algebra is the following.

Theorem 16.1.3 (Baer). *Every module is (isomorphic to) a submodule of an injective module.*

Proof. A proof can be found, for example, in [33, p. 627]. □

Proposition 16.1.4. *Let $X = \mathrm{Spec}_m R$ be an affine scheme. If I is an injective R-module, then $\tilde{I}$ is flasque.*

Proof. A complete proof can be found in [60, III, 3.4]. We indicate one step in the argument to illustrate the role of injectivity of I. We have to prove that $\tilde{I}(X) \to \tilde{I}(U)$ is surjective for any open U. We do this when $U = D(f)$. Then this amounts to the surjectivity of the canonical map $\kappa : I \to I[1/f]$. Let $J_r = \{g \in R \mid gf^r = 0\}$ denote the annihilator of f^r. We have $J_1 \subseteq J_2 \subseteq \cdots$. Since R is Noetherian by Hilbert's basis theorem [8], the sequence stabilizes. So there is an r such that $J_r = J_{r+1} = \cdots$. An element of $I[1/f]$ can be written as a fraction e/f^n with $e \in I$. We have to show that this lies in the image of κ. There is an isomorphism between the ideal $K = (f^{n+r})$ and R/J_{n+r}. Since the element $f^r e \in I$ is annihilated by $J_r = J_{n+r}$, we obtain a map $\phi : K \to I$ satisfying $\phi(f^{n+r}) = f^r e$. Since I is injective, ϕ extends to a map $\psi : R \to I$. Let $e' = \psi(1) \in I$. We have $f^{n+r} e' = f^r e$. Therefore $e/f^n = f^r e / f^{n+r} = e'$. □

Theorem 16.1.5 (Grothendieck–Serre). *Let R be an affine ring, and let $Y = \mathrm{Spec}_m R$. Then for any R-module M,*

$$H^i(Y, \tilde{M}) = 0.$$

Remark 16.1.6. The proof given here works for arbitrary Noetherian affine schemes. That is why we give the dual attribution.

Proof. The module M embeds into an injective module I. Let $N = I/M$. By Corollary 15.2.5, the functor $M \mapsto \tilde{M}$ is exact. Therefore we have an exact sequence of sheaves

$$0 \to \tilde{M} \to \tilde{I} \to \tilde{N} \to 0.$$

Since $H^i(Y, \tilde{I}) = 0$ for $i > 0$, we obtain

$$I \to N \to H^1(Y, \tilde{M}) \to 0$$

and

$$H^{i+1}(Y, \tilde{M}) \cong H^i(Y, \tilde{N}).$$

Since $I \to N$ is surjective, $H^1(Y, \tilde{M}) = 0$. This proves the theorem for $i = 1$. Applying this case to N shows that

$$H^2(Y, \tilde{M}) \cong H^1(Y, \tilde{N}) = 0.$$

We can kill all higher cohomology groups in the same fashion. □

We get a new proof of (a strengthened form of) Lemma 2.3.4.

Corollary 16.1.7. *If $X \subset Y$ is a closed subscheme of an affine scheme defined by an ideal sheaf $\mathscr{I}$, then $\mathscr{O}(X) \cong \mathscr{O}(Y)/\mathscr{I}(Y)$.*

Proof. The sequence

$$0 \to \mathscr{I}(Y) \to \mathscr{O}(Y) \to \mathscr{O}(X) \to 0$$

is exact. □

Corollary 16.1.8. *An open affine cover of a variety is Leray with respect to any coherent sheaf.*

Exercises

16.1.9. Let $\mathscr{U} = \{D(f_i)\}$ be a finite cover of $X = \mathrm{Spec}_m R$ with R affine. Prove that $\check{H}^1(\mathscr{U}, \mathscr{O}_X) = 0$ as follows.

(a) Show that for every $n > 0$ there exists $r_i \in R$ such that $\sum r_i f_i^n = 1$.
(b) Given an alternating cocycle $g_{ij} \in C^1_{alt}(\mathscr{U}, \mathscr{O}_X)$ (Exercise 7.3.8), choose n such that $f_i^n f_j^n g_{ij} \in R$ for all i, j. Show that

$$h_i = \sum_\ell r_\ell f_\ell^n g_{\ell i} \in R[1/f_i].$$

(c) Check that $g = \pm \partial h$.

16.1.10. Generalize the argument from the previous exercise to prove that $\check{H}^i(\mathscr{U}, \mathscr{O}_X) = 0$ for all $i > 0$.

16.2 Cohomology of Coherent Sheaves on $\mathbb{P}^n$

In this section, we do the important computation of the cohomology of line bundles over projective space. We use this to extract some more general statements regarding cohomology of coherent sheaves. Let us be clear about the notation. We will write $(\mathbb{P}^n = \mathbb{P}^n_k, \mathscr{O}_{\mathbb{P}^n})$ for projective space over a fixed field k, viewed as an algebraic variety. This convention applies to the case of $k = \mathbb{C}$ as well. We recall that the line bundles are given by $\mathscr{O}_{\mathbb{P}}(d)$ with $d \in \mathbb{Z}$. Let $S = k[x_0, \ldots, x_n]$ be the homogeneous coordinate ring. Let $S_i \subset S$ be the space of homogeneous polynomials of degree i. It is convenient to allow i to be negative, in which case $S_i = 0$. The degree of a ratio of homogeneous polynomials is the difference of the degrees of the numerator and denominator. Consequently, the localization $S[1/f]$ carries a $\mathbb{Z}$-grading for any homogeneous polynomial f.

Theorem 16.2.1 (Serre).

(a) $H^0(\mathbb{P}^n, \mathscr{O}_{\mathbb{P}}(d)) \cong S_d$.
(b) $H^i(\mathbb{P}^n, \mathscr{O}_{\mathbb{P}}(d)) = 0$ *if* $i \notin \{0, n\}$.
(c) $H^n(\mathbb{P}^n, \mathscr{O}_{\mathbb{P}}(d)) \cong S_{-d-n-1}$.

Proof. We use Čech cohomology with respect to the standard open affine cover $\{U_i\}$, where $U_i = \{x_i \neq 0\}$. This is a Leray cover by Corollary 16.1.8. Our first task is to identify the Čech complex with a complex of polynomials. As explained above, the localization

$$S\left[\frac{1}{x_i}\right] = \bigoplus_d S\left[\frac{1}{x_i}\right]_d \subset K = k(x_0,\ldots,x_n)$$

has a natural $\mathbb{Z}$-grading such that the degree-0 piece of $S[1/x_i]$ is exactly $k[x_0/x_i, \ldots, x_n/x_i] = \mathscr{O}(U_i)$; recall that U_i can be identified with affine n-space with coordinates

$$\frac{x_0}{x_i},\ldots,\frac{\widehat{x_i}}{x_i},\ldots,\frac{x_n}{x_i}.$$

Similar statements apply to localizations $S[1/x_{i_1}x_{i_2}\cdots]$ along monomials.

We start with the description of $\mathscr{O}(d)$ given in Section 7.3, where we identify a section of $\mathscr{O}(d)(U)$ with a collection of rational functions $f_i \in \mathscr{O}(U \cap U_i) \subset K$ satisfying

$$f_i = \left(\frac{x_j}{x_i}\right)^d f_j,$$

or equivalently

$$x_i^d f_i = x_j^{dd} f_j. \tag{16.2.1}$$

Now identify $\mathscr{O}(d)(U_i)$ with $S[1/x_i]_d$ under the bijection that sends f_i to $x_i^d f_i$. Equation (16.2.1) just says that these elements agree in K. By carrying out similar identifications for elements of $\mathscr{O}(d)(U_i \cap U_j)$ etc., we can realize the Čech complex for $\mathscr{O}(d)$ as the degree-d piece of the complex

$$\bigoplus_i S\left[\frac{1}{x_i}\right] \to \bigoplus_{i<j} S\left[\frac{1}{x_i x_j}\right] \to \cdots, \tag{16.2.2}$$

where the differentials are defined as alternating sums of the inclusions along the lines of Section 7.3.

An element of $H^0(\mathbb{P}^n, \mathscr{O}(d))$ can be represented by an $(n+1)$-tuple $(p_0/x_0^N, p_1/x_1^N, \ldots)$ of homogeneous rational expressions of degree d, where the p_i's are polynomials satisfying

$$\frac{p_0}{x_0^N} = \frac{p_1}{x_1^N} = \frac{p_2}{x_2^N} = \cdots.$$

This forces the polynomials p_i to be divisible by x_i^N. Thus $H^0(\mathbb{P}^n, \mathscr{O}(d))$ can be identified with S_d as claimed in (a).

From (16.2.2), $\bigoplus_d H^n(\mathbb{P}^n, \mathscr{O}(d))$ is isomorphic to $S[\frac{1}{x_0\cdots x_n}]$ modulo the the space of coboundaries

$$B_n = \sum_i S\left[\frac{1}{x_0\cdots\hat{x}_i\cdots x_n}\right].$$

$S[\frac{1}{x_0\cdots x_n}]$ is spanned by monomials $x_0^{i_0}\cdots x_n^{i_n}$ with arbitrary integer exponents. The image B is the space spanned by those monomials where at least one of the exponents is nonnegative. Therefore the quotient can be identified with the complementary submodule spanned by monomials with negative exponents. In particular,

$$H^n(\mathbb{P}^n,\mathscr{O}(d)) \cong \bigoplus_{\substack{i_0+\cdots+i_n=d\\ i_1,\ldots,i_n<0}} kx_0^{i_0}\cdots x_n^{i_n}.$$

This is isomorphic to S_{-d-n-1} via

$$x_0^{i_0}\cdots x_n^{i_n} \mapsto x_0^{-i_0-1}\cdots x_n^{-i_n-1}.$$

This proves (c).

It remains to prove (b). Part of this is easy. Since (16.2.2) has length n, $H^i(\mathbb{P}^n,\mathscr{O}(d))$ is automatically zero when $i>n$. For the case $0<i<n$, it is convenient to do all the degrees simultaneously by showing that $H^i(\mathbb{P}^n,\mathscr{F})=0$, where $\mathscr{F}=\bigoplus_d \mathscr{O}(d)$. The advantage here is that the sheaf $\mathscr{F}$ has the structure of a graded S-module. Thus $H^i(\mathbb{P}^n,\mathscr{F})$ inherits the structure of a graded S-module. We claim that $H^i(\mathbb{P}^n,\mathscr{F})[1/x_n]=0$ when $i>0$. Since localization is exact, this can be computed as the cohomology of the localization of (16.2.2) with respect to x_n. After localizing, we obtain the Čech complex for $\mathscr{F}$ with respect to the cover $\{U_i\cap U_n\}$ of U_n. This complex is necessarily acyclic, because $H^i(U_n,\mathscr{F})=0$ for $i>0$. Therefore the claim is proved, and it implies that every element of $H^i(\mathbb{P}^n,\mathscr{F})$ is annihilated by a power of x_n. So to prove (b), it suffices to prove that multiplication by x_n is bijective on $H^i(\mathbb{P}^n,\mathscr{F})$ for $0<i<n$. We do this by induction on n. Let H be the hyperplane defined by $x_n=0$. We can identify the ideal of H with $\mathscr{O}_{\mathbb{P}^n}(-1)$. Thus we have an exact sequence

$$0\to \mathscr{O}_{\mathbb{P}^n}(-1)\to \mathscr{O}_{\mathbb{P}^n}\to \mathscr{O}_H\to 0.$$

After tensoring this with $\mathscr{F}$, we obtain

$$0\to \mathscr{F}(-1)\to \mathscr{F}\to \mathscr{F}_H\to 0.$$

Therefore we get an exact sequence

$$\cdots\to H^i(\mathbb{P}^n,\mathscr{F}(-1))\xrightarrow{m} H^i(\mathbb{P}^n,\mathscr{F})\to H^i(H,\mathscr{F}_H)\to\cdots, \tag{16.2.3}$$

where $H^i(\mathbb{P}^n,\mathscr{F}(-1))\cong H^i(\mathbb{P}^n,\mathscr{F})$ with a shift in grading, and the arrow labeled m can be identified with multiplication by x_n. By induction, $H^i(H,\mathscr{F}_H)=0$ for $0<i<n-1$, which proves (b) for $1<i<n-1$. The remaining cases can be obtained by proving exactness of (16.2.3) at the two ends:

$$0\to H^0(\mathbb{P}^n,\mathscr{F}(-1))\xrightarrow{m} H^0(\mathbb{P}^n,\mathscr{F})\xrightarrow{\alpha} H^0(H,\mathscr{F}_H)\to 0,$$

$$0\to H^{n-1}(H,\mathscr{F}_H)\xrightarrow{\beta} H^n(\mathbb{P}^n,\mathscr{F}(-1))\xrightarrow{m} H^n(\mathbb{P}^n,\mathscr{F})\to 0.$$

This can be done by identifying these with

$$0 \to S \xrightarrow{x_n} S \xrightarrow{\alpha} k[x_0,\ldots,x_{n-1}] \to 0$$

and

$$0 \to S\left[\frac{1}{x_0\cdots x_{n-1}}\right]/B_{n-1} \xrightarrow{\beta} S\left[\frac{1}{x_0\cdots x_n}\right]/B_n \xrightarrow{x_n} S\left[\frac{1}{x_0\cdots x_n}\right]/B_n \to 0$$

respectively. □

We can now deduce finite-dimensionality of cohomology.

Theorem 16.2.2. *If $\mathscr{E}$ is a coherent sheaf on $\mathbb{P}^n$, then $H^i(\mathbb{P}^n,\mathscr{E})$ is finite-dimensional for each i, and is zero for $i>n$. Furthermore, there exist d_0 such that $H^i(\mathbb{P}^n,\mathscr{E}(d)) = 0$ for $d \geq d_0$ and $i>0$.*

Proof. We prove this by induction on r, where r is the length of the shortest syzygy, i.e., resolution of the type given by Theorem 15.3.9. If $r=0$, we are done by Theorem 16.2.1. Suppose that the theorem holds for all $r'<r$. Choose a resolution

$$0 \to \mathscr{E}_r \to \mathscr{E}_{r-1} \to \cdots \to \mathscr{E}_0 \to \mathscr{E} \to 0$$

and let $\mathscr{E}' = \ker[\mathscr{E}_0 \to \mathscr{E}]$. We have exact sequences

$$0 \to \mathscr{E}_r \to \mathscr{E}_{r-1} \to \cdots \to \mathscr{E}_1 \to \mathscr{E}' \to 0$$

and

$$0 \to \mathscr{E}' \to \mathscr{E}_0 \to \mathscr{E} \to 0.$$

Since $\mathscr{E}'$ has syzygy of length $r-1$, $H^*(\mathbb{P}^n,\mathscr{E}')$ is finite-dimensional by the induction hypothesis. Therefore the long exact sequence

$$\cdots \to H^i(\mathbb{P}^n,\mathscr{E}_0) \to H^i(\mathbb{P}^n,\mathscr{E}) \to H^{i+1}(\mathbb{P}^n,\mathscr{E}') \to \cdots$$

implies the finite-dimensionality of $H^i(\mathbb{P}^n,\mathscr{E})$. The vanishing of $H^i(\mathbb{P}^n,\mathscr{E}(d))$ for $i>0$ and $d \gg 0$ can be proved by induction in a similar manner. Details are left as an exercise. □

Corollary 16.2.3. *If $\mathscr{E}$ is a coherent sheaf on a projective variety X, then the cohomology groups $H^i(X,\mathscr{E})$ are finite-dimensional.*

Proof. Embed $\iota: X \hookrightarrow \mathbb{P}^n$. Then $\iota_*\mathscr{E}$ is a coherent $\mathscr{O}_{\mathbb{P}}$-module, so it has finite-dimensional cohomology. Since ι_* is exact,

$$H^i(X,\mathscr{E}) \cong H^i(\mathbb{P}^n,\iota_*\mathscr{E}).$$ □

The Euler characteristic of a sheaf $\mathscr{F}$ of k-modules on a space X is

$$\chi(\mathscr{F}) = \sum(-1)^i \dim_k H^i(X,\mathscr{F}),$$

provided that the sum is finite. From Lemma 11.3.2, we easily obtain the following result:

Lemma 16.2.4. *If* $0 \to \mathscr{F}_r \to \cdots \to \mathscr{F}_0 \to \mathscr{F} \to 0$ *is an exact sequence of sheaves with finite-dimensional cohomology, then*

$$\chi(\mathscr{F}) = \sum_j (-1)^j \chi(\mathscr{F}_j).$$

Proof. Break this up into short exact sequences and use induction. □

Theorem 16.2.5. *If* $\mathscr{E}$ *is a coherent sheaf on* $\mathbb{P}^n_k$*, then* $i \mapsto \chi(\mathscr{E}(i))$ *is a polynomial in* i *of degree at most* n*. If* X *is a subvariety of* $\mathbb{P}^n_k$ *and* $\mathscr{E} = \mathscr{O}_X$*, then this polynomial has degree* $\dim X$.

Proof. From Theorem 16.2.1,

$$\chi(\mathscr{O}_{\mathbb{P}^n}(i)) = \begin{cases} \dim S_i & \text{if } i \geq 0, \\ (-1)^n \dim S_{-d-n-1} & \text{otherwise}, \end{cases}$$

so that

$$\chi(\mathscr{O}_{\mathbb{P}^n}(i)) = \binom{n+i}{n}, \tag{16.2.4}$$

which is a polynomial of degree n. This implies that $\chi(\mathscr{E}(i))$ is a polynomial of degree $\leq n$ when $\mathscr{E}$ is a sum of line bundles $\mathscr{O}_{\mathbb{P}^n}(j)$. In general, Theorem 15.3.9 and the above corollary imply that for any coherent $\mathscr{E}$,

$$\chi(\mathscr{E}(i)) = \sum_{j=0}^{r} (-1)^j \chi(\mathscr{E}_j(i)),$$

where $\mathscr{E}_j$ are sums of line bundles. This shows that $\chi(\mathscr{E}(i))$ is a polynomial with degree at most n.

To get the sharper bound on the degree when $\mathscr{E} = \mathscr{O}_X$, we use induction on $\dim X$ and the relation

$$\chi(\mathscr{O}_{H\cap X}(i)) = \chi(\mathscr{O}_X(i)) - \chi(\mathscr{O}_X(i-1)),$$

which follows from the sequence

$$0 \to \mathscr{O}_X(-1) \to \mathscr{O}_X \to \mathscr{O}_{H\cap X} \to 0,$$

where H is a general hyperplane. □

The polynomial $\chi(\mathscr{E}(i))$ is called the *Hilbert polynomial* of $\mathscr{E}$. It has an elementary algebraic interpretation:

Corollary 16.2.6 (Hilbert). *Let* M *be a finitely generated graded* S*-module. Then* $i \mapsto \dim M_i$ *is a polynomial for* $i \gg 0$*, and this coincides with the Hilbert polynomial* $\chi(\tilde{M}(i))$.

Proof. This is obtained by combining the last theorem with Theorems 15.3.8 and 16.2.2. □

Exercises

16.2.7. Compute $H^i(\mathbb{P}^1, \mathcal{O}_{\mathbb{P}^1}(d))$ directly using Mayer–Vietoris with respect to $\{U_0, U_1\}$ without referring to the above arguments.

16.2.8. Give a direct proof that $H^1(\mathbb{P}^2, \mathcal{O}_{\mathbb{P}^2}(d)) = 0$ by solving the equation

$$\frac{p_{ij}}{(x_i x_j)^n} = \frac{q_i}{x_i^n} - \frac{q_j}{x_j^n}$$

for any 1-cocycle $\{p_{ij}/(x_i x_j)^n\} \in C^1(\{U_i\}, \mathcal{O}(d))$ on $\mathbb{P}^2$.

16.2.9. Let f be a homogeneous polynomial of degree d in $S = k[x_0, x_1, x_2]$. Then corresponding to the exact sequence of graded modules

$$0 \to S(-d) \cong Sf \to S \to S/(f) \to 0$$

there is an exact sequence of sheaves

$$0 \to \mathcal{O}_{\mathbb{P}^2}(-d) \to \mathcal{O}_{\mathbb{P}^2} \to \mathcal{O}_X \to 0.$$

Prove that

$$\dim H^1(X, \mathcal{O}_X) = \frac{(d-1)(d-2)}{2}.$$

16.2.10. Let f be a homogeneous polynomial of degree d in four variables. Repeat the above exercise for $H^1(X, \mathcal{O}_X)$ and $H^2(X, \mathcal{O}_X)$.

16.2.11. Prove the remaining parts of Theorem 16.2.2.

16.2.12. If $\mathcal{E}$ is a nonzero coherent sheaf on $\mathbb{P}^n$, prove that the degree of $\chi(\mathcal{E}(i))$ coincides with the dimension of its support $\mathrm{supp}(\mathcal{E}) = \{x \in \mathbb{P}^n \mid \mathcal{E}_x \neq 0\}$.

16.3 Cohomology of Analytic Sheaves

As a preparation for the next section, we summarize some analogues of earlier results in the analytic setting. We start with *Stein* manifolds, which can be regarded as analogues of affine varieties. The characteristic feature of these manifolds is the abundance of global holomorphic functions. A complex manifold X is called Stein if the following three conditions hold:

(1) Global homolomorphic functions separate points, i.e., given $x_1 \neq x_2$, there exists $f \in \mathcal{O}(X)$ such that $f(x_1) \neq f(x_2)$.

(2) Global holomorphic functions separate tangent vectors, i.e., for any $x_0 \in X$, there exist global holomorphic functions $f_1, \dots, f_n$ that generate the maximal ideal at x_0.
(3) X is holomorphically convex, i.e., for any sequence $x_i \in X$ without accumulation points, there exists $f \in \mathscr{O}(X)$ such that $|f(x_i)| \to \infty$.

The third condition is a bit less intuitive than the others. It is best illustrated by an example where it does not hold. Hartogs' theorem [66] tells us that any holomorphic function on $X = \mathbb{C}^2 - \{0\}$ extends to $\mathbb{C}^2$. Thus (3) will fail for any sequence $x_i \to 0$. The first two conditions do hold for this X. Here are a few key examples:

Example 16.3.1. The ball in $\mathbb{C}^n$ is Stein.

Example 16.3.2. Closed submanifolds of $\mathbb{C}^n$ (hence in particular, nonsingular affine varieties) are Stein.

The following is an analogue of Theorem 16.1.5, although it predates it.

Theorem 16.3.3 (Cartan's Theorem B). *Let $\mathscr{E}$ be a coherent sheaf on a Stein manifold. Then $H^i(X, \mathscr{E}) = 0$ for all $i > 0$.*

Proof. See [66, Theorem 7.4.3] □

Corollary 16.3.4. *An open cover $\{U_i\}$ is Leray for a coherent analytic sheaf if each intersection $U_{i_1 \dots i_r}$ is Stein. In particular, this applies to open affine covers of nonsingular varieties.*

We also have the following finiteness theorem, which is parallel to Corollary 16.2.3.

Theorem 16.3.5 (Cartan–Serre). *If $\mathscr{E}$ is a coherent sheaf on a compact complex manifold X, then the cohomology groups $H^i(X, \mathscr{E})$ are finite-dimensional.*

Proof. A proof can be found in [18, 47] in general. When $\mathscr{E}$ is a vector bundle, a proof due to Kodaira can be given using Hodge theory for the $\bar{\partial}$-operator [49, 120].
□

Exercises

16.3.6. Check that a product of Stein manifolds is Stein.

16.3.7. Check that $\mathbb{C}^n$ is Stein.

16.3.8. Check that a closed submanifold of a Stein manifold is Stein.

16.3.9. Prove the converse to Cartan's Theorem B. Hint: Consider exact sequences $0 \to \mathscr{I} \to \mathscr{O}_X \to \mathscr{O}_X/\mathscr{I} \to 0$ for appropriate ideal sheaves.

16.4 GAGA, Part II

We complete our discussion of the GAGA theorems. So we assume $k = \mathbb{C}$. We already stated Theorem 15.4.3, which gives an equivalence between the categories of coherent algebraic and analytic sheaves on projective space. This will be deduced from another GAGA theorem that gives comparison of their cohomologies.

Theorem 16.4.1 (Serre). *Let $\mathscr{E}$ be a coherent $\mathscr{O}_{\mathbb{P}^n}$-module. Then there is an isomorphism*

$$H^i(\mathbb{P}^n, \mathscr{E}) \cong H^i(\mathbb{P}^n_{an}, \mathscr{E}_{an}).$$

We start with a special case.

Proposition 16.4.2. *The map*

$$\iota^* : H^i(\mathbb{P}^n, \mathscr{O}_{\mathbb{P}^n}(d)) \to H^i(\mathbb{P}^n_{an}, \mathscr{O}_{\mathbb{P}^n_{an}}(d))$$

is an isomorphism for all n and d.

Proof. We first check the result for $d = 0$. We have that $H^i(\mathbb{P}^n_{an}, \mathscr{O}_{\mathbb{P}^n,an}) = H^{(0,i)}(\mathbb{P}^n_{an})$. From Theorem 7.2.2, it follows that $H^i(\mathbb{P}^n_{an}, \mathbb{C})$ is zero if i is odd, and generated by the fundamental class of a linear subspace $L \subset \mathbb{P}^n$ otherwise. It follows that the Hodge structure on $H^i(\mathbb{P}^n_{an}, \mathbb{C})$ is of type $(i/2, i/2)$ when i is even. Therefore $H^{(0,i)}(X)$ is $\mathbb{C}$ if $i = 0$ and zero otherwise.

Let $H \subset \mathbb{P}^n$ be a hyperplane. By induction on n, we can assume that the result holds for $H \cong \mathbb{P}^{n-1}$. We have an exact sequence

$$0 \to \mathscr{O}_{\mathbb{P}^n}(d-1) \to \mathscr{O}_{\mathbb{P}^n}(d) \to \mathscr{O}_H(d) \to 0,$$

which yields a diagram

$$\begin{array}{ccccccc} \cdots & H^i(\mathscr{O}_{\mathbb{P}}(d-1)) & \to & H^i(\mathscr{O}_{\mathbb{P}}(d)) & \to & H^i(\mathscr{O}_H(d)) & \cdots \\ & \downarrow & & \downarrow & & \downarrow & \\ & H^i(\mathscr{O}_{\mathbb{P}_{an}}(d-1)) & \to & H^i(\mathscr{O}_{\mathbb{P}_{an}}(d)) & \to & H^i(\mathscr{O}_{H_{an}}(d)) & \end{array}$$

By the induction assumption and the five lemma, we see that the vertical map ι^* is an isomorphism for $\mathscr{O}_{\mathbb{P}^n}(d)$ if and only if it is for $\mathscr{O}_{\mathbb{P}^n}(d-1)$. Since the isomorphism holds for $d = 0$ by the previous step, it holds for all d. □

We can now prove Theorem 16.4.1.

Proof. By Theorem 15.3.9, there is a resolution

$$0 \to \mathscr{E}_r \to \mathscr{E}_{r-1} \to \cdots \to \mathscr{E}_0 \to \mathscr{E} \to 0,$$

where $r \le n$ and each $\mathscr{E}_i$ is a sum of a finite number of line bundles $\mathscr{O}_{\mathbb{P}^n}(j)$. Let $r(\mathscr{E})$ denote the length of the shortest resolution of this type. Proposition 16.4.2 implies the theorem when $r(\mathscr{E}) = 0$. If $r(\mathscr{E}) > 0$, then we can find an exact sequence

$$0 \to \mathscr{R} \to \mathscr{F} \to \mathscr{E} \to 0,$$

where $\mathscr{F}$ is a direct sum of line bundles and $r(\mathscr{R}) < r(\mathscr{E})$. Then the theorem holds for $\mathscr{R}$ and $\mathscr{F}$ by induction on the length. Therefore it holds for $\mathscr{E}$ by the 5-lemma. □

The GAGA theorem can fail for nonprojective varieties. For example, $H^0(\mathscr{O}_{\mathbb{C}^n_{an}})$ is the space of holomorphic functions on $\mathbb{C}^n$ which is much bigger than the space $H^0(\mathscr{O}_{\mathbb{C}^n})$ of polynomials.

Corollary 16.4.3. *If X is a smooth projective algebraic variety, then* $\dim H^q(X, \Omega^p_X)$ *coincides with the Hodge number h^{pq} of the manifold X_{an}.*

We are now ready to outline a proof of Theorem 15.4.3.

Proof. One part of the theorem follows immediately from the previous theorem. Suppose that $\mathscr{E}, \mathscr{F}$ are coherent algebraic on $\mathbb{P}^n$. Then we claim that $\mathrm{Hom}(\mathscr{E}, \mathscr{F}) \cong \mathrm{Hom}(\mathscr{E}_{an}, \mathscr{F}_{an})$. The left- and right-hand sides are the spaces of global sections of $\mathscr{H}om(\mathscr{F}, \mathscr{E})$ and

$$\mathscr{H}om(\mathscr{E}, \mathscr{F})_{an} = \mathscr{H}om(\mathscr{E}_{an}, \mathscr{F}_{an}),$$

respectively. These spaces of global sections are isomorphic by Theorem 16.4.1.

In order to make the remainder of the proof clear, we will break it up into a series of assertions. The last assertion is precisely what we need to obtain the theorem:

(A_n) *Given a coherent analytic sheaf $\mathscr{E}$ on $\mathbb{P}^n$, there exists a constant d_0 such that $\mathscr{E}(d)$ is generated by global sections for $d \geq d_0$.*
(B_n) *Given a coherent analytic sheaf $\mathscr{E}$ on $\mathbb{P}^n$, there exists a constant d_0 such that $H^i(\mathbb{P}^n, \mathscr{E}(d)) = 0$ for $i > 0$ and $d \geq d_0$.*
(C_n) *Every coherent analytic sheaf $\mathscr{E}$ on $\mathbb{P}^n$ arises from a coherent algebraic sheaf.*

The logic of the proof is as follows:

$$A_{n-1} \& B_{n-1} \Rightarrow A_n, \quad A_n \Rightarrow B_n, \quad A_n \Rightarrow C_n.$$

We carry out these steps below.

A: We assume for simplicity that $\mathscr{E}$ is a torsion-free coherent analytic sheaf on $\mathbb{P}^n$. (See the exercises for the general case). Given $x \in \mathbb{P}^n$, choose a hyperplane H passing through x. Consider the sequence

$$0 \to \mathscr{O}_{\mathbb{P}^n}(d-1) \to \mathscr{O}_{\mathbb{P}^n}(d) \to \mathscr{O}_H \to 0.$$

Tensoring this with $\mathscr{E}$ yields

$$0 \to \mathscr{E}(d-1) \to \mathscr{E}(d) \to \mathscr{E}|_H(d) \to 0$$

(the injectivity of the first map follows from the torsion-freeness of $\mathscr{E}$). By induction, we may assume assertion (B) for H. Therefore we have that

$H^1(\mathscr{E}|_H(d)) = 0$ for $d \geq d_1$. Therefore, there is a d_1 such that for $d \geq d_1$ we get a sequence of surjections

$$H^1(\mathscr{E}(d-1)) \twoheadrightarrow H^1(\mathscr{E}(d)) \twoheadrightarrow H^1(\mathscr{E}(d+1)) \twoheadrightarrow \cdots.$$

Since these spaces are finite-dimensional by Theorem 16.3.5, these maps must stabilize to isomorphisms for all $d \geq d_2$ for some $d_2 \geq d_1$. Thus

$$H^0(\mathscr{E}(d)) \to H^0(\mathscr{E}|_H(d))$$

is surjective for $d \geq d_2$. By induction, we can assume that $\mathscr{E}|_H(d)$ is generated by global sections $H^0(\mathscr{E}(d)|_H)$ for $d \geq d_0 \geq d_2$. These can be lifted to sections $H^0(\mathscr{E}(d))$, which will span the stalk $\mathscr{E}(d)_x$. This proves (A_n).

B: (B_n) will be proved by descending induction on i. To start the process, observe that $H^i(\mathbb{P}^n, \mathscr{E}(d))$ can be computed using the Čech complex with respect to the standard cover by Corollary 16.3.4. Therefore $H^i(\mathbb{P}^n, \mathscr{E}(d)) = 0$ for $i > n$ and any d. From (A_n), we have that $\mathscr{E}(a)$ is generated by global sections $s_1, \ldots, s_N$ for some $a \gg 0$. These sections can be viewed as maps $\mathscr{O}(-a) \to \mathscr{E}$. Thus we have an exact sequence

$$0 \to \mathscr{R} \to \mathscr{F} \xrightarrow{\sum s_j} \mathscr{E} \to 0,$$

where $\mathscr{F} = \mathscr{O}(-a)^N$, and $\mathscr{R}$ is the kernel. Since $\mathscr{F}$ is a direct sum of line bundles, by Theorems 16.4.1 and 16.2.1, we have $H^i(\mathscr{F}(d)) = 0$ for $i > 0$ and $d \gg 0$. Therefore

$$H^i(\mathscr{E}(d)) \cong H^{i+1}(\mathscr{R}(d)).$$

Thus this step follows by descending induction on i.

C: Since $\mathscr{E}(d)$ is globally generated for $d \gg 0$ by (A_n), we can argue as in B to show that $\mathscr{E}$ fits into an exact sequence

$$0 \to \mathscr{R} \to \mathscr{F}_2 \to \mathscr{E} \to 0,$$

where $\mathscr{F}_2$ is a direct sum of line bundles. Repeating this yields an epimorphism

$$\mathscr{F}_1 \to \mathscr{R} \to 0,$$

where $\mathscr{F}_1$ is a direct sum of line bundles. Thus we obtain a presentation

$$\mathscr{F}_1 \xrightarrow{f} \mathscr{F}_2 \to \mathscr{E} \to 0.$$

In particular, note that the $\mathscr{F}_i$ are algebraic. It follows from the previous step that f is a morphism of the underlying algebraic sheaves. Consequently, $\mathscr{E}$ is the cokernel of an algebraic map, and hence also algebraic. □

Exercises

16.4.4. If X is a smooth projective variety, show that $H^1(X, \mathcal{O}_X^*) \cong H^1(X_{an}, \mathcal{O}_{X_{an}}^*)$.

16.4.5. Consider the setup in as step A of the proof of Theorem 15.4.3, but without the torsion-freeness assumption for $\mathcal{E}$. Let $\mathcal{K} = \ker[\mathcal{E}(-1) \to \mathcal{E}]$, where the map is given by tensoring $\mathcal{E}$ with the inclusion of the ideal sheaf of H.

(a) Show that K is supported on H. Thus we can apply the induction assumption to conclude that $H^i(\mathcal{K}(d)) = 0$ for $i > 0$ and $d \gg 0$.
(b) By breaking up

$$0 \to \mathcal{K}(d) \to \mathcal{E}(d-1) \to \mathcal{E}(d) \to \mathcal{E}(d)|_H \to 0$$

into short exact sequences, show that $H^0(\mathcal{E}(d)) \to H^0(\mathcal{E}(d)|_H)$ is surjective for $d \gg 0$.

Chapter 17
Computation of Some Hodge Numbers

The Hodge numbers of a smooth projective algebraic variety are very useful invariants. By Hodge theory, these determine the Betti numbers. In this chapter, we turn to the practical matter of actually computing these for a number of examples such as projective spaces, hypersurfaces, and double covers. The GAGA theorem, Theorem 16.4.1, allows us to do this by working in the algebraic setting, where we may employ some of the tools developed in the earlier chapters.

17.1 Hodge Numbers of $\mathbb{P}^n$

Let $S = k[x_0,\dots,x_n]$ and $\mathbb{P} = \mathbb{P}^n_k$ for some field k. We first need to determine the sheaf of differentials.

Proposition 17.1.1. *There is an exact sequence*

$$0 \to \Omega^1_{\mathbb{P}} \to \mathscr{O}_{\mathbb{P}}(-1)^{n+1} \to \mathscr{O}_{\mathbb{P}} \to 0.$$

Proof. Let $\Omega_S = \oplus S dx_i \cong S^{n+1}$ be the module of Kähler differentials of S. Construct the graded S-module

$$M = \Gamma_*(\Omega^1_{\mathbb{P}}) = \Gamma(\mathbb{A}^{n+1} - \{0\}, \pi^*\Omega^1_{\mathbb{P}}),$$

where $\pi : \mathbb{A}^{n+1} - \{0\} \to \mathbb{P}$ is the projection. This can be realized as the submodule Ω_S consisting of those forms that annhilate the tangent spaces of the fibers of π. The tangent space of the fiber over $[x_0,\dots,x_n]$ is generated by the Euler vector field $\sum x_i \frac{\partial}{\partial x_i}$. Thus a 1-form $\sum f_i dx_i$ lies in M if and only if $\sum f_i x_i = 0$.

Next, we have to check the gradings. Ω_S has a grading such that the dx_i lie in degree 0. Under the natural grading of $M = \Gamma_*(\Omega^1_{\mathbb{P}})$, sections of $\Omega^1_{\mathbb{P}}(U_i)$ that are generated by

$$d\left(\frac{x_j}{x_i}\right) = \frac{x_i dx_j - x_j dx_i}{x_i^2}$$

should have degree 0. Thus the gradings on Ω_S and M are off by a shift of one.

To conclude, we have an exact sequence of graded modules

$$0 \to M \to \Omega_S(-1) \to m \to 0, \tag{17.1.1}$$

where $m = (x_0, \ldots, x_n)$ and the first map sends dx_i to x_i. Since $m \sim S$, (17.1.1) implies the result. □

Dualizing yields the sequence of Example 15.3.10.

Proposition 17.1.2. *Suppose we are given an exact sequence of locally free sheaves*

$$0 \to \mathscr{A} \to \mathscr{B} \to \mathscr{C} \to 0.$$

If $\mathscr{A}$ *has rank one, then*

$$0 \to \mathscr{A} \otimes \wedge^{p-1}\mathscr{C} \to \wedge^p \mathscr{B} \to \wedge^p \mathscr{C} \to 0$$

is exact for any $p \geq 1$. *If* $\mathscr{C}$ *has rank one, then*

$$0 \to \wedge^p \mathscr{A} \to \wedge^p \mathscr{B} \to \wedge^{p-1}\mathscr{A} \otimes \mathscr{C} \to 0$$

is exact for any $p \geq 1$.

Proof. We prove the first statement, where $\operatorname{rank}(\mathscr{A}) = 1$, by induction, leaving the second as an exercise. When $p = 1$ we have the original sequence. In general, the maps in the putative exact sequence need to be explained. The last map $\lambda : \wedge^p \mathscr{B} \to \wedge^p \mathscr{C}$ is the natural one. The multiplication $\mathscr{B} \otimes \wedge^{p-1}\mathscr{B} \to \wedge^p \mathscr{B}$ restricts to give $\mu : \mathscr{A} \otimes \wedge^{p-1}\mathscr{B} \to \wedge^p \mathscr{B}$. We claim that μ factors through a map $\alpha : \mathscr{A} \otimes \wedge^{p-1}\mathscr{C} \to \wedge^p \mathscr{B}$. For this we can, by induction, appeal to the exactness of

$$0 \to \mathscr{A} \otimes \wedge^{p-2}\mathscr{C} \to \wedge^{p-1}\mathscr{B} \to \wedge^{p-1}\mathscr{C} \to 0.$$

Since $\mathscr{A}$ has rank one,

$$\mu|_{\mathscr{A} \otimes (\mathscr{A} \otimes \wedge^{p-2}\mathscr{C})} = 0,$$

so μ factors as claimed. Therefore all the maps in the sequence are defined.

Exactness can be checked on stalks. For this the sheaves can be replaced by free modules. Let $\{b_0, b_1, \ldots\}$ be a basis for $\mathscr{B}$ with b_0 spanning $\mathscr{A}$. Then the images $\bar{b}_1, \bar{b}_2, \ldots$ give a basis for $\mathscr{C}$. Then the above maps are given by

$$\alpha : b_0 \otimes \bar{b}_{i_1} \wedge \cdots \wedge \bar{b}_{i_{p-1}} \mapsto b_0 \otimes b_{i_1} \wedge \cdots \wedge b_{i_{p-1}},$$
$$\lambda : b_{i_1} \wedge \cdots \wedge b_{i_p} \mapsto \bar{b}_{i_1} \wedge \cdots \wedge \bar{b}_{i_p},$$

where $\cdots > i_2 > i_1 > 0$. The exactness is immediate. □

Corollary 17.1.3. *There is an exact sequence*

$$0 \to \Omega^p_{\mathbb{P}} \to \mathcal{O}_{\mathbb{P}}(-p)^{\binom{n+1}{p}} \to \Omega^{p-1}_{\mathbb{P}} \to 0$$

and in particular,

$$\Omega^n_{\mathbb{P}} \cong \mathcal{O}_{\mathbb{P}}(-n-1).$$

Proof. This follows from the above proposition and Proposition 17.1.1, together with the isomorphism

$$\wedge^p[\mathcal{O}_{\mathbb{P}}(-1)^{n+1}] \cong \mathcal{O}_{\mathbb{P}}(-p)^{\binom{n+1}{p}}.$$ □

This corollary can be understood from another point of view. Using the notation introduced in the proof of Proposition 17.1.1, we can extend the map $\Omega_S(-1) \to m$ to an exact sequence

$$0 \to [\wedge^{n+1}\Omega_S](-n-1) \xrightarrow{\delta} \cdots \to [\wedge^2\Omega_S](-2) \xrightarrow{\delta} \Omega_S(-1) \xrightarrow{\delta} m \to 0, \quad (17.1.2)$$

where

$$\delta(dx_{i_1} \wedge \cdots \wedge dx_{i_p}) = \sum (-1)^p x_{i_j} dx_{i_1} \wedge \cdots \wedge \hat{dx}_{i_j} \wedge \cdots \wedge dx_{i_p}$$

is contraction with the Euler vector field. The sequence is called the Koszul complex, and it is one of the basic workhorses of homological algebra [33, Chapter 17]. The associated sequence of sheaves is

$$0 \to [\wedge^{n+1}\mathcal{O}^{n+1}_{\mathbb{P}}](-n-1) \to \cdots \to [\wedge^2\mathcal{O}^{n+1}_{\mathbb{P}}](-2) \to [\mathcal{O}^{n+1}_{\mathbb{P}}](-1) \to \mathcal{O}_{\mathbb{P}} \to 0.$$

If we break this up into short exact sequences, then we obtain exactly the sequences in Corollary 17.1.3.

Proposition 17.1.4.

$$H^q(\mathbb{P}, \Omega^p_{\mathbb{P}}) = \begin{cases} k & \text{if } p = q \leq n, \\ 0 & \text{otherwise.} \end{cases}$$

Proof. When $p = 0$, this follows from Theorem 16.2.1. In general, the same theorem together with Corollary 17.1.3 implies

$$H^q(\Omega^p_{\mathbb{P}}) \cong H^{q-1}(\Omega^{p-1}_{\mathbb{P}}).$$

Therefore, we get the result by induction. □

When $k = \mathbb{C}$, this gives a new proof of the formula for Betti numbers of $\mathbb{P}^n$ given in Section 7.2. By a somewhat more involved induction we can obtain the following theorem of Raoul Bott:

Theorem 17.1.5 (Bott). $H^q(\mathbb{P}^n, \Omega^p_{\mathbb{P}^n}(r)) = 0$ *unless*

(a) $p = q$, $r = 0$,
(b) $q = 0$, $r > p$,
(c) or $q = n$, $r < -n + p$.

Proof. A complete proof will be left for the exercises. We give the proof for $p \leq 1$. For $p = 0$, this is a consequence of Theorem 16.2.1. We now turn to $p = 1$. Corollary 17.1.3 implies that

$$H^{q-1}(\mathscr{O}(r)) \to H^q(\Omega^1(r)) \to H^q(\mathscr{O}(r-1))^e \tag{17.1.3}$$

is exact. We use constants $e, e', \dots$ for exponents whose exact values are immaterial for the argument. The sequence (17.1.3), along with Theorem 16.2.1, forces $H^q(\Omega^1(r)) = 0$ in the following four cases: $q = 0, r < 1$; $q = 1, r < 0$; $1 < q < n$; $q = n, r \geq -n+1$. The remaining cases are $q = 0, r = 1$ and $q = 1, r > 0$. The trick is to apply Corollary 17.1.3 with other values of p. This yields exact sequences

$$H^q(\mathscr{O}(r-2))^e \to H^q(\Omega^1(r)) \to H^{q+1}(\Omega^2(r)) \to H^{q+1}(\mathscr{O}(r-2))^e,$$
$$H^{q+1}(\mathscr{O}(r-3))^{e'} \to H^{q+1}(\Omega^2(r)) \to H^{q+2}(\Omega^3(r)) \to H^{q+2}(\mathscr{O}(r-3))^{e'},$$
$$\dots$$

leading to isomorphisms

$$H^0(\Omega^1(1)) \cong H^1(\Omega^2(1)) \cong \cdots \cong H^{n-1}(\Omega^n(1)) = H^{n-1}(\mathscr{O}(-n)) = 0.$$

Likewise, for $r > 0$,

$$H^1(\Omega^1(r)) \cong H^2(\Omega^2(r)) \cong \cdots \cong H^n(\Omega^n(r)) = H^n(\mathscr{O}(-n-1+r)) = 0. \qquad \square$$

Exercises

17.1.6. Finish the proof of Proposition 17.1.2.

17.1.7. Given an exact sequence $0 \to \mathscr{A} \to \mathscr{B} \to \mathscr{C} \to 0$ of locally free sheaves, prove that the top exterior power $\det \mathscr{B}$ is isomorphic to $(\det \mathscr{A}) \otimes (\det \mathscr{C})$. Use this to rederive the formula for $\Omega^n_{\mathbb{P}^n}$.

17.1.8. Finish the proof of Theorem 17.1.5.

17.2 Hodge Numbers of a Hypersurface

We now let $X \subset \mathbb{P} = \mathbb{P}^n_k$ be a nonsingular hypersurface defined by a degree-d polynomial.

Proposition 17.2.1. *The restriction map*

$$H^q(\mathbb{P}^n, \Omega^p_{\mathbb{P}}) \to H^q(X, \Omega^p_X)$$

is an isomorphism when $p+q < n-1$.

We give two proofs, one now, over $\mathbb{C}$, and another later for general k.

Proof. Let $k = \mathbb{C}$. The weak Lefschetz theorem, Theorem 14.3.1, implies that the restriction map $H^i(X_{an}, \mathbb{C}) \to H^i(Y_{an}, \mathbb{C})$ is an isomorphism for $i < n-1$. The proposition is a consequence of this together with the the canonical Hodge decomposition (Theorem 12.2.4) and GAGA (Theorem 16.4.1). □

As a corollary, we can calculate many of the Hodge numbers of X.

Corollary 17.2.2. *The Hodge numbers* $h^{pq}(X)$ *equal* δ_{pq}, *where* δ_{pq} *is the Kronecker symbol, when* $n-1 \neq p+q < 2n-2$.

Proof. We give a proof when $k = \mathbb{C}$. For $p+q < n-1$, this follows from the above proposition and Proposition 17.1.4. For $p+q > n-1$, this follows from GAGA and Corollary 10.2.3. □

We prepare for the second proof by establishing a few key lemmas.

Lemma 17.2.3. *There is an exact sequence*

$$0 \to \Omega^p_{\mathbb{P}}(-d) \to \Omega^p_{\mathbb{P}} \to \Omega^p_{\mathbb{P}}|_X \to 0.$$

(Recall that $\Omega^p_{\mathbb{P}}|_X$ *is shorthand for* $i_* i^* \Omega^p_{\mathbb{P}}$, *where* $i : X \to \mathbb{P}$ *is the inclusion.)*

Proof. Tensor

$$0 \to \mathcal{O}_{\mathbb{P}}(-d) \to \mathcal{O}_{\mathbb{P}} \to \mathcal{O}_X \to 0$$

with $\Omega^p_{\mathbb{P}}$ to get

$$\begin{array}{ccccccccc}
0 & \longrightarrow & \Omega^p_{\mathbb{P}} \otimes \mathcal{O}_{\mathbb{P}}(-d) & \longrightarrow & \Omega^p_{\mathbb{P}} \otimes \mathcal{O}_{\mathbb{P}} & \longrightarrow & \Omega^p_{\mathbb{P}} \otimes \mathcal{O}_X & \longrightarrow & 0 \\
 & & \downarrow \cong & & \downarrow \cong & & \downarrow \cong & & \\
0 & \longrightarrow & \Omega^p_{\mathbb{P}}(-d) & \longrightarrow & \Omega^p_{\mathbb{P}} & \longrightarrow & \Omega^p_{\mathbb{P}}|_X & \longrightarrow & 0
\end{array}$$

For the last isomorphism, it is simply a matter of expanding the notation. Observe that if $\iota : X \to \mathbb{P}$ is the inclusion, then

$$\Omega^p_{\mathbb{P}}|_X = \iota_* \iota^* \Omega^p_{\mathbb{P}} = \iota_*(\iota^{-1} \Omega^p_{\mathbb{P}} \otimes \mathcal{O}_X) \cong \Omega^p_{\mathbb{P}} \otimes \mathcal{O}_X. \qquad \square$$

Lemma 17.2.4. $0 \to \mathcal{O}_X(-d) \to \Omega^1_{\mathbb{P}}|_X \to \Omega^1_X \to 0$ *is exact.*

Proof. We have a natural epimorphism $\Omega^1_{\mathbb{P}}|_X \to \Omega^1_X$ corresponding to restriction of 1-forms. We just have to determine the kernel. Let f be a defining polynomial of X, and let $M = \Gamma_*(\Omega^1_{\mathbb{P}})$ and $\overline{M} = \Gamma_*(\Omega^1_X)$. We embed M as a submodule of $\Omega_S(-1)$ as in the proof of Proposition 17.1.1. In particular, the symbols dx_i have degree 1. Then $\ker[M/fM \to \overline{M}]$ is a free $S/(f)$-module generated by

$$df = \sum_i \frac{\partial f}{\partial x_i} dx_i.$$

Thus it is isomorphic to $S/(f)(-d)$. □

Corollary 17.2.5. $0 \to \Omega^{p-1}_X(-d) \to \Omega^p_{\mathbb{P}}|_X \to \Omega^p_X \to 0$.

Proof. Apply Proposition 17.1.2 to the lemma. □

For the second proof of Proposition 17.2.1, it is convenient to prove something stronger.

Proposition 17.2.6. *If $p+q < n-1$, then*

$$H^q(X, \Omega^p_X(-r)) = \begin{cases} H^q(\mathbb{P}^n, \Omega^p_{\mathbb{P}}) & \text{if } r = 0, \\ 0 & \text{if } r > 0. \end{cases}$$

Proof. We prove this by induction on p. For $p = 0$, this follows from the long exact sequences associated to

$$0 \to \mathcal{O}_{\mathbb{P}}(-d-r) \to \mathcal{O}_{\mathbb{P}}(-r) \to \mathcal{O}_X(-r) \to 0$$

and Theorem 16.2.1.

In general, by induction and Corollary 17.2.5 we deduce

$$H^q(X, \Omega^p_X(-r)) = H^q(\Omega^p_{\mathbb{P}}(-r)|_X)$$

for $r \geq 0$ and $p+q < n-1$. Lemma 17.2.3 and Theorem 17.1.5 give

$$H^q(\Omega^p_{\mathbb{P}}(-r)|_X) = \begin{cases} H^q(\mathbb{P}^n, \Omega^p_{\mathbb{P}}) & \text{if } r = 0, \\ 0 & \text{if } r > 0. \end{cases}$$ □

Exercises

17.2.7. Using Exercise 17.1.7, deduce a version of the adjunction formula $\Omega^{n-1}_X \cong \mathcal{O}_X(d-n-1)$ with X as above.

17.2.8. Compute $H^q(X, \Omega^{n-1}_X)$ for all q.

17.3 Hodge Numbers of a Hypersurface II

As in the previous section, $X \subset \mathbb{P}^n$ is a nonsingular degree-d hypersurface. By Corollary 17.2.2, the Hodge numbers $h^{pq}(X)$ equal δ_{pq} when $n-1 \neq p+q < 2(n-1)$. So the only thing left to compute are the Hodge numbers in the middle. The formulas simplify a bit by setting $h_0^{pq}(X) = h^{pq}(X) - \delta_{pq}$. These can be expressed by the Euler characteristics:

Lemma 17.3.1. $h_0^{p,n-1-p}(X) = (-1)^{n-1-p}\chi(\Omega_X^p) + (-1)^n$.

We can calculate these Hodge numbers by hand using the following recurrence formulas.

Proposition 17.3.2.

(a)

$$\chi(\Omega_{\mathbb{P}}^p(i)) = \sum_{j=0}^{p}(-1)^j\binom{n+1}{p-j}\binom{i-p+j+n}{n}.$$

(b)

$$\chi(\mathscr{O}_X(i)) = \binom{i+n}{n} - \binom{i+n-d}{n}.$$

(c)

$$\chi(\Omega_X^p(i)) = \chi(\Omega_{\mathbb{P}}^p(i)) - \chi(\Omega_{\mathbb{P}}^p(i-d)) - \chi(\Omega_X^{p-1}(i-d)).$$

Proof. Corollary 17.1.3 yields the recurrence

$$\chi(\Omega_{\mathbb{P}}^p(i)) = \binom{n+1}{p}\chi(\mathscr{O}_{\mathbb{P}}(i-p)) - \chi(\Omega_{\mathbb{P}}^{p-1}(i)).$$

Therefore (a) follows by induction on p. The base case was obtained previously in (16.2.4).

Lemma 17.2.3 and Corollary 17.2.5 imply

$$\begin{aligned}\chi(\Omega_X^p(i)) &= \chi(\Omega_{\mathbb{P}}^p(i)|_X) - \chi(\Omega_X^{p-1}(i-d)) \\ &= \chi(\Omega_{\mathbb{P}}^p(i)) - \chi(\Omega_{\mathbb{P}}^p(i-d)) - \chi(\Omega_X^{p-1}(i-d)).\end{aligned}$$

When $p=0$, the right side can be evaluated explicitly to obtain (b). □

Corollary 17.3.3. *The Hodge numbers of X depend only on d and n, and are given by polynomials in these variables.*

In principle, formulas for all the Hodge numbers can be calculated using the above recurrence formulas. For example,

$$h^{0,n-1}(X) = (-1)^n\binom{n-d}{n} = \binom{d-1}{n}, \tag{17.3.1}$$

$$h_0^{1,n-2}(X) = (-1)^n\left[(n+1)\binom{n-1}{n} - (n+1)\binom{n-d-1}{n} + \binom{n-2d}{n}\right].$$

But this gets quite messy as p increases. So, instead, we give a closed form for the generating function below. Let $h^{pq}(d)$ denote the pqth Hodge number of a smooth hypersurface of degree d in $\mathbb{P}^{p+q+1}$. Define the formal power series

$$H(d) = \sum_{pq}(h^{pq}(d) - \delta_{pq})x^p y^q$$

in x and y.

Theorem 17.3.4 (Hirzebruch).

$$H(d) = \frac{(1+y)^{d-1} - (1+x)^{d-1}}{(1+x)^d y - (1+y)^d x}.$$

Corollary 17.3.5. *Hodge symmetry $h^{pq} = h^{qp}$ holds for smooth hypersurfaces in projective space over arbitrary fields.*

Remark 17.3.6. Hodge symmetry can fail for arbitrary smooth projective varieties in positive characteristic [90].

Corollary 17.3.7. *If $X \subset \mathbb{P}^{n+1}_{\mathbb{C}}$ has degree 2, then $b_n(X) = 0$ if n is odd; otherwise, $b_n(X) = h^{n/2,n/2}(X) = 2$.*

Proof.

$$H(2) = \frac{1}{1-xy}.$$

□

By expanding the series $H(3)$ for a few terms, we obtain the following corollary:

Corollary 17.3.8. *If $X \subset \mathbb{P}^{n+1}$ has degree 3, the middle hodge numbers are*

$$1,1$$

$$0,7,0$$

$$0,5,5,0$$

$$0,1,21,1,0$$

$$0,0,21,21,0,0$$

for $n \leq 5$.

Although this result can be deduced from the previous formulas in principle, Hirzebruch [63, 22.1.1] obtained this from his general Riemann–Roch theorem. His original formula gave a generating function for $\chi(\Omega_X^p)$; the above form can be obtained by a change of variables, cf. [30, Example XI Corollary 2.4]. Similar formulas are available for complete intersections. We will be content to work out the case $y = 0$. On one side we have the generating function $\sum h^{n0}(d)x^n$. By (17.3.1), this equals

$$\sum \binom{d-1}{n+1} x^n = \frac{(1+x)^{d-1} - 1}{x},$$

which is what one gets by substituting $y = 0$ into Hirzebruch's formula.

Exercises

17.3.9. Prove Lemma 16.2.4.

17.3.10. Calculate the Hodge numbers of a degree-d surface in $\mathbb{P}^3$ (a) using the recurrence formulas, (b) using the generating function. Compare the expressions.

17.3.11. Prove that for every fixed d and p, there exists q_0 such $h^{pq}(d) = 0$ for $q \geq q_0$.

17.4 Double Covers

Our goal is to compute the Hodge numbers for another natural class of examples that generalize hyperelliptic curves. Let $f(x_0,\dots,x_n) \in \mathbb{C}[x_0,\dots,x_n]$ be a homogeneous polynomial of degree $2d$ such that the hypersurface $D \subset \mathbb{P}^n = \mathbb{P}$ defined by $f = 0$ is nonsingular. Let $\pi : X \to \mathbb{P}$ be the double cover branched along D (Example 3.4.9). By construction, this is gotten by gluing the affine varieties defined by $y_i^2 = f(x_0,\dots,1,\dots,x_n)$ over U_i. It follows that X is nonsingular. Using these coordinates, it is also clear that the local coordinate ring $\mathscr{O}_X(\pi^{-1}U_i)$ is a free $\mathscr{O}(U_i)$-module generated by 1 and y_i. Globally, we have

$$\pi_*\mathscr{O}_X \cong \mathscr{O}_{\mathbb{P}} \oplus L,$$

where L is the line bundle locally generated by y_i. The ratios y_i/y_j give a cocycle for L, from which it easily follows that $L = \mathscr{O}(\pm d)$. To get the correct sign, we need to observe that L is a nontrivial ideal in $\pi_*\mathscr{O}_X$, so it has no nonzero global sections. Therefore we obtain the following:

Lemma 17.4.1.

$$\pi_*\mathscr{O}_X \cong \mathscr{O}_{\mathbb{P}} \oplus \mathscr{O}_{\mathbb{P}}(-d).$$

It is worth observing that the summands $\mathscr{O}_{\mathbb{P}}$ and $\mathscr{O}_{\mathbb{P}}(-d)$ are exactly the invariant and anti-invariant parts under the action of the Galois group, which is generated by the involution $\sigma : y_i \mapsto -y_i$. A more abstract, but less ad hoc, argument involves observing that $\mathscr{O}_{\mathbb{P}} \oplus \mathscr{O}_{\mathbb{P}}(-d)$ is a sheaf of algebras, and defining X as its relative spectrum [35]. Then the lemma becomes a tautology. We need an extension of the previous lemma to forms:

Lemma 17.4.2 (Esnault–Viehweg). *There are isomorphisms*

$$\pi_*\Omega_X^p \cong \Omega_{\mathbb{P}}^p \oplus (\Omega_{\mathbb{P}}^p(\log D) \otimes \mathscr{O}(-d))$$

for every p.

Proof. We sketch the proof. See [35, pp. 6–7] for further details. The Galois group acts on $\pi_*\Omega_X^p$. We check that the invariant and anti-invariant parts correspond to

$\Omega^p_{\mathbb{P}}$ and $\Omega^p_{\mathbb{P}}(\log D)\otimes\mathcal{O}(-d)$ respectively. It is enough to do this for the associated analytic sheaves. By the implicit function theorem, we can choose new analytic local coordinates such that X is given locally by $y^2 = x_1$. Then $y, x_2, \ldots, x_n$ are coordinates on X, so that their derivatives locally span Ω^1_X. It follows that a local basis for $\pi_*\Omega^1_X$ is given by

$$\underbrace{ydy = \frac{1}{2}dx_1, dx_2, \ldots, dx_n}_{\text{invariant}}, \quad \underbrace{dy = y\frac{dx_1}{2x_1}, ydx_2, \ldots, ydx_n}_{\text{anti-invariant}}.$$

The forms in the first group are invariant and give a local basis for $\Omega^1_{\mathbb{P}}$. The remainder are anti-invariant and form a local basis for $\Omega^1_{\mathbb{P}}(\log D)\otimes\mathcal{O}(-d)$. By taking wedge products, we get a similar decomposition for p-forms. □

Corollary 17.4.3.

$$H^q(X,\Omega^p_X) \cong H^q(\mathbb{P},\Omega^p_{\mathbb{P}}) \oplus H^q(\mathbb{P},\Omega^p_{\mathbb{P}}(\log D)\otimes\mathcal{O}(-d)).$$

Proof. Let $\{U_i\}$ be the standard affine cover of $\mathbb{P}^n$. Then $\tilde{U}_i = \pi^{-1}U_i$ gives an affine cover of X. We can compute $H^q(X,\Omega^p_X)$ using the Čech complex

$$\check{C}(\{\tilde{U}_i\},\Omega^p_X) = \check{C}(\{U_i\},\Omega^p_{\mathbb{P}}) \oplus \check{C}(\{U_i\},\Omega^p_{\mathbb{P}}(\log D)),$$

which decomposes into a sum. This decomposition passes to cohomology. □

Corollary 17.4.4. *We have*

$$\begin{aligned} h^{pq}(X) = \delta_{pq} &+ \dim\operatorname{coker}[H^{q-1}(\Omega^{p-1}_D(-d)) \to H^q(\Omega^p_{\mathbb{P}}(-d))] \\ &+ \dim\ker[H^q(\Omega^p_D(-d)) \to H^{q+1}(\Omega^p_{\mathbb{P}}(-d))] \end{aligned}$$

in general, and $h^{pq}(X) = \delta_{pq}$ *if* $p+q<n$.

Proof. This follows from (12.6.4) from §12.6 together with Bott's vanishing theorem, Theorem 17.1.5. □

We can obtain more explicit formulas by combining this with earlier results.

Exercises

17.4.5. When $n=2$, check that $h^{02}(X) = \frac{(d-1)(d-2)}{2}$ and $h^{11}(X) = 3d^2-3d+2$.

17.4.6. Verify that $h^{pq}(X) = \delta_{pq}$ also holds when $p+q>n$.

17.5 Griffiths Residues*

In this section, we describe an alternative method for computing the Hodge numbers of a hypersurface due to Griffiths [48], although the point is really that the method gives more, namely a method for computing the Hodge structure (or more precisely the part one gets by ignoring the lattice). Further details and applications can be found in books of Carlson, Peters, Müller-Stach [17, §3.2] and Voisin [116, Chapter 6] in addition to Griffiths' paper. We work over $\mathbb{C}$ in this section.

Suppose that $X \subset \mathbb{P} = \mathbb{P}^{n+1}$ is a smooth hypersurface defined by a polynomial $f \in \mathbb{C}[x_0, \dots x_{n+1}]$ of degree d. Let $U = \mathbb{P} - X$. The exact sequence (12.6.5) yields

$$H^{n-1}(X) \to H^n(\mathbb{P}) \to H^{n+1}(U) \to H^n(X) \to H^{n+2}(\mathbb{P}).$$

The first map is an isomorphism by weak Lefschetz. Therefore $H^{n+1}(U)$ maps isomorphically onto the primitive cohomology $P^n(X) = \ker[H^n(X) \to H^{n+2}(\mathbb{P})]$. This is the same as $H^n(X)$ if n is odd, and has dimension one less if n is even. The Hodge filtration on $F^pH^{n+1}(U)$ maps onto the Hodge filtration on X with a shift $F^{p-1}P^n(X)$. We refer to Section 12.6 for the definition of this and of the pole filtration $\mathrm{Pole}^pH^{n+1}(U)$. The key step is to compare these filtrations.

Theorem 17.5.1 (Griffiths). *The Hodge filtration $F^pP^n(X)$ coincides with the shifted pole filtration $\mathrm{Pole}^{p+1}H^{n+1}(U)$. This can, in turn, be identified with the quotient*

$$\frac{H^0(\Omega_{\mathbb{P}}^{n+1}((n-p+1)X))}{dH^0(\Omega_{\mathbb{P}}^{n}((n-p)X))}.$$

Proof. For $p = n$, this is immediate because

$$F^nP^n(X) = F^{n+1}H^{n+1}(U) = H^0(\Omega_{\mathbb{P}}^{n+1}(\log X)) = H^0(\Omega_{\mathbb{P}}^{n+1}(X)).$$

For $p = n-1$, we use the exact sequence

$$0 \to \Omega_{\mathbb{P},cl}^{n}(\log X) \to \Omega_{\mathbb{P}}^{n}(X) \xrightarrow{d} \Omega_{\mathbb{P}}^{n+1}(2X) \to 0,$$

where $\Omega_{\mathbb{P},cl}(\dots)$ is the subsheaf of closed forms. Then

$$H^0(\Omega_{\mathbb{P}}^{n}(X)) \xrightarrow{d} H^0(\Omega_{\mathbb{P}}^{n+1}(2X)) \to H^1(\Omega_{\mathbb{P},cl}^{n}(\log X)) \to H^1(\Omega_{\mathbb{P}}^{n}(X)).$$

On the right, the group $H^1(\Omega_{\mathbb{P}}(X))$ is equal to $H^1(\Omega_{\mathbb{P}}^{n}(d)) = 0$ by Bott's Theorem 17.1.5. Thus $F^nH^{n+1}(U) = H^1(\Omega_{\mathbb{P},cl}^{n}(\log X))$ is isomorphic to the cokernel of the first map labeled by d. This proves the theorem for this case. The remaining p's can be handled by a similar argument, which is left for the exercises. □

This can be made explicit using the following lemma:

Lemma 17.5.2. *Let*

$$\omega = \sum (-1)^i x_i dx_0 \wedge \cdots \wedge \widehat{dx_i} \wedge \cdots \wedge dx_{n+1}.$$

Then

$$H^0(\Omega_{\mathbb{P}}^{n+1}(kX)) = \left\{ \frac{g\omega}{f^k} \mid g \text{ homogeneous of } \deg = kd - (n+2) \right\}.$$

With these identifications, elements of $P^n(X)$ can be represented by homogeneous rational differential forms $g\omega/f^k$ modulo exact forms. Set

$$R = \frac{\mathbb{C}[x_0, \ldots, x_{n+1}]}{(\partial f/\partial x_1, \ldots, \partial f/\partial x_{n+1})}.$$

The ring inherits a grading $R = \oplus R_i$ from the polynomial ring. We define a map $P^n(X) \to R$ by sending the class of $g\omega/f^k$ to the class of g.

Theorem 17.5.3 (Griffiths). *Under this map, the intersection of $H^{n-p}(X, \Omega_X^p)$ with $P^n(X)$ maps isomorphically to $R_{\tau(p)}$, where $\tau(p) = (n-p+1)d - (n+2)$.*

This leads to an alternative method for computing the Hodge numbers of a degree-d hypersurface.

Corollary 17.5.4. *$h^{pq}(d) - \delta_{pq}$ is the coefficient of $t^{\tau(p)}$ in $(1+t+\cdots+t^{d-2})^{n+2}$.*

Proof. We can assume that $f = x_0^d + x_1^d + \cdots + x_{n+1}^d$ is the Fermat equation. It suffices to prove that the Poincaré series of R, which is the generating function $p(t) = \sum \dim R_i t^i$, is given by

$$p(t) = (1+t+\cdots+t^{d-2})^{n+2}.$$

Note that

$$R = \frac{\mathbb{C}[x_0, \ldots, x_{n+1}]}{(x_0^{d-1}, x_1^{d-1}, \ldots)} \cong \frac{\mathbb{C}[x]}{(x^{d-1})} \otimes \frac{\mathbb{C}[x]}{(x^{d-1})} \otimes \cdots \quad (n+2 \text{ times}).$$

Since Poincaré series for graded rings are multiplicative for tensor products, $p(t)$ is the $(n+2)$ power of the Poincaré series of $\mathbb{C}[x]/(x^{d-1})$, and this is given by the above formula. □

Exercises

17.5.5. Using exact sequences

$$0 \to \Omega^{n-i}_{\mathbb{P},cl}((j-1)X) \to \Omega^{n-i}_{\mathbb{P}}((j-1)X) \to \Omega^{n-i+1}_{\mathbb{P}}(jX) \to 0$$

and identifications

$$F^p H^{n+1}(U) \cong H^{n+1-p}(\Omega^p_{\mathbb{P},cl}(\log X)) \cong H^{n+1-p}(\Omega^p_{\mathbb{P},cl}(X)),$$

finish the proof of Theorem 17.5.1.

Chapter 18
Deformations and Hodge Theory

We have now come to the penultimate chapter. Over the course of this book, we have introduced a number of invariants, such as Betti numbers, Hodge numbers, and Picard numbers, that can be used to distinguish (complex smooth projective) varieties from one another. Varieties tend to occur in families. For example, we have encountered the Legendre family $y^2 = x(x-1)(x-\lambda)$ of elliptic curves. We have, at least implicitly, considered the family of all nonsingular hypersurfaces $\sum a_{d_0\cdots d_n} x_0^{d_0}\dots x_n^{d_n} = 0$ of fixed degree. So the question is, what happens to these invariants as the coefficients vary? Or in more geometric language, what happens as the variety deforms? For Betti numbers of complex smooth projective varieties, we have already seen that as a consequence of Ehresmann's theorem, they will not change in a family, because all the fibers are diffeomorphic. However, as algebraic varieties, or as complex manifolds, they can be very different. So it is perhaps surprising that the Hodge numbers will not change either. This is a theorem of Kodaira and Spencer, whose proof we outline. The proof will make use of pretty much everything we have done, plus one more thing. We will need a basic result due to Grothendieck in the algebraic setting, and Grauert in the analytic, that under the appropriate assumptions, the dimensions of coherent cohomology are upper semicontinuous. This means that the Hodge numbers could theoretically jump upward at special values of the parameters. On the other hand, by the Hodge decomposition, their sums, which are the Betti numbers, cannot. That is the basic idea. In the last section, we look at the behavior of the Picard number. Here the results are much less definitive. We end with the Noether–Lefschetz theorem, which explains what happens for general surfaces in $\mathbb{P}^3_{\mathbb{C}}$.

18.1 Families of Varieties via Schemes

Fix an algebraically closed field k. Let $R = \mathscr{O}(Y)$ be the coordinate ring of an affine variety Y over k. Given a collection of polynomials $f_j(x_1,\dots,x_n;y) \in R[x_1,\dots,x_n]$, the collection of subschemes

$$V(f_j(x,b)) = \{a \in \mathbb{A}_k^n \mid f_j(a,b) = 0\}, \quad b \in Y, \tag{18.1.1}$$

is what we mean by a *family of subschemes* of $\mathbb{A}_k^n$ parameterized by Y. This can be formulated more geometrically as follows. Affine space over Y or R is

$$\mathbb{A}_Y^n = \mathbb{A}_R^n = \mathrm{Spec}_m R[x_1,\ldots,x_n] \cong \mathbb{A}_k^n \times Y.$$

Let $I \subset R[x_1,\ldots,x_n]$ be the ideal generated by the f_j's. We get a morphism

$$\mathrm{Spec}_m(R[x_1,\ldots,x_n]/I) \to \mathrm{Spec}_m R = Y,$$

and the members of the above family are precisely the fibers

$$\mathrm{Spec}_m(R[x_1,\ldots,x_n]/I \otimes R/m_b) = \mathrm{Spec}_m k[x_1,\ldots,x_n]/(f_j(x,b))$$

over the points $b \in Y$.

Given a collection of polynomials $f_j(x_0,\ldots,x_n;y) \in S = R[x_0,\ldots,x_n]$ homogenous in the x's, the above setup (18.1.1) can be easily modified to define a family of subschemes of $\mathbb{P}_k^n$. In more coordinate-free terms, let $I \subset S$ be the homogeneous ideal generated by the f_j's. Then we can construct an ideal sheaf $\mathscr{I} = \widetilde{I}$ on projective space over Y,

$$\mathbb{P}_Y^n = \mathbb{P}_R^n = \mathbb{P}_k^n \times Y,$$

by mimicking the procedure in Section 15.3. We then get the associated closed subscheme $X = V(\mathscr{I}) \subset \mathbb{P}_Y^n$, which fits into the diagram

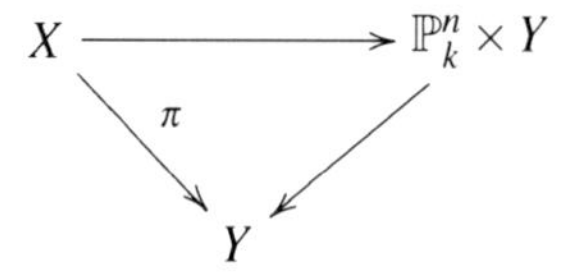

The family is the set of fibers of π. An equivalent alternative is to use the Proj functor from graded rings to schemes given in [60]. Then $V(\mathscr{I}) = \mathrm{Proj}(S/I)$. In general, a family of (projective) schemes over a scheme Y is nothing more than a morphism of schemes $f : X \to Y$ (fitting into a diagram as above).

Example 18.1.1. Fix d,n, and let $V = V_{d,n}$ be the vector space of all degree-d homogeneous polynomials in $n+1$ variables. Then

$$\mathscr{U}_{n,d} = \{([a_0,\ldots,a_n],[F]) \in \mathbb{P}_k^n \times \mathbb{P}(V) \mid F(a_0,\ldots,a_n) = 0\}$$

with its projection to $\mathbb{P}(V_{d,n})$ is the universal family of degree-d hypersurfaces.

Our notion of a family of schemes needs some fine tuning. At least for a "good" family, we would like have to some uniformity among fibers. For example, we would like the dimensions of the fibers to be the same. However, the example of a constant map $\mathbb{A}_k^n \to \mathbb{A}_k^m$ shows that our present notion is too general. We can certainly avoid this sort of example by requiring surjectivity, but this is not enough either:

Example 18.1.2. Consider the morphism

$$B = \{(x,y,t) \in \mathbb{A}^3_k \mid y - xt = 0\} \to \mathbb{A}^2_k$$

given by projection to (x,y). The fiber over $(0,0)$ is one-dimensional, but the other fibers are zero-dimensional.

However, on the positive side, we have the following:

Proposition 18.1.3. *If $f : X \to Y$ is a surjective morphism of a variety to a nonsingular curve, then all fibers have dimension equal to* $\dim X - 1$.

Proof. Fix $p \in Y$. We can assume that X and Y are affine. By assumption, Y is the spectrum of a Dedekind domain [8, Chapter 9]. Therefore, we can assume (after shrinking Y further) that the maximal ideal m_p is principal. Therefore the fiber $f^{-1}p$ is also defined by a principal ideal, and the result follows from Krull's principal ideal theorem [8, Corollary 11.17]. □

Over more general bases, the right notion, due to Grothendieck, of a good family is more subtle. To partially motivate it, let us look at a case in which things could go wrong. Suppose $Y = \mathrm{Spec}_m R$ is affine, and that X were reducible and had a component X_1 that mapped to a proper subvariety of Y. Then fibers need not have the same dimension. Choosing a nonzero regular function $r \in R$ vanishing along $f(X_1)$, r would annihilate any function vanishing along $X - X_1$. So this is something to avoid; in good cases, $T = \mathcal{O}(U)$ should be a torsion-free R-module for any nonempty open $U \subset X$. Example 18.1.2 shows that torsion-freeness is still too weak, but at least we are heading in the right direction. Torsion-freeness is equivalent to the injectivity of $I \otimes T \to T$ for every principal ideal $I \subset R$. Requiring this for every ideal is equivalent to the flatness of T, and this is the key.

Definition 18.1.4. Let R be a commutative ring. An R-module T is flat if $T \otimes_R N \to T \otimes_R M$ is injective whenever $N \to M$ is an injective map of R-modules. If $Y = \mathrm{Spec}_m R$, a morphism or family $f : X \to Y$ of schemes is flat if $\mathcal{O}(U)$ is a flat R-module for every open set $U \subseteq X$. For general Y, f is flat if $f^{-1}U_i \to U_i$ are flat for some affine open cover $\{U_i\}$ of Y.

Standard properties of flat modules can be found in [15, 33], and these show that the last part of the definition is well defined. Some intuition behind the geometric meaning of flatness can be found in [34, 60, 92]. Example 18.1.1 is flat. We point out one aspect that makes this property attractive. We will deduce a special case of this from a stronger result in the next section.

Proposition 18.1.5. *The fibers of a flat surjective morphism of varieties $f : X \to Y$ all have the same dimension* $\dim X - \dim Y$.

Proof. See [60, Chapter III, Proposition 9.5]. □

We can recover Proposition 18.1.3 from this by the following observation:

Lemma 18.1.6. *If $f : X \to Y$ is a surjective morphism of a variety to a nonsingular curve, then it is flat.*

Proof. The lemma is a consequence of the well-known fact that a torsion-free module over a Dedekind domain is flat. This fact can be proved by observing that torsion-free modules over Dedekind domains are direct limits of finitely generated locally free modules, and applying Exercise 18.1.12. □

We encountered smoothness very briefly already as the algebraic analogue of a submersion, which is to say that the maps on Zariski tangent spaces were required to be surjective. This definition does not work well if the varieties are singular. Instead, we can formulate the definition directly using a Jacobian criterion below.

Definition 18.1.7. A morphism $f : X \to Y$ of schemes is smooth (of relative dimension m) at $p \in X$ if it is locally given by

$$\mathrm{Spec}_m R[x_1, \ldots, x_{n+m}]/(f_1, \ldots, f_n) \to \mathrm{Spec}_m R,$$

where

$$\mathrm{rank}\left(\frac{\partial f_i}{\partial x_j}(p)\right) = n.$$

The morphism is smooth if this holds for all $p \in X$.

To be clear, "locally" above means that both X and Y are allowed to be replaced by neighborhoods of p and $f(p)$ respectively.

Example 18.1.8. Let $V_{d,n}$ and $\mathscr{U}_{d,n}$ be as in Example 18.1.1. There is a hypersurface $\Delta \subset \mathbb{P}(V_{d,n})$ parameterizing singular hypersurfaces. The restriction of $\mathscr{U}_{d,n} \to \mathbb{P}(V_{d,n})$ to the complement of Δ is smooth.

Lemma 18.1.9. *When $Y = \mathrm{Spec}_m k$ is a point, $p \in X$ is smooth in this sense if and only if it is nonsingular or smooth in the sense of Definition 2.5.14.*

Proof. Locally, we have

$$X = \mathrm{Spec}_m k[x_1, \ldots, x_{n+m}]/(f_1, \ldots, f_n),$$

so that $\dim X \geq m$ by Krull's theorem [8, Corollary 11.6]. On the other hand, the Jacobian at p has rank n, which implies that $\dim T_{X,p} = m$. Since $m \leq \dim \mathscr{O}_{X,p} \leq \dim T_{X,p}$ by standard commutative algebra, this forces $p \in X$ to be nonsingular or smooth in the sense of Definition 2.5.14. The converse is also true by [92, Chapter III, §4, Theorem 4], so these are equivalent conditions. □

In general, we have the following characterizations. In particular, smooth maps between nonsingular varieties are indeed the same as submersions:

Theorem 18.1.10. *Let $f : X \to Y$ be a morphism. Then*

(a) The morphism f is smooth of relative dimension m if and only if f is flat and all the fibers are nonsingular of dimension m.
(b) If X and Y are nonsingular, then f is smooth if and only if it induces a surjection on all tangent spaces $T_x \to T_{f(x)}$.

Proof. See [92, Chapter III, §10, Theorem 3′] for (a).

Now suppose that the spaces are nonsingular. From our definition of smoothness, it is clear that the map on tangent spaces given by the Jacobian has maximal rank. Conversely, we may apply [60, Chapter III Propositions 10.1, 10.4] to see that conditions (a) hold. □

This can be modified to prove that Definition 18.1.7 is equivalent to the usual definition of smoothness given in [60].

Exercises

18.1.11. Let $R = \mathscr{O}(Y)$ be the coordinate ring of an affine variety, and let $S = R[x_0,\dots,x_n]$. Given a graded S-module M, construct a sheaf of modules $\widetilde{M}$ on $\mathbb{P}^n_Y$ using (15.3.1) and whatever refinements are needed. Show that this is coherent if M is finitely generated.

18.1.12. Prove that the following examples are flat R-modules:

(a) Free modules.
(b) Direct summands of flat modules, hence locally free modules.
(c) Unions of a directed families of locally free modules. A family of subsets is directed, or filtered, if the union of any pair of members of the family is contained in another member.
(d) $R[x_1,\dots,x_n]$.

18.1.13. Show that Example 18.1.2 is not flat directly without appealing to Proposition 18.1.5.

18.1.14. Check the smoothness of Example 18.1.8.

18.1.15. Given a morphism $f : X \to Y$, the relative differentials are defined by $\Omega^1_{X/Y} = \text{coker}[\Omega^1_Y \to \Omega^1_X]$. Show that if f is smooth of relative dimension n, then $\Omega^1_{X/Y}$ is locally free of rank n.

18.2 Semicontinuity of Coherent Cohomology

In the last section, we defined the notion of a family of subschemes of $\mathbb{P}^n_k$, which amounted to a family of ideal sheaves. It is convenient to extend this a bit. We can regard a coherent sheaf $\mathscr{M}$ on $\mathbb{P}^n_Y$ as a family of coherent sheaves $\mathscr{M}_y = \mathscr{M}|_{\mathbb{P}^n \times \{y\}}$ on $\mathbb{P}^n_k$ parameterized by Y. In more explicit terms, when Y is affine with coordinate ring $R = \mathscr{O}(Y)$, a finitely generated graded module M over

$$S = R[x_0, \dots, x_n] = \bigoplus S_i$$

gives rise to a coherent sheaf $\mathscr{M} = \widetilde{M}$ on $\mathbb{P}^n_R = \mathbb{P}^n_k \times Y$ by Exercise 18.1.11. Then $\mathscr{M}_y = \widetilde{M \otimes_R R/m_y}$. The basic question that we want to address is, *how do the dimensions of the cohomology groups*

$$h^i(\mathscr{M}_y) = \dim H^i(\mathscr{M}_y)$$

vary with $y \in Y$? Let us consider some examples.

Example 18.2.1. Let M be a finitely generated graded $k[x_0, \dots, x_n]$-module. Then $R \otimes_k M$ is a finitely generated graded S-module. $\mathscr{M} = \widetilde{R \otimes M}$ is a "constant" family of sheaves; we have $\mathscr{M}|_y = \widetilde{M}$. This can be constructed geometrically. Let $\pi : \mathbb{P}^n_k \times Y \to Y$ be the projection. Then $\mathscr{M} = \pi^*(\tilde{M})$. Clearly $y \mapsto h^i(\mathscr{M}_y)$ is a constant function of y.

In general, this is not constant:

Example 18.2.2. Set $R = k[s,t]$, and choose three points

$$p_1 = [0,0,1], \quad p_2 = [s,0,1], \quad p_3 = [0,t,1]$$

in $\mathbb{P}^2_k$ with s,t variable. Let $\mathscr{I}$ be the ideal sheaf of the union of these points in $\mathbb{P}^2_R = \mathbb{P}^2_k \times \mathbb{A}^2_k$. This can also be described as $\widetilde{I}$, where I is the product

$$(x,y,z-1)(x-s,y,z-1)(x,y-t,z-1) \subset R[x,y,z].$$

Consider $\mathscr{I}(1) = \widetilde{I(1)}$. The global sections of this sheaf correspond to the space of linear forms vanishing at p_1, p_2, p_3. Such a form can exist only when the points are collinear, and it is unique up to scalars unless the points coincide. Thus

$$h^0(\mathscr{I}(1)_{(s,t)}) = \begin{cases} 2 \text{ if } s = t = 0, \\ 1 \text{ if } s = 0 \text{ or } t = 0 \text{ but not both,} \\ 0 \text{ if } s \neq 0 \text{ and } t \neq 0. \end{cases}$$

One can also see that $\chi(\mathscr{I}(1)_{(s,t)}) = \chi(\mathscr{O}_{\mathbb{P}^2}(1)) - 3$ is constant, so h^1 has similar jumping behavior.

In the above example, the sets where the cohomology jumps are Zariski closed. This is typical:

Theorem 18.2.3 (Grothendieck). *Let R be an affine algebra, and $Y = \operatorname{Spec}_m R$. Suppose that $\mathscr{M} = \widetilde{M}$, where M is a finitely generated graded $S = R[x_0, \dots, x_n]$-module that is flat as an R-module. Then*

$$y \mapsto h^i(\mathcal{M}_y)$$

is upper semicontinuous, i.e., the sets $\{y \mid h^i(\mathcal{M}_y) \geq r\}$ *are Zariski closed. The Euler characteristic*

$$y \mapsto \chi(\mathcal{M}_y) = \sum(-1)^i h^i(\mathcal{M}_y)$$

is locally constant.

By GAGA, we deduce semicontinuity of $h^i(\mathcal{M}_{y,an})$ under the same conditions. In fact, Grauert has established more general statements in the analytic setting where GAGA is not available. Precise statements and proofs can be found in [47, 46]. The following important corollary is actually a characterization of flatness [60, Chapter III, 9.9].

Corollary 18.2.4. *If* $X \to Y$ *is a flat family of projective schemes, then the Hilbert polynomial* $\chi(\mathcal{O}_{X_y}(d))$ *is locally constant. In particular, the fibers have the same dimension over connected components.*

We now set up the preliminaries for the proof of Theorem 18.2.3. Let

$$C^\bullet(M) = \bigoplus_i M\left[\frac{1}{x_i}\right]_0 \to \bigoplus_{i<j} M\left[\frac{1}{x_i x_j}\right]_0 \to \cdots$$

denote the Čech complex of $\mathcal{M} = \widetilde{M}$ with respect to the standard affine cover of $\mathbb{P}^n_R$. This is a complex of flat R-modules by assumption. The Čech complex of the fiber $\mathcal{M}_y$ can be identified with $C^\bullet(M \otimes R/m_y) \cong C^\bullet(M) \otimes R/m_y$. Theorem 7.4.5 and Corollary 16.1.8 imply the following:

Lemma 18.2.5.

(a) $\mathcal{H}^i(C^\bullet(M)) \cong H^i(\mathbb{P}^n_R, \mathcal{M})$.
(b) $\mathcal{H}^i(C^\bullet(M) \otimes R/m_y) \cong H^i(\mathbb{P}^n_R, \mathcal{M}_y)$.

Most of the arguments of Chapter 16 can be reprised to prove that these cohomologies are finitely generated as R-modules (see [60]). Therefore the flat complex $C^\bullet(M)$ can be replaced by a complex $D^\bullet$ of finitely generated locally free R-modules with the same property (b) by the following lemma:

Lemma 18.2.6. *Suppose that* $C^\bullet$ *is a finite complex of flat R-modules with finitely generated cohomology. Then there exist a finite complex* $D^\bullet$ *of finitely generated locally free modules and a map of complexes* $D^\bullet \to C^\bullet$ *such that*

$$\mathcal{H}^i(D^\bullet \otimes N) \cong \mathcal{H}^i(C^\bullet \otimes N)$$

for any R-module N.

Proof. See [93, pp. 47–49]. □

We now prove Theorem 18.2.3.

Proof. We assume that a complex $D^\bullet$ of finitely generated locally free modules has been constructed as above. By shrinking Y if necessary, there is no loss in assuming that these are free modules. Thus the differentials $D^i \to D^{i+1}$ can be represented by matrices A^i over $R = \mathscr{O}(Y)$. The differentials $A^\bullet(y)$ of $D^\bullet \otimes R/m_y$ are obtained by evaluating these matrices at y. We note that

$$h^i(\tilde{M}_y) = \mathrm{rank}(D^i) - \mathrm{rank}(A^i(y)) - \mathrm{rank}(A^{i-1}(y)).$$

Thus the inequality $h^i(\tilde{M}_y) \geq r$ can be expressed in terms of vanishing of the appropriate minors of these matrices, and this defines a Zariski closed set. By Lemma 11.3.3, $\chi(\tilde{M}_y)$ is the alternating sum of ranks of the D^i. This is independent of y. □

Exercises

18.2.7. The arithmetic genus of a possibly singular curve X is $1 - \chi(\mathscr{O}_X)$. Calculate it for a plane curve.

18.2.8. Show that Theorem 16.1.5 fails for $j_!\mathscr{O}_U$ as defined in the Exercises of §15.1.

18.2.9. Let $\pi : \mathbb{P}^n_R \to Y$ be the projection. Show that $R^i\pi_*(\mathscr{M}) \cong \widetilde{\mathscr{H}^i(D^\bullet)}$ with the notation used above in the proof of the theorem.

18.2.10. Show that the functions $h^i(\mathscr{M}_y)$ are all locally constant for all i if and only if the sheaves $R^i\pi_*(\mathscr{M})$ are all locally free for all i.

18.3 Deformation Invariance of Hodge Numbers

In this section, we revert to working over $\mathbb{C}$.

Theorem 18.3.1 (Kodaira–Spencer). *If $f : X \to Y$ is a smooth projective morphism of nonsingular varieties, then the Betti and Hodge numbers of all the fibers are the same.*

Proof. The map f is a proper submersion of the associated C^∞-manifolds. Therefore any two fibers X_t and X_s are diffeomorphic by Theorem 13.1.3. Thus they have the same Betti numbers. By Theorem 18.2.3, for all p, q, there are constants g^{pq} such that

$$h^{pq}(X_t) \geq g^{qp} \tag{18.3.1}$$

for all $t \in X$ with equality on a nonempty open set U. Suppose that (18.3.1) is strict for at least one pair p, q and $t \notin U$. Choose a point $s \in U$. Then we would have

$$\sum_i b_i(X_t) = \sum_{p,q} h^{pq}(X_t) > \sum_{p,q} g^{pq} = \sum_i b_i(X_s),$$

which is impossible. □

This result relies on the Hodge theorem in an essential way, and generalizes to compact Kähler manifolds. However, it is known to fail when $\mathbb{C}$ is replaced by a field of positive characteristic. Also it can fail for many of the other invariants considered so far, even over $\mathbb{C}$. We consider some examples below.

We want to discuss one more positive result. For a smooth projective variety X, the *canonical bundle* $\omega_X = \Omega_X^{\dim X}$ and the nth *plurigenus* is defined by

$$P_n(X) = h^0(X, \omega_X^{\otimes n}).$$

These invariants are of fundamental importance in the classification theory of surfaces, which we briefly touched on, and also in the study of higher-dimensional varieties. Gentle introductions to the latter highly technical subject can be found in [73, 83]. There are examples of surfaces that can be distinguished by plurigenera, but not by Hodge numbers alone. Curves have three distinct types with different behaviors: $\mathbb{P}^1$, elliptic curves, and everything else. A similar division can be made in higher dimensions using growth rates of the plurigenera as a function of n, which is called the *Kodaira dimension*. From this point of view, the analogues of $\mathbb{P}^1$ are the varieties for which all the plurigenera are zero. The "everything else" category consists of the varieties of *general type*. These are the X for which $P_n(X)$ grows at the maximum possible rate, which is to say like a polynomial of degree $\dim X$. For example, X has general type if ω_X is ample. The first plurigenus $P_1(X) = h^{1,0}(X)$ is deformation-invariant by the above theorem, and the result was long conjectured for all n. Siu [107] proved this for the class of varieties of general type, and later for all varieties. A simplified presentation of the proof can be found in Lazarsfeld's book [78, §11.5]. We can get a very special case quite cheaply from what we have proved so far.

Theorem 18.3.2. *Suppose that $f : X \to Y$ is a smooth projective morphism of complex varieties such that ω_{X_0} is ample for some $0 \in Y$. Then $P_n(X_y) = P_n(X_0)$ for all $n > 0$ and y in a Zariski open neighborhood of 0.*

Proof. We may assume that $n > 1$. Using standard ampleness criteria [60], one can see that there is an open neighborhood of 0 such that ω_{X_y} is ample for all $y \in U$. By Kodaira's vanishing theorem (Corollary 12.6.4),

$$H^i(X_y, \omega_{X_y}^{\otimes n}) = 0$$

for $y \in U$, $i > 0$ (and the standing assumption $n > 1$). Thus

$$P_n(X_y) = \chi(\omega_{X_y}^{\otimes n})$$

is constant. □

Exercises

18.3.3. Give a formula for the nth plurigenus of a smooth projective curve using Riemann–Roch.

18.3.4. There exist examples of surfaces X for which $\omega_X^{\otimes 2} \cong \mathscr{O}_X$ but with ω_X nontrivial. Check that $P_n(X) = 0$ when n is odd, and 1 if even.

18.4 Noether–Lefschetz*

The Picard number is an important and subtle invariant that is not stable under deformation. For example, we saw in Section 11.2 that the Picard number of $E \times E$ is not constant. But any two (products of) elliptic curves are deformation equivalent, since any elliptic curve can be realized as a smooth cubic in $\mathbb{P}^2$. Let us now look at surfaces in $\mathbb{P}^3$.

Theorem 18.4.1 (Noether–Lefschetz). *Let $d \geq 4$. Then there exists a surface $X \subset \mathbb{P}^3_{\mathbb{C}}$ of degree d with Picard number* 1.

Remark 18.4.2. Noether refers not to Emmy, but her father Max, who was also a mathematician.

Remark 18.4.3. The result is true "for almost all" surfaces. See the exercises.

Before getting to the proof, we need to explain one key ingredient, which is the space of all curves in $\mathbb{P}^3$. In order to motivate this, consider the corresponding problem for curves in $\mathbb{P}^2$. A curve in the plane is determined by a single homogeneous polynomial in $\mathbb{C}[x_0,x_1,x_2]$ unique up to scalars. Thus a plane curve is determined by its degree d and a point in the projective space of polynomials of degree d. For a curve $C \subset \mathbb{P}^3$ (which we understand to mean a 1-dimensional subscheme), things are more complicated. We first fix the discrete invariants, which are the degree and genus. These can be identified, up to a change of variables, with coefficients of the Hilbert polynomial $\chi(\mathscr{O}_C(m))$. The curve C is determined by a homogeneous ideal $I \subset S = \mathbb{C}[x_0,x_1,x_2,x_3]$, and the set of these cannot be parameterized directly by any reasonable algebrogeometric object. The key technical result is the following:

Proposition 18.4.4 (Grothendieck). *Fix a linear polynomial $p(m) \in \mathbb{Q}[m]$. Then there exists a constant d depending only on $p(m)$ such that for any curve $C \subset \mathbb{P}^3$ with $\chi(\mathscr{O}_C(m)) = p(m)$, the ideal $I_C = \Gamma_*(\mathscr{I}_C)$ is generated by elements of degree d. Furthermore, $e = \dim(I_C \cap S_d)$ is determined by $p(m)$.*

Proof. The most natural proof can be found in [91, Lecture 14] using what is now referred to as Castelnuovo–Mumford regularity. □

Thus we see that C is determined by the subspace $I_C \cap S_d \subset S_d$. So the set of curves with Hilbert polynomial p is parameterized by a subset of the Grassmannian of e-dimensional subspaces of S_d. This is an important first step, but there is more to do. Then end result is the following theorem:

Theorem 18.4.5 (Grothendieck). *Fix a linear polynomial $p \in \mathbb{Q}[m]$. There is a projective scheme $H = \mathrm{Hilb}^p_{\mathbb{P}^3}$, called the Hilbert scheme, with a universal flat family $\mathscr{C} \to H$ of curves in $\mathbb{P}^3_{\mathbb{C}}$ with Hilbert polynomial p. Universality implies that every curve with Hilbert polynomial p occurs as a fiber of $\mathscr{C}$ exactly once.*

Proof. An intuitive account can be found in [34, Chapter VI §2.2]. For more precise details, see [37]. □

Remark 18.4.6. Although we have stated only a special case, the result is much more general. It holds for arbitrary subschemes of $\mathbb{P}^n_{\mathbb{Z}}$.

We can now prove Theorem 18.4.1

Proof. Let $X \subset \mathbb{P}^3$ be a smooth surface of degree $d \geq 4$. From the exact sequence

$$0 \to \mathscr{O}_{\mathbb{P}^3}(-d) \to \mathscr{O}_{\mathbb{P}^3} \to \mathscr{O}_X \to 0$$

and Theorem 16.2.1, we can deduce that $H^1(X, \mathscr{O}_X) = 0$ and $H^2(X, \mathscr{O}_X) \neq 0$. From the exponential sequence

$$H^1(X, \mathscr{O}_X) \to \mathrm{Pic}(X) \to H^2(X, \mathbb{Z}) \to H^2(X, \mathscr{O}_X)$$

we can conclude that $c_1 : \mathrm{Pic}(X) \to H^2(X, \mathbb{Z})$ is injective but not surjective. In fact, it is not surjective after tensoring with $\mathbb{Q}$.

Now we will apply the language and results of §14.4. Choose a Lefschetz pencil $\{X_t\}_{t \in \mathbb{P}^1}$ of surfaces, let $\tilde{X} \to \mathbb{P}^1$ be the incidence variety, and let $U \subset \mathbb{P}^1$ be $\{t | X_t$ is smooth$\}$. Call $t \in \mathbb{P}^1$ special if for some $p \in \mathbb{Q}[m]$, all the curves parameterized by Hilb^p lie on X_t. There are only countably many special values. So we can choose a nonspecial $t_0 \in U$. Therefore any curve lying on X_{t_0} will propagate to all the members X_t of the pencil. In particular, it can be "pushed along" any path covering a loop in U. Consequently, the group $\mathrm{NS} = c_1(\mathrm{Pic}(X_{t_0}))$ generated by curves on X_{t_0} is stable under the action of $\pi_1(U, t_0)$. (A less fishy argument is outlined in Exercise 18.4.10.) By Proposition 14.4.10, $H^2(X_{t_0}, \mathbb{Q}) = \mathrm{im}(H^2(\mathbb{P}^3)) \oplus V$, where V is generated by vanishing cycles. The fundamental group $\pi_1(U, t_0)$ acts irreducibly on V. Since $\mathrm{NS} \otimes \mathbb{Q}$ contains $\mathrm{im}(H^2(\mathbb{P}^3))$, it follows that either $\mathrm{NS} \otimes \mathbb{Q}$ equals $\mathrm{im}(H^2(\mathbb{P}^3))$ or it equals $H^2(X_{t_0})$. The last case is impossible, because $c_1 \otimes \mathbb{Q}$ was not surjective. Therefore $\mathrm{NS} \otimes \mathbb{Q} = \mathrm{im}(H^2(\mathbb{P}^3)) = \mathbb{Q}$. □

Exercises

18.4.7. Is the restriction $d \geq 4$ in Noether–Lefschetz necessary?

18.4.8. Let $X \subset \mathbb{P}^3$ be a smooth quartic containing a line ℓ (refer back to the exercises of §11.1). Prove that the Picard number of X is at least 2.

18.4.9. Using the Baire category theorem (that a countable union of nowhere-dense sets in a complete metric space is nowhere dense), show that the subset of surfaces for which the conclusion of Noether–Lefschetz fails is nowhere dense in the projective space of all surfaces of degree $d \geq 4$.

18.4.10. Let

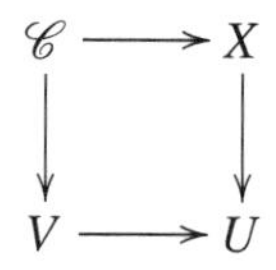

be a commutative diagram, where V/U is a finite unbranched Galois cover with Galois group G, and $\mathscr{C}$ is a family of curves in the fiber product $X \times_U V$. Choose $s_0 \in V$ lying over t_0. These assumptions ensure that we have an exact sequence

$$1 \to \pi_1(V, s_0) \to \pi_1(U, t_0) \to G \to 1.$$

(a) Show that the restriction $\alpha = c_1(\mathscr{O}_X(\mathscr{C}))|_{X_{t_0}}$ belongs to $H^2(X_{t_0})^{\pi_1(V)}$, where we identify the fiber of $X \times_U V$ over s_0 with X_{t_0}.
(b) Show that the span of the G-orbit of α lies in $\mathrm{NS}(X_{t_0})$ and is stable under $\pi_1(U, t_0)$.

For X_{t_0} as in the proof of Theorem 18.4.1, a Hilbert scheme argument shows that any curve $C \subset X_{t_0}$ can be extended to such a family $\mathscr{C}$ after possiblly shrinking U.

Part V
Analogies and Conjectures*

Chapter 19
Analogies and Conjectures

In this final chapter, we end our story by beginning another. Although we have mostly worked over $\mathbb{C}$, and occasionally over a general algebraically closed field, algebraic geometry can be done over any field. Each field has its own character: transcendental over $\mathbb{C}$, and arithmetic over fields such as $\mathbb{Q}, \mathbb{F}_p, \ldots$. It may seem that aside from a few formal similarities, the arithmetic and transcendental sides would have very little to do with each other. But in fact they are related in deep and mysterious ways. We start by briefly summarizing the results of Weil, Grothendieck, and Deligne for finite fields. Then we return to complex geometry and prove Serre's analogue of the Weil conjecture. This result inspired Grothendieck to formulate his standard conjectures. We explain some of these along with the closely related Hodge conjecture. These are among the deepest open problems in algebraic geometry.

19.1 Counting Points and Euler Characteristics

Let $\mathbb{F}_q$ be the field with $q = p^r$ elements, where p is a prime number. Consider the algebraic closure $k = \bar{\mathbb{F}}_p = \bigcup \mathbb{F}_{q^n}$. Suppose that $X \subseteq \mathbb{P}^d_k$ is a quasiprojective variety defined over $\mathbb{F}_p$, that is, assume that the coefficients of the defining equations lie in $\mathbb{F}_p$. Let $X(\mathbb{F}_{p^n})$ be the set of points of $\mathbb{P}^N_{\mathbb{F}_{p^n}}$ satisfying the equations defining X. Let $N_n(X)$ be the number of points of $X(\mathbb{F}_{p^n})$. Here are a few simple computations:

Example 19.1.1. $N_n(\mathbb{A}^m_{\mathbb{F}_p}) = p^{nm}$.

Example 19.1.2. Expressing $\mathbb{P}^m = \mathbb{A}^m \cup \mathbb{A}^{m-1} \cup \cdots$ as a disjoint union yields

$$N_n(\mathbb{P}^m_{\mathbb{F}_p}) = p^{nm} + p^{n(m-1)} + \cdots + p^n + 1.$$

Example 19.1.3. $N_n((\mathbb{P}^1_{\mathbb{F}_p})^m) = (p^n+1)^m$.

The last two computations are based on the following obvious properties.

(1) Additivity:

$$N_n(X) = N_n(X - Z) + N_n(Z)$$

whenever $Z \subset X$ is closed.

(2) Multiplicativity:

$$N_n(X \times Y) = N_n(X)N_n(Y).$$

Now let us return to complex geometry, so that $k = \mathbb{C}$. The Euler characteristic with respect to compactly supported cohomology is

$$\chi_c(X) = \sum(-1)^i \dim H^i_c(X, \mathbb{R}).$$

It is not difficult to compute this number in the above examples using the techniques from the earlier chapters:

$$\chi_c(\mathbb{A}^m_{\mathbb{C}}) = 1, \quad \chi_c(\mathbb{P}^m_{\mathbb{C}}) = m, \quad \chi_c((\mathbb{P}_{\mathbb{C}})^m) = 2^m.$$

This leads to the following curious observation that *if we set $p = 1$ in the above formulas, then we get χ_c*. Is there a deeper reason for this? First note that although we defined compactly supported cohomology using differential forms in Section 5.4, there is a purely topological definition that works for any locally compact Hausdorff space. We can set

$$H^i_c(X) = H^i(\bar{X}, \bar{X} - X)$$

for any (or some) compactification $\bar{X}$. Then (7.2.1) and a little diagram chasing yields the long exact sequence

$$\cdots \to H^i_c(X) \to H^i_c(X) \to H^i_c(X - Z) \to H^{i+1}_c(X) \to \cdots. \tag{19.1.1}$$

The first clue that there is a deeper relation between N_n and χ_c is the following.

Lemma 19.1.4. *The invariant χ_c is additive and multiplicative i.e.,*

$$\chi_c(X) = \chi_c(X - Z) + \chi_c(Z),$$
$$\chi_c(X \times Y) = \chi_c(X)\chi_c(Y)$$

holds.

Proof. The additivity follows immediately from (19.1.1). The multiplicativity follows from the Künneth formula

$$H^i_c(X \times Y, \mathbb{R}) = \bigoplus_{j+l=i} H^j_c(X, \mathbb{R}) \otimes H^l_c(Y, \mathbb{R}).$$ □

Suppose we start with a complex quasiprojective algebraic variety X with a fixed embedding into $\mathbb{P}^N_{\mathbb{C}}$. If the defining equations (and inequalities) have integer coefficients, then we can reduce these modulo a prime p to get a quasiprojective variety (or more accurately scheme) X_p defined over the finite field $\mathbb{F}_p$. In less prosaic terms, we have a scheme over $\operatorname{Spec}\mathbb{Z}$, and X_p is the fiber over p. (In practice, $\mathbb{Z}$ might be replaced by something bigger such as the ring of integers of a number field, but the essential ideas are the same.) Then we can count points on X_p and compare it to $\chi_c(X)$. To avoid certain pathologies, we should take p sufficiently large.

Lemma 19.1.5. *If X is expressible as a disjoint union of affine spaces, then $N_n(X_p)$ is a polynomial in p^n for sufficiently large p. Substituting $p = 1$ yields $\chi_c(X)$.*

Proof. If $X = \bigcup \mathbb{A}^{m_i}$ is a disjoint union, then we get a similar decomposition for X_p with $p \gg 0$. Therefore $N_n(X_p) = \sum p^{nm_i}$ and $\chi_c(X) = \sum 1^{nm_i}$. □

The lemma applies to the above examples of course, as well as to the larger class of toric varieties [41, p. 103], Grassmannians, and more generally flag varieties [42, 19.1.11]. Nevertheless, most varieties do not admit such decompositions (e.g., a curve of positive genus does not). So this is of limited use.

It is worth pointing out that these days, the material of this section is usually embedded into the framework of motivic integration. A succinct introduction to this is given in [81].

Exercises

19.1.6. Let $G = \mathbb{G}(2,4)$ be the Grassmannian of two-dimensional subspaces of k^4. Calculate $N_n(G)$ over $\mathbb{F}_p$ and $\chi(G)$ over $\mathbb{C}$ and compare.

19.1.7. Generalize this to $\mathbb{G}(2,n)$.

19.2 The Weil Conjectures

We may ask whether something like Lemma 19.1.5 holds for arbitrary varieties. We start by looking at an elliptic curve, which is the simplest example where the lemma does not apply.

Example 19.2.1. Let X be the elliptic curve given by the affine equation $y^2 = x^3 - 1$. This defines a smooth curve X_p over $\mathbb{F}_p$ when $p \geq 5$. So let us analyze what happens when $p = 5$. When n is odd, $5^n - 1$ is not divisible by 3. This implies that $x \mapsto x^3$ is an automorphism of $\mathbb{F}_{5^n}^*$. Therefore $y^2 + 1$ has a unique cube root. Thus $N_n(X_5) = 1 + 5^n$ if n is odd. When n is even, we can compute a few values by brute force on a machine.

p^n	N_n
5^2	$36 = 1 + 5^2 + 2 \cdot 5$
5^4	$576 = 1 + 5^4 - 2 \cdot 5^2$
5^6	$15876 = 1 + 5^6 + 2 \cdot 5^3$
5^8	$389376 = 1 + 5^8 - 2 \cdot 5^4$
5^{10}	$9771876 = 1 + 5^{10} + 2 \cdot 5^5$

So at least empirically, we have the formula

$$N_n(X_5) = \begin{cases} 1+5^n & \text{if } n \text{ odd}, \\ 1+5^n-2(-5)^{n/2} & \text{if } n \text{ even}, \end{cases}$$
$$= 1+5^n-(\sqrt{-5})^n-(-\sqrt{-5})^n.$$

Example 19.2.2. Calculating the number of points for the elliptic curve defined by $y^2 = x^3 - x$ with $p = 3$, we get

p^n	N_n
3	$4 = 1+3$
3^2	$16 = 1+3^2+2\cdot 3$
3^3	$28 = 1+3^3$
3^4	$64 = 1+3^4-2\cdot 3^2$
3^5	$244 = 1+3^5$

Then

$$N_n(X_3) = 1+3^n-(\sqrt{-3})^n-(-\sqrt{-3})^n$$

fits this data.

From these and additional examples, we observe a pattern that on an elliptic curve, $N_n = 1+p^n-\lambda_1^n-\lambda_2^n$ for appropriate constants λ_i with order of magnitude $\sqrt{p}$. We can generate more examples by taking products of these with the previous ones. Based on this data, we may guess that in general $N_n(X_p)$ is a linear combination of powers λ_i^n, and setting $\lambda_i = 1$ yields the Euler characteristic. This turns out to be correct, but it seems to come out of nowhere. We need some guiding principle to explain these formulas. The basic insight goes back to Weil [119] (who proved a number of cases). Suppose that $X \subset \mathbb{P}^d_{\mathbb{C}}$ is a nonsingular projective variety with equations defined over $\mathbb{Z}$ as above. Let us denote by $\bar{X}_p$ the variety over the algebraic closure $\bar{\mathbb{F}}_p$ determined by reducing the equations modulo p. The Frobenius morphism $F_p : \bar{X}_p \to \bar{X}_p$ is the map that raises the coordinates to the pth power (see [60, p. 301] for a more precise description). Then $X_p(\mathbb{F}_{p^n})$ are the points of $\bar{X}_p$ fixed by F_p^n. If this were a manifold with a self-map F (satisfying appropriate transversallity conditions), then we could calculate this number using the Lefschetz trace formula. Weil conjectured that this sort of argument could be carried out in the present setting for some suitable cohomology theory. He made some additional conjectures that will be discussed a bit later. Grothendieck eventually constructed such a Weil cohomology theory—in fact several. For each prime $\ell \neq p$, he constructed functors $H^i_{et}(-,\mathbb{Q}_\ell)$ called ℓ-adic cohomology such that:

(1) $H^i_{et}(\bar{X}_p,\mathbb{Q}_\ell)$ is a vector space over the field of ℓ-adic numbers $\mathbb{Q}_\ell = (\varprojlim \mathbb{Z}/\ell^n)\otimes \mathbb{Q}$.
(2) $\dim H^i_{et}(\bar{X}_p,\mathbb{Q}_\ell)$ is the usual ith Betti number of X.
(3) F_p acts on these spaces. The action on $H^0_{et}(\bar{X}_p,\mathbb{Q}_\ell)$ is trivial, but nontrivial in general.

(4) There is a Lefschetz trace formula that implies that

$$N_n(X_p) = \sum_i (-1)^i \,\mathrm{trace}[F_p^{n*} : H^i_{et}(\bar{X}_p, \mathbb{Q}_\ell) \to H^i_{et}(\bar{X}_p, \mathbb{Q}_\ell)].$$

Grothendieck constructed this by generalizing sheaf cohomology. The details, which are quite involved, can be found in the books by Freitag and Kiehl [39] or Milne [85]. The last formula can be rewritten as

$$N_n(X_p) = \sum_i (-1)^i \sum \lambda_{ij}^n,$$

where λ_{ij} are the generalized eigenvalues of F_p^* on $H^i_{et}(\bar{X}_p, \mathbb{Q}_\ell)$. In the previous examples, the numbers $\pm\sqrt{-5}, \pm\sqrt{-3}$ above were precisely the eigenvalues of F_p acting on H^1. Although this is overkill, we can also use this formalism to re-prove the formulas of the last section. For example,

$$H^i_{et}(\mathbb{P}^m_{\bar{\mathbb{F}}_p}, \mathbb{Q}_\ell) = \begin{cases} \mathbb{Q}_\ell \text{ with } F_p \text{ acting by } p^{i/2} & \text{if } i < 2m \text{ is even,} \\ 0 & \text{otherwise,} \end{cases} \tag{19.2.1}$$

gives the formula for $N_n(\mathbb{P}^m)$. Notice that the absolute values of the eigenvalues in these examples have very specific sizes. This is consistent with a deep theorem of Deligne proving the last of Weil's conjectures on the analogue of the Riemann hypothesis. (For more background and in particular what this has to do with the Riemann hypothesis, see [39], [60, Appendix C], [69], [85] and of course [25].)

Theorem 19.2.3 (Deligne). *Let X be a smooth and projective variety defined over $\mathbb{F}_p$, and $\bar{X}$ its extension to $\bar{\mathbb{F}}_p$. Then the eigenvalues of the Frobenius action on $H^i_{et}(\bar{X}, \mathbb{Q}_\ell)$ are algebraic numbers λ all of whose absolute values satisfy $|\lambda| = p^{i/2}$.*

Remark 19.2.4. This is valid over any finite field. The word "eigenvalue" above really means generalized eigenvalue, although in fact the action of F_p has been conjectured to be diagonalizable. This is still wide open.

This abstract theorem has concrete consequences. The first goes from topology to number theory, and the second goes in the opposite direction.

Corollary 19.2.5. *Let $X \subset \mathbb{P}^{N+1}$ be a smooth degree-d hypersurface defined over $\mathbb{F}_p$. Then*

$$|N_n(X) - (1 + p + \cdots + p^N)| \le b_N \cdot p^{N/2},$$

where b_N is the Nth Betti number of a smooth degree-d hypersurface in $\mathbb{P}^n_{\mathbb{C}}$.

Proof. We can assume that X comes from a hypersurface over $\mathbb{C}$ by reducing modulo a prime. By the weak Lefschetz theorem, $H^i(X, \mathbb{Q}) \cong H^i(\mathbb{P}^{N+1}, \mathbb{Q})$ for $i \in [0, 2N] - \{N\}$. So the Betti numbers of X and $\mathbb{P}^{N+1}$ are the same in this range. In fact, the action of F_p would be compatible with this isomorphism. Therefore eigenvalues would be the same in both spaces for $i \in [0, 2N] - \{N\}$. For $\mathbb{P}^{N+1}$, these are given in (19.2.1). Let λ_{jN} denote the eigenvalues on the Nth cohomology of X. Then

$$|N_1(X) - (1 + p + \cdots + p^N)| \leq |\sum_{j=1}^{b_N} \lambda_{jN}| \leq b_N \cdot p^{N/2}. \qquad \square$$

Corollary 19.2.6. *If X is determined by reducing a complex smooth projective variety mod $p \gg 0$ as above, the Betti numbers of the complex variety can be determined from $N_n(X)$.*

In most cases, the Betti numbers are easier to calculate than N_n. A nontrivial example in which the last corollary was usefully applied was given by Harder and Narasimhan [57].

When X is singular or open, then the above theorem is no longer true. Deligne [27] has shown that the eigenvalues can have varying sizes or weights independent of cohomological degree. Surprisingly, this has Hodge-theoretic meaning. If one counts the number of eigenvalues on $H^i_{et}(\bar{X}_p, \mathbb{Q}_\ell)$ of a given absolute value $p^{k/2}$, then this is the dimension of the weight-k quotient of the mixed Hodge structure on $H^i(X)$ that we touched on in §12.6. See [26] for a more precise summary of these results.

Exercises

19.2.7. Assuming Theorem 19.2.3, deduce the Hasse–Weil bound that if X is a smooth projective genus-g curve over $\mathbb{F}_p$, then $|N_n(X) - 1 - p^n| \leq 2gp^{n/2}$. (Of course this bound came first.)

19.3 A Transcendental Analogue of Weil's Conjecture

After this excursion into arithmetic, let us return to Hodge theory and prove an analogue of the Weil–Riemann hypothesis found by Serre [102]. To set up the analogy let us replace $\bar{X}$ above by a smooth complex projective variety Y, and F_p by an endomorphism $f : Y \to Y$. As for p, if we consider the effect of the Frobenius on $\mathbb{P}^N_{\bar{\mathbb{F}}_p}$, the pullback of $\mathscr{O}(1)$ under this map is $\mathscr{O}(p)$. To complete the analogy, we require the existence of a very ample line bundle $\mathscr{O}_Y(1)$ on Y, so that $f^*\mathscr{O}_Y(1) \cong \mathscr{O}_Y(1)^{\otimes q}$. We can take $c_1(\mathscr{O}_Y(1))$ to be the Kähler class ω. Then we have $f^*\omega = q\omega$.

Theorem 19.3.1 (Serre). *If $f : Y \to Y$ is a holomomorphic endomorphism of a compact Kähler manifold with Kähler class ω such that $f^*\omega = q\omega$ for some $q \in \mathbb{R}$, then q is an algebraic integer, $f^* : H^i(Y, \mathbb{Q}) \to H^i(Y, \mathbb{Q})$ is diagonalizable, and its eigenvalues are algebraic integers with absolute value $q^{i/2}$.*

Proof. The theorem holds for $H^{2n}(Y)$, since ω^n generates it. Note that q^n is the degree of f, which is necessarily a (rational) integer. Therefore q is an algebraic integer. By hypothesis, f^* preserves the Lefschetz decomposition (Theorem 14.1.1). Thus we can replace $H^i(Y)$ by primitive cohomology $P^i(Y)$. Recall from Corollary 14.1.4 that

$$\tilde{Q}(\alpha,\beta) = Q(\alpha, C\bar{\beta})$$

is a positive definite Hermitian form on $P^i(Y)$, where

$$Q(\alpha,\beta) = (-1)^{i(i-1)/2}\int \alpha\wedge\beta\wedge\omega^{n-i}.$$

Consider the endomorphism $F = q^{-i/2}f^*$ of $P^i(Y)$. We have

$$Q(F(\alpha),F(\beta)) = (-1)^{i(i-1)/2}q^{-n}\int f^*(\alpha\wedge\beta\wedge\omega^{n-i}) = Q(\alpha,\beta).$$

Moreover, since f^* is a morphism of Hodge structures, it preserves the Weil operator C. Therefore F is unitary with respect to $\tilde{Q}$, so its eigenvalues have norm 1. This gives the desired estimate on absolute values of the eigenvalues of f^*. □

Since f^* can be represented by an integer matrix, the set of its eigenvalues is a Galois-invariant set of algebraic integers. So we get a stronger conclusion that all Galois conjugates have absolute value $q^{i/2}$. This would imply that when $q = 1$ (e.g., if f is an automorphism) then these are roots of unity.

Exercises

19.3.2. Verify the above theorem for Y a complex torus, and $f: Y \to Y$ multiplication by a nonzero integer n, by direct calculation.

19.3.3. Show, by example, that if $f^*\omega$ is not a multiple of ω, then the eigenvalues of f^* on $H^i(Y)$ can have different absolute values.

19.4 Conjectures of Grothendieck and Hodge

Prior to Deligne's proof, Grothendieck [54] had suggested a strategy for carrying out a proof of the Weil–Riemann hypothesis similar to Serre's proof of the transcendental version. This required first establishing his *standard conjectures* [54, 71]. All but one of these conjectures are open over $\mathbb{C}$. The exception follows from the Hodge index theorem. For general fields, they are essentially all wide open. Grothendieck had also formulated his conjectures in order to construct his theory of *motives*, which gives a deeper explanation for some of the analogies between the worlds of arithmetic and complex geometry. So even though Deligne managed to prove the last of Weil's conjectures by another method, the problem of solving these conjectures is fundamental.

We want to spell out some of these conjectures in the complex case, and indicate their relation to a better-known Hodge conjecture [65]. Let X be an n-dimensional nonsingular complex projective variety. A codimension-p algebraic cycle is a finite formal sum $\sum n_i Z_i$, where $n_i \in \mathbb{Z}$ and $Z_i \subset X$ are codimension-p closed subvarieties. These form an abelian group $Z^p(X)$ of infinite rank. The first task is to cut it down to a more manageable size. Given a nonsingular $\iota : Z \hookrightarrow X$, we defined its fundamental class $[Z] = \iota_!(1) \in H^{2p}(X,\mathbb{Z})$. The fundamental class can be defined even when Z has singularities. This can be done in several ways (see [7]). A quick but nonelementary method is to use Hironaka's famous theorem [62] on resolution of singularities. This implies that there exists a smooth projective variety $\tilde{Z}$ with a birational map $\pi : \tilde{Z} \to Z$. Let $\tilde{i} : \tilde{Z} \to X$ denote the composition of π and the inclusion. Then set $[Z] = i_!(1) \in H^{2p}(X,\mathbb{Z})$.

Lemma 19.4.1. *This class is independent of the choice of resolution of singularities.*

Proof. Let $\tilde{Z}' \to Z$ be another resolution. Then by applying Hironaka's theorem to the fibered product, we see that there exists a third resolution $\tilde{Z}'' \to Z$ fitting into a commutative diagram

$$\begin{array}{ccc} \tilde{Z}'' & \xrightarrow{\psi} & \tilde{Z} \\ \downarrow & & \downarrow \pi \\ \tilde{Z}' & \xrightarrow{\pi'} & Z \end{array}$$

Then $i_!(1) = i_!(\psi_! 1) = (i \circ \psi)_!(1)$. Therefore $\tilde{Z}$ and $\tilde{Z}''$ give the same class. By symmetry, $\tilde{Z}'$ and $\tilde{Z}''$ also give the same class. □

We thus get a homomorphism $[] : Z^p(X) \to H^{2p}(X,\mathbb{Z})$ by sending $\sum n_i Z_i \mapsto \sum n_i [Z_i]$. The space of algebraic cohomology classes is given by

$$H^{2p}_{\text{alg}}(X,\mathbb{Z}) = \text{im}[Z^p(X) \to H^{2p}(X,\mathbb{Z})],$$

$$H^{2p}_{\text{alg}}(X,\mathbb{Q}) = \text{im}[Z^p(X) \otimes \mathbb{Q} \to H^{2p}(X,\mathbb{Q})].$$

We define the space of codimension-p Hodge cycles on X to be

$$H^{2p}_{\text{hodge}}(X) = H^{2p}(X,\mathbb{Q}) \cap H^{pp}(X)$$

and let $H^{2p}_{\text{hodge}}(X,\mathbb{Z})$ denote the preimage of $H^{pp}(X)$ in $H^{2p}(X,\mathbb{Z})$.

Lemma 19.4.2. $H^{2p}_{\text{alg}}(X,\mathbb{Z}) \subseteq H^{2p}_{\text{hodge}}(X,\mathbb{Z})$

Proof. It is enough to prove that the fundamental class $[Z]$ of a codimension-p subvariety is a Hodge class. Let $\tilde{Z} \to Z$ be a resolution of singularities, and let $i : \tilde{Z} \to X$ be the natural map. By Corollary 12.2.10, the map

$$i_! : H^0(\tilde{Z}) = \mathbb{Z} \to H^{2p}(X,\mathbb{Z})(-p)$$

is a morphism of Hodge structures. Therefore it takes 1 to a Hodge class.

In more down-to-earth terms, this amounts to the fact that for any form α of type $(r, 2n-2p-r)$ on X,

$$\int_{\tilde{Z}} i^* \alpha = 0$$

unless $r = n - p$. □

The *Hodge conjecture* asserts the converse.

Conjecture 19.4.3 (Hodge). $H^{2p}_{\text{alg}}(X, \mathbb{Q})$ and $H^{2p}_{\text{hodge}}(X, \mathbb{Q})$ coincide.

Note that in the original formulation, $\mathbb{Z}$ was used in place of $\mathbb{Q}$, but Atiyah and Hirzebruch have shown that this version is false [7]. It is also worth pointing out that Voisin [114] has shown that the Hodge conjecture (in various formulations) can fail for compact Kähler manifolds. On the positive side, we should mention that for $p = 1$, the conjecture is true — even over $\mathbb{Z}$ — by the Lefschetz $(1,1)$ theorem, Theorem 10.3.3. We prove that it holds for $p = \dim X - 1$.

Proposition 19.4.4. *If the Hodge conjecture holds for X in degree $2p$ (i.e., if $H^{2p}_{\text{alg}}(X, \mathbb{Q}) = H^{2p}_{\text{hodge}}(X, \mathbb{Q})$) with $p < n = \dim X$, then it holds in degree $2n - 2p$.*

Proof. Let L be the Lefschetz operator corresponding to a projective embedding $X \subset \mathbb{P}^N$. Then for any subvariety Y, we have $L[Y] = [Y \cap H]$, where H is a hyperplane section chosen in general position. It follows that L^{n-2p} takes $H^{2p}_{\text{alg}}(X)$ to $H^{2n-2p}_{\text{alg}}(X)$. Moreover, the map is injective by hard Lefschetz. Thus

$$\dim H^{2p}_{\text{hodge}}(X) = \dim H^{2p}_{\text{alg}}(X) \leq \dim H^{2n-2p}_{\text{alg}}(X) \leq \dim H^{2n-2p}_{\text{hodge}}(X).$$

On the other hand, L^{n-2p} induces an isomorphism of Hodge structures $H^{2p}(X, \mathbb{Q})(p-n) \cong H^{2n-2p}(X, \mathbb{Q})$, and therefore an isomorphism $H^{2p}_{\text{hodge}}(X) \cong H^{2n-2p}_{\text{hodge}}(X)$. This forces equality of the above dimensions. □

Corollary 19.4.5. *The Hodge conjecture holds in degree $2n - 2$. In particular, it holds for three-dimensional varieties.*

Given a cycle $Y \in Z^{n-p}(X)$, define the intersection number

$$Z \cdot Y = \int_X [Z] \cup [Y] \in \mathbb{Z}.$$

This can be defined by purely algebrogeometric methods over any field [42].

Definition 19.4.6. A cycle $Z \in Z^p(X)$ is said to be homologically equivalent to 0 if $[Z] = 0$. It is numerically equivalent to 0 if for any $Y \in Z^{n-p}(X)$ we have $Z \cdot Y = 0$. Two cycles are homologically (respectively numerically) equivalent if their difference is homologically (respectively numerically) equivalent to 0.

Numerical equivalence is a purely algebrogeometric notion, independent of any cohomology theory. (This is clearly an issue in positive characteristic where one has several equally good cohomology theories, such as the various ℓ-adic theories.) On the other hand, it is usually easier to prove things about homological equivalence. For example, $H^{2p}_{\text{alg}}(X)$, which is $Z^p(X)$ modulo homological equivalence, is finitely generated, since it is sits in the finitely generated group $H^{2p}(X,\mathbb{Z})$. Therefore $Z^p(X)$ modulo numerical equivalence is also finitely generated, because clearly homological equivalence implies numerical equivalence. The converse is one of Grothendieck's standard conjectures.

Conjecture 19.4.7 ("Conjecture D"). Numerical equivalence coincides with homological equivalence.

In order to explain the relationship to Hodge, we state another of Grothendieck's conjectures.

Conjecture 19.4.8 ("Conjecture A"). The Lefschetz operator induces an isomorphism on the spaces of algebraic cycles

$$L^i : H^{n-i}_{\text{alg}}(X,\mathbb{Q}) \to H^{n+i}_{\text{alg}}(X,\mathbb{Q})$$

There are several other conjectures, which we will not state. One of them, which is a version of the Hodge index theorem, is true over $\mathbb{C}$. The remaining conjectures are known to follow if Conjecture A is true for all X [54, 71]. These conjectures are weaker than Hodge's and are known in many more cases. For example, they are known for all abelian varieties, while Hodge is still open for this class.

Proposition 19.4.9. *If the Hodge conjecture holds for X, then Conjecture A will also hold for it. If Conjecture A holds for X, then conjecture D holds for it.*

Proof. By the hard Lefschetz theorem,

$$L^i : H^{n-i}(X,\mathbb{Q}) \to H^{n+i}(X,\mathbb{Q})$$

is an isomorphism of vector spaces. It follows that L^i gives an isomorphism $H^{n-i}(X) \cong H^{n+i}(X)(i)$ of Hodge structures. Therefore it induces an isomorphism on the spaces of Hodge cycles

$$H^{n-i}_{\text{hodge}}(X,\mathbb{Q}) \to H^{n+i}_{\text{hodge}}(X,\mathbb{Q}).$$

This is an isomorphism of the spaces of algebraic cycles, since we are assuming the Hodge conjecture.

By the Hodge index theorem, Theorem 14.1.4, we get a positive bilinear form Q on $H^i(X)$ given by

$$Q(\alpha,\beta) = \int_X \alpha \cup \beta',$$

where

$$\beta' = \sum \pm L^{n-i+j}\beta_j$$

with $\beta = \sum L^j \beta_j$ the decomposition into primitive parts in Theorem 14.1.1. Suppose that Conjecture A holds for X. Then we have a Lefschetz decomposition on the space of algebraic cycles

$$H^{2p}_{\text{alg}}(X) = \bigoplus L^j(P^{2p-2j}(X) \cap H^{2p-2j}_{\text{alg}}(X)).$$

Therefore $\beta' \in H^{2n-2p}_{\text{alg}}(X)$ whenever $\beta \in H^{2p}_{\text{alg}}(X)$. Suppose that $Z \in Z^p(X)$ is numerically equivalent to 0. Then $[Z] = 0$, since otherwise we get a contradiction $Z \cdot [Z]' = Q([Z],[Z]) > 0$. □

Further information about the Hodge conjecture, and related conjectures, can be found in Lewis [80]. See André [3] for an introduction to motives, which have been lurking behind the scenes. The historically minded reader will find a fascinating glimpse into how these ideas evolved in the letters of Grothendieck and Serre [21].

Exercises

19.4.10. Prove that the Hodge conjecture holds for products of projective spaces. (Hint: the Hodge conjecture is trivially true for a variety whose cohomology is spanned by algebraic cycles.)

19.4.11. Let X be an n-dimensional smooth projective variety. Another of Grothendieck's standard conjectures asserts that the components in $H^i(X,\mathbb{Q}) \otimes H^{2n-i}(X,\mathbb{Q})$ of the class of the diagonal $[\Delta] \in H^*(X \times X, \mathbb{Q})$ under the Künneth decomposition are algebraic. Show that this follows from the Hodge conjecture.

19.5 Problem of Computability

As we saw in this book, it is relatively straightforward to compute Hodge numbers. For things like hypersurfaces, we obtained formulas. More generally, given explicit equations for a subvariety $X \subset \mathbb{P}^n$, we may use the following strategy for computing Hodge numbers.

- View the sheaves Ω^p_X as coherent sheaves on $\mathbb{P}^n$. Specifically:

$$\Omega^p_X = \Omega^p_{\mathbb{P}^n}/(\mathscr{I}\Omega^p_{\mathbb{P}^n} + d\mathscr{I} \wedge \Omega^{p-1}_{\mathbb{P}^n}),$$

  where $\mathscr{I}$ is the ideal sheaf of X. These sheaves can be given an explicit presentation by combining this formula with the presentation

$$\mathscr{O}_{\mathbb{P}^n}(-p-2)^{\binom{n+1}{p+2}} \to \mathscr{O}_{\mathbb{P}^n}(-p-1)^{\binom{n+1}{p+1}} \to \Omega^p_{\mathbb{P}^n}$$

  coming from Corollary 17.1.3.

- Resolve these as in Theorem 15.3.9.
- Calculate cohomology using the resolution.

This can be turned into an algorithm using standard Gröbner basis techniques. (We are using the term "algorithm" somewhat loosely, but to really get one we should assume that the coefficients are given to us in some computable subfield of $\mathbb{C}$ such as $\mathbb{Q}$ or $\mathbb{Q}(\sqrt{2},\pi)$.)

If one wanted to verify (or disprove) the Hodge conjecture in an example, one would run up against the following problem, which by contrast to the case of Hodge numbers seems extremely difficult:

Problem 19.5.1. Find an algorithm to compute the dimensions of $H^*_{\text{alg}}(X,\mathbb{Q})$ and $H^*_{\text{hodge}}(X,\mathbb{Q})$ given the equations (or some other explicit representation).

The following special case of the problem would already be interesting and probably very hard.

Problem 19.5.2. Find an algorithm for computing the Picard number of a surface in $\mathbb{P}^3$ or of a product of two curves.

We end with a few comments. Given a variety X, the Hodge structure on $H^k(X)$ is determined by the period matrices

$$P^p = \left(\int_{\gamma_i} \omega_j\right),$$

where γ_i is a basis of $H_k(X,\mathbb{Z})$ and ω_j a basis of $H^{k-p}(X,\Omega^p_X)$. When X is defined over $\overline{\mathbb{Q}}$, we should in principle be able to compute the entries of these matrices to any desired degree of accuracy, by combing symbolic methods with numerical ones. But this does not (appear to) help. However, Kontsevich and Zagier [74] propose that there may be an algorithm to determine whether such a number (which they call a period) is rational. More generally, we can ask for an algorithm for deciding whether any finite set of periods is linearly dependent over $\mathbb{Q}$. Such an algorithm would be instrumental in finding an algorithm for computing $\dim H^*_{\text{hodge}}(X,\mathbb{Q})$.

Regarding $H^*_{\text{alg}}(X)$, Tate [113] made a conjecture that can be loosely viewed as an arithmetic version of the Hodge conjecture, although there is even less evidence for it. Suppose that X is defined over $\overline{\mathbb{Q}}$ and in fact for simplicity $\mathbb{Z}$. We obtain varieties X_p defined by reducing X mod p. These are smooth for all but finitely many p. Given an algebraic cycle $Z \in Z^i(X)$ on X defined over $\mathbb{Z}$ (but this is not essential), we get induced cycles $Z_p \in Z^i(X_p)$. We can form a fundamental class in $[Z_p] \in H^{2i}_{et}(\bar{X}_p,\mathbb{Q}_\ell) \cong H^{2i}_{et}(\bar{X},\mathbb{Q}_\ell)$, and it will be an eigenvector for the Frobenius F_p with eigenvalue exactly p^i for $p \gg 0$. Tate conjectured conversely that the dimension of the intersection of the p^ith eigenspaces of F_p, as p varies, is precisely $H^{2i}_{\text{alg}}(X,\mathbb{Q})$. Even without assuming the conjecture, this should give some sort of bound on $H^{2i}_{\text{alg}}(X,\mathbb{Q})$. The challenge would be to make this effective.

19.6 Hodge Theory without Analysis

As we have seen, Hodge theory has a number of important consequences for the cohomology of a smooth complex projective variety X:

(1) Hodge decomposition
$$\sum_{p+q=i} h^{pq}(X) = b_i(X).$$

(2) Hodge symmetry
$$h^{pq}(X) = h^{qp}(X).$$

(3) Kodaira vanishing
$$H^i(X, \Omega_X^n \otimes L) = 0$$
for $i > 0$, $n = \dim X$, and L an ample line bundle on X.

Thanks to GAGA, we can replace (the dimensions of) the above analytic cohomology groups by their algebraic counterparts. To be precise, for $b_i(X)$ we can use the dimension of either the hypercohomology $\dim \mathbb{H}^i(X, \Omega_X^\bullet)$ or some suitable Weil cohomology such as ℓ-adic theory. A rather natural question, which occurs for example in [53], is whether these consequences can be established directly without analysis. In particular, can these be extended to arbitrary fields? First, the bad news: the answer to the second question is in general no. Counterexamples have been constructed in positive characteristic by Mumford [90], Raynaud [96], and others. In spite of this, the first question has a positive answer. Faltings [36] gave the first entirely algebraic proof of (1). This was soon followed by an easier algebraic proof of (1) and (3) by Deligne and Illusie [29], which made surprising use of characteristic-p techniques. An explanation of their proof can be found in Esnault and Viehweg's book [35].

The only thing left is see how to prove (2) without harmonic forms. In outline, first apply the decomposition (1) and hard Lefschetz (which also has an algebraic proof [27]) to get
$$h^{pq}(X) = h^{n-q,n-p}(X).$$
Now combine this with Serre duality [60, Chapter III, Corollary 7.13],
$$h^{n-q,n-p}(X) = h^{qp}(X).$$

At this point, we should remind ourselves that Hodge theory gives much more than the items (1), (2), and (3). For instance, we have seen how to associate a canonical Hodge structure to every smooth projective variety over $\mathbb{C}$. As far as the author knows, there is no purely algebraic substitute for this. Nevertheless, we can devise the following test to see how close we can get. Suppose that X is a smooth complex projective variety defined by equations $\sum a_{i_0\ldots i_n} x_0^{i_0}\cdots x_n^{i_n} = 0$ with coefficients in $\overline{\mathbb{Q}} \subset \mathbb{C}$ (or some other algebraically closed subfield). Given $\sigma \in \mathrm{Gal}(\overline{\mathbb{Q}}/\mathbb{Q})$, we get a new complex variety X_σ defined by $\sum \sigma(a_{i_0\ldots i_n}) x_0^{i_0}\cdots x_n^{i_n} = 0$.

Problem 19.6.1. Now suppose that Y is another smooth projective variety defined over $\overline{\mathbb{Q}}$ such that $H^i(X,\mathbb{Q}) \cong H^i(Y,\mathbb{Q})$ as Hodge structures. Show that $H^i(X_\sigma,\mathbb{Q}) \cong H^i(Y_\sigma,\mathbb{Q})$ as Hodge structures for every $\sigma \in \mathrm{Gal}(\overline{\mathbb{Q}}/\mathbb{Q})$.

If we assumed the Hodge conjecture, we would get a solution as follows. The isomorphism $H^i(X) \cong H^i(Y)$ would give a class in $H^*_{\mathrm{hodge}}(X \times Y)$, which would be an algebraic cycle α, necessarily defined over a field $\overline{\mathbb{Q}}(t_1,\dots,t_N)$. After specializing the t_i, we can assume that α is defined over $\overline{\mathbb{Q}}$. Then $\sigma^*(\alpha)$ would induce the desired isomorphism $H^i(X_\sigma,\mathbb{Q}) \cong H^i(Y_\sigma,\mathbb{Q})$. To make this work, we really need only the following weak form of the Hodge conjecture due to Deligne [31]:

Conjecture 19.6.2 ("Hodge cycles are absolute"). If $\alpha \in \mathbb{H}^{2p}(X,\Omega^\bullet_X)$ is a rational (p,p) class, then $\sigma^*\alpha$ is a rational (p,p) class on X_σ.

This conjecture is known in many more cases than the Hodge conjecture. Although it looks rather technical, it does have some down-to-earth applications to showing that certain natural constants are algebraic numbers.

References

1. Ahlfors, L.V.: Complex analysis, third edn. McGraw-Hill Book Co., New York (1978). An introduction to the theory of analytic functions of one complex variable, International Series in Pure and Applied Mathematics
2. Amorós, J., Burger, M., Corlette, K., Kotschick, D., Toledo, D.: Fundamental groups of compact Kähler manifolds, *Mathematical Surveys and Monographs*, vol. 44. American Mathematical Society, Providence, RI (1996)
3. André, Y.: Une introduction aux motifs (motifs purs, motifs mixtes, périodes), *Panoramas et Synthèses [Panoramas and Syntheses]*, vol. 17. Société Mathématique de France, Paris (2004)
4. Arapura, D.: Fundamental groups of smooth projective varieties. In: Current topics in complex algebraic geometry (Berkeley, CA, 1992/93), *Math. Sci. Res. Inst. Publ.*, vol. 28, pp. 1–16. Cambridge Univ. Press, Cambridge (1995)
5. Arbarello, E., Cornalba, M., Griffiths, P.A., Harris, J.: Geometry of algebraic curves. Vol. I, *Grundlehren der Mathematischen Wissenschaften [Fundamental Principles of Mathematical Sciences]*, vol. 267. Springer-Verlag, New York (1985)
6. Arnold, V.I., Guseĭn-Zade, S.M., Varchenko, A.N.: Singularities of differentiable maps. Vol. II, *Monographs in Mathematics*, vol. 83. Birkhäuser Boston Inc., Boston, MA (1988). Monodromy and asymptotics of integrals, Translated from the Russian by Hugh Porteous, Translation revised by the authors and James Montaldi
7. Atiyah, M.F., Hirzebruch, F.: Analytic cycles on complex manifolds. Topology **1**, 25–45 (1962)
8. Atiyah, M.F., Macdonald, I.G.: Introduction to commutative algebra. Addison-Wesley Publishing Co., Reading, Mass.-London-Don Mills, Ont. (1969)
9. Barth, W.P., Hulek, K., Peters, C.A.M., Van de Ven, A.: Compact complex surfaces, *Ergebnisse der Mathematik und ihrer Grenzgebiete. 3. Folge. A Series of Modern Surveys in Mathematics [Results in Mathematics and Related Areas. 3rd Series. A Series of Modern Surveys in Mathematics]*, vol. 4, second edn. Springer-Verlag, Berlin (2004)
10. Beauville, A.: Complex algebraic surfaces, *London Mathematical Society Lecture Note Series*, vol. 68. Cambridge University Press, Cambridge (1983). Translated from the French by R. Barlow, N. I. Shepherd-Barron and M. Reid
11. Berline, N., Getzler, E., Vergne, M.: Heat kernels and Dirac operators. Grundlehren Text Editions. Springer-Verlag, Berlin (2004). Corrected reprint of the 1992 original
12. Bidal, P., de Rham, G.: Les formes différentielles harmoniques. Comment. Math. Helv. **19**, 1–49 (1946)
13. Bloch, S., Gieseker, D.: The positivity of the Chern classes of an ample vector bundle. Invent. Math. **12**, 112–117 (1971)
14. Bott, R., Tu, L.W.: Differential forms in algebraic topology, *Graduate Texts in Mathematics*, vol. 82. Springer-Verlag, New York (1982)

15. Bourbaki, N.: Commutative algebra. Chapters 1–7. Elements of Mathematics (Berlin). Springer-Verlag, Berlin (1998). Translated from the French, Reprint of the 1989 English translation
16. Carlson, J.: Extensions of mixed Hodge structures. In: Journées de Géometrie Algébrique d'Angers, Juillet 1979/Algebraic Geometry, Angers, 1979, pp. 107–127. Sijthoff & Noordhoff, Alphen aan den Rijn (1980)
17. Carlson, J., Müller-Stach, S., Peters, C.: Period mappings and period domains, *Cambridge Studies in Advanced Mathematics*, vol. 85. Cambridge University Press, Cambridge (2003)
18. Cartan, H., Serre, J.P.: Un théorème de finitude concernant les variétés analytiques compactes. C. R. Acad. Sci. Paris **237**, 128–130 (1953)
19. Chavel, I.: Eigenvalues in Riemannian geometry, *Pure and Applied Mathematics*, vol. 115. Academic Press Inc., Orlando, FL (1984). Including a chapter by Burton Randol, With an appendix by Jozef Dodziuk
20. Clemens, C.H.: A scrapbook of complex curve theory, *Graduate Studies in Mathematics*, vol. 55, second edn. American Mathematical Society, Providence, RI (2003)
21. Colmez, P., Serre, J.P. (eds.): Correspondance Grothendieck-Serre. Documents Mathématiques (Paris) [Mathematical Documents (Paris)], 2. Société Mathématique de France, Paris (2001)
22. Cox, D., Little, J., O'Shea, D.: Ideals, varieties, and algorithms, third edn. Undergraduate Texts in Mathematics. Springer, New York (2007). DOI 10.1007/978-0-387-35651-8. URL http://dx.doi.org/10.1007/978-0-387-35651-8. An introduction to computational algebraic geometry and commutative algebra
23. Deligne, P.: Théorème de Lefschetz et critères de dégénérescence de suites spectrales. Inst. Hautes Études Sci. Publ. Math. (35), 259–278 (1968)
24. Deligne, P.: Théorie de Hodge. II. Inst. Hautes Études Sci. Publ. Math. (40), 5–57 (1971)
25. Deligne, P.: La conjecture de Weil. I. Inst. Hautes Études Sci. Publ. Math. (43), 273–307 (1974)
26. Deligne, P.: Poids dans la cohomologie des variétés algébriques. In: Proceedings of the International Congress of Mathematicians (Vancouver, B. C., 1974), Vol. 1, pp. 79–85. Canad. Math. Congress, Montreal, Que. (1975)
27. Deligne, P.: La conjecture de Weil. II. Inst. Hautes Études Sci. Publ. Math. (52), 137–252 (1980). URL http://www.numdam.org/item?id=PMIHES_1980__52__137_0
28. Deligne, P., Griffiths, P., Morgan, J., Sullivan, D.: Real homotopy theory of Kähler manifolds. Invent. Math. **29**(3), 245–274 (1975)
29. Deligne, P., Illusie, L.: Relèvements modulo p^2 et décomposition du complexe de de Rham. Invent. Math. **89**(2), 247–270 (1987). DOI 10.1007/BF01389078. URL http://dx.doi.org/10.1007/BF01389078
30. Deligne, P., Katz, N.: Groupes de monodromie en géométrie algébrique. II. Lecture Notes in Mathematics, Vol. 340. Springer-Verlag, Berlin (1973). Séminaire de Géométrie Algébrique du Bois-Marie 1967–1969 (SGA 7 II), Dirigé par P. Deligne et N. Katz
31. Deligne, P., Milne, J.S., Ogus, A., Shih, K.y.: Hodge cycles, motives, and Shimura varieties, *Lecture Notes in Mathematics*, vol. 900. Springer-Verlag, Berlin (1982)
32. Diamond, F., Shurman, J.: A first course in modular forms, *Graduate Texts in Mathematics*, vol. 228. Springer-Verlag, New York (2005)
33. Eisenbud, D.: Commutative algebra, *Graduate Texts in Mathematics*, vol. 150. Springer-Verlag, New York (1995). With a view toward algebraic geometry
34. Eisenbud, D., Harris, J.: The geometry of schemes, *Graduate Texts in Mathematics*, vol. 197. Springer-Verlag, New York (2000)
35. Esnault, H., Viehweg, E.: Lectures on vanishing theorems, *DMV Seminar*, vol. 20. Birkhäuser Verlag, Basel (1992)
36. Faltings, G.: p-adic Hodge theory. J. Amer. Math. Soc. **1**(1), 255–299 (1988). DOI 10.2307/1990970. URL http://dx.doi.org/10.2307/1990970
37. Fantechi, B., Göttsche, L., Illusie, L., Kleiman, S.L., Nitsure, N., Vistoli, A.: Fundamental algebraic geometry, *Mathematical Surveys and Monographs*, vol. 123. American Mathematical Society, Providence, RI (2005). Grothendieck's FGA explained

38. Forster, O.: Lectures on Riemann surfaces, *Graduate Texts in Mathematics*, vol. 81. Springer-Verlag, New York (1991). Translated from the 1977 German original by Bruce Gilligan, Reprint of the 1981 English translation
39. Freitag, E., Kiehl, R.: Étale cohomology and the Weil conjecture, *Ergebnisse der Mathematik und ihrer Grenzgebiete (3) [Results in Mathematics and Related Areas (3)]*, vol. 13. Springer-Verlag, Berlin (1988). Translated from the German by Betty S. Waterhouse and William C. Waterhouse, With an historical introduction by J. A. Dieudonné
40. Fulton, W.: Algebraic curves. Advanced Book Classics. Addison-Wesley Publishing Company Advanced Book Program, Redwood City, CA (1989). An introduction to algebraic geometry, Notes written with the collaboration of Richard Weiss, Reprint of 1969 original
41. Fulton, W.: Introduction to toric varieties, *Annals of Mathematics Studies*, vol. 131. Princeton University Press, Princeton, NJ (1993). The William H. Roever Lectures in Geometry
42. Fulton, W.: Intersection theory, *Ergebnisse der Mathematik und ihrer Grenzgebiete. 3. Folge. A Series of Modern Surveys in Mathematics [Results in Mathematics and Related Areas. 3rd Series. A Series of Modern Surveys in Mathematics]*, vol. 2, second edn. Springer-Verlag, Berlin (1998)
43. Fulton, W., Harris, J.: Representation theory, *Graduate Texts in Mathematics*, vol. 129. Springer-Verlag, New York (1991). A first course, Readings in Mathematics
44. Gelfand, S.I., Manin, Y.I.: Methods of homological algebra, second edn. Springer Monographs in Mathematics. Springer-Verlag, Berlin (2003)
45. Godement, R.: Topologie algébrique et théorie des faisceaux. Hermann, Paris (1973). Troisième édition revue et corrigée, Publications de l'Institut de Mathématique de l'Université de Strasbourg, XIII, Actualités Scientifiques et Industrielles, No. 1252
46. Grauert, H., Peternell, T., Remmert, R. (eds.): Several complex variables. VII, *Encyclopaedia of Mathematical Sciences*, vol. 74. Springer-Verlag, Berlin (1994). Sheaf-theoretical methods in complex analysis, Current problems in mathematics. Fundamental directions. Vol. 74 (Russian), Vseross. Inst. Nauchn. i Tekhn. Inform. (VINITI), Moscow
47. Grauert, H., Remmert, R.: Coherent analytic sheaves, *Grundlehren der Mathematischen Wissenschaften [Fundamental Principles of Mathematical Sciences]*, vol. 265. Springer-Verlag, Berlin (1984)
48. Griffiths, P.: On the periods of certain rational integrals. I, II. Ann. of Math. (2) 90 (1969), 460-495; ibid (2) **90**, 465–541 (1969)
49. Griffiths, P., Harris, J.: Principles of algebraic geometry. Wiley Classics Library. John Wiley & Sons Inc., New York (1994). Reprint of the 1978 original
50. Griffiths, P., Schmid, W.: Recent developments in Hodge theory: a discussion of techniques and results. In: Discrete subgroups of Lie groups and applicatons to moduli (Internat. Colloq., Bombay, 1973), pp. 31–127. Oxford Univ. Press, Bombay (1975)
51. Grothendieck, A.: Sur quelques points d'algèbre homologique. Tôhoku Math. J. (2) **9**, 119–221 (1957)
52. Grothendieck, A.: La théorie des classes de Chern. Bull. Soc. Math. France **86**, 137–154 (1958)
53. Grothendieck, A.: The cohomology theory of abstract algebraic varieties. In: Proc. Internat. Congress Math. (Edinburgh, 1958), pp. 103–118. Cambridge Univ. Press, New York (1960)
54. Grothendieck, A.: Standard conjectures on algebraic cycles. In: Algebraic Geometry (Internat. Colloq., Tata Inst. Fund. Res., Bombay, 1968), pp. 193–199. Oxford Univ. Press, London (1969)
55. Grothendieck, A., Dieudonné, J.: Éléments de géométrie algébrique. I, II, III, IV. Inst. Hautes Études Sci. Publ. Math. (4, 8, 11, 17, 24, 28, 32) (1960-1967)
56. Guillemin, V., Pollack, A.: Differential topology. AMS Chelsea Publishing, Providence, RI (2010). Reprint of the 1974 original
57. Harder, G., Narasimhan, M.S.: On the cohomology groups of moduli spaces of vector bundles on curves. Math. Ann. **212**, 215–248 (1974/75)
58. Harris, J.: Algebraic geometry, *Graduate Texts in Mathematics*, vol. 133. Springer-Verlag, New York (1995). A first course, Corrected reprint of the 1992 original

59. Hartshorne, R.: Varieties of small codimension in projective space. Bull. Amer. Math. Soc. **80**, 1017–1032 (1974)
60. Hartshorne, R.: Algebraic geometry. Springer-Verlag, New York (1977). Graduate Texts in Mathematics, No. 52
61. Hatcher, A.: Algebraic topology. Cambridge University Press, Cambridge (2002)
62. Hironaka, H.: Resolution of singularities of an algebraic variety over a field of characteristic zero. I, II. Ann. of Math. (2) 79 (1964), 109–203; ibid. (2) **79**, 205–326 (1964)
63. Hirzebruch, F.: Topological methods in algebraic geometry. Classics in Mathematics. Springer-Verlag, Berlin (1995). Translated from the German and Appendix One by R. L. E. Schwarzenberger, With a preface to the third English edition by the author and Schwarzenberger, Appendix Two by A. Borel, Reprint of the 1978 edition
64. Hodge, W.V.D.: The theory and applications of harmonic integrals. Cambridge, at the University Press (1952). 2d ed
65. Hodge, W.V.D.: The topological invariants of algebraic varieties. In: Proceedings of the International Congress of Mathematicians, Cambridge, Mass., 1950, vol. 1, pp. 182–192. Amer. Math. Soc., Providence, R. I. (1952)
66. Hörmander, L.: An introduction to complex analysis in several variables, *North-Holland Mathematical Library*, vol. 7, third edn. North-Holland Publishing Co., Amsterdam (1990)
67. Huybrechts, D.: Complex geometry. Universitext. Springer-Verlag, Berlin (2005). An introduction
68. Iversen, B.: Cohomology of sheaves. Universitext. Springer-Verlag, Berlin (1986)
69. Katz, N.M.: An overview of Deligne's proof of the Riemann hypothesis for varieties over finite fields. In: Mathematical developments arising from Hilbert problems (Proc. Sympos. Pure Math., Vol. XXVIII, Northern Illinois Univ., De Kalb, Ill., 1974), pp. 275–305. Amer. Math. Soc., Providence, R.I. (1976)
70. Kelley, J.L.: General topology. Springer-Verlag, New York (1975). Reprint of the 1955 edition [Van Nostrand, Toronto, Ont.], Graduate Texts in Mathematics, No. 27
71. Kleiman, S.L.: Algebraic cycles and the Weil conjectures. In: Dix esposés sur la cohomologie des schémas, pp. 359–386. North-Holland, Amsterdam (1968)
72. Kodaira, K.: Collected works. Vol. I, II, III. Princeton University Press (1975)
73. Kollár, J.: The structure of algebraic threefolds: an introduction to Mori's program. Bull. Amer. Math. Soc. (N.S.) **17**(2), 211–273 (1987). DOI 10.1090/S0273-0979-1987-15548-0. URL http://dx.doi.org/10.1090/S0273-0979-1987-15548-0
74. Kontsevich, M., Zagier, D.: Periods. In: Mathematics unlimited—2001 and beyond, pp. 771–808. Springer, Berlin (2001)
75. Lamotke, K.: The topology of complex projective varieties after S. Lefschetz. Topology **20**(1), 15–51 (1981). DOI 10.1016/0040-9383(81)90013-6. URL http://dx.doi.org/10.1016/0040-9383(81)90013-6
76. Lang, S.: Algebra, *Graduate Texts in Mathematics*, vol. 211, third edn. Springer-Verlag, New York (2002)
77. Lazarsfeld, R.: Positivity in algebraic geometry. I, *Ergebnisse der Mathematik und ihrer Grenzgebiete. 3. Folge. A Series of Modern Surveys in Mathematics [Results in Mathematics and Related Areas. 3rd Series. A Series of Modern Surveys in Mathematics]*, vol. 48, 49. Springer-Verlag, Berlin (2004). Classical setting: line bundles and linear series
78. Lazarsfeld, R.: Positivity in algebraic geometry. II, *Ergebnisse der Mathematik und ihrer Grenzgebiete. 3. Folge. A Series of Modern Surveys in Mathematics [Results in Mathematics and Related Areas. 3rd Series. A Series of Modern Surveys in Mathematics]*, vol. 49. Springer-Verlag, Berlin (2004). Positivity for vector bundles, and multiplier ideals
79. Lefschetz, S.: L'analysis situs et la géométrie algébrique. Gauthier-Villars, Paris (1950)
80. Lewis, J.D.: A survey of the Hodge conjecture, *CRM Monograph Series*, vol. 10, second edn. American Mathematical Society, Providence, RI (1999). Appendix B by B. Brent Gordon
81. Looijenga, E.: Motivic measures. Astérisque (276), 267–297 (2002). Séminaire Bourbaki, Vol. 1999/2000
82. Mac Lane, S.: Categories for the working mathematician, *Graduate Texts in Mathematics*, vol. 5, second edn. Springer-Verlag, New York (1998)

83. Matsuki, K.: Introduction to the Mori program. Universitext. Springer-Verlag, New York (2002)
84. Milgram, A.N., Rosenbloom, P.C.: Harmonic forms and heat conduction. I. Closed Riemannian manifolds. Proc. Nat. Acad. Sci. U. S. A. **37**, 180–184 (1951)
85. Milne, J.S.: Étale cohomology, *Princeton Mathematical Series*, vol. 33. Princeton University Press, Princeton, N.J. (1980)
86. Milnor, J.: Morse theory. Based on lecture notes by M. Spivak and R. Wells. Annals of Mathematics Studies, No. 51. Princeton University Press, Princeton, N.J. (1963)
87. Milnor, J.: Singular points of complex hypersurfaces. Annals of Mathematics Studies, No. 61. Princeton University Press, Princeton, N.J. (1968)
88. Moishezon, B.: On n-dimensional compact complex varieties with n algebraically independent meromorphic functions. I-III. Am. Math. Soc., Transl., II. Ser. **63**, 51–177 (1967)
89. Morrow, J., Kodaira, K.: Complex manifolds. AMS Chelsea Publishing, Providence, RI (2006). Reprint of the 1971 edition with errata
90. Mumford, D.: Pathologies of modular algebraic surfaces. Amer. J. Math. **83**, 339–342 (1961)
91. Mumford, D.: Lectures on curves on an algebraic surface. With a section by G. M. Bergman. Annals of Mathematics Studies, No. 59. Princeton University Press, Princeton, N.J. (1966)
92. Mumford, D.: The red book of varieties and schemes, *Lecture Notes in Mathematics*, vol. 1358, expanded edn. Springer-Verlag, Berlin (1999). Includes the Michigan lectures (1974) on curves and their Jacobians, With contributions by Enrico Arbarello
93. Mumford, D.: Abelian varieties, *Tata Institute of Fundamental Research Studies in Mathematics*, vol. 5. Published for the Tata Institute of Fundamental Research, Bombay (2008). With appendices by C. P. Ramanujam and Yuri Manin, Corrected reprint of the second (1974) edition
94. Patodi, V.K.: Curvature and the eigenforms of the Laplace operator. J. Differential Geometry **5**, 233–249 (1971)
95. Peters, C.A.M., Steenbrink, J.H.M.: Mixed Hodge structures, *Ergebnisse der Mathematik und ihrer Grenzgebiete. 3. Folge. A Series of Modern Surveys in Mathematics [Results in Mathematics and Related Areas. 3rd Series. A Series of Modern Surveys in Mathematics]*, vol. 52. Springer-Verlag, Berlin (2008)
96. Raynaud, M.: Contre-exemple au "vanishing theorem" en caractéristique $p > 0$. In: C. P. Ramanujam—a tribute, *Tata Inst. Fund. Res. Studies in Math.*, vol. 8, pp. 273–278. Springer, Berlin (1978)
97. Reid, M.: Undergraduate algebraic geometry, *London Mathematical Society Student Texts*, vol. 12. Cambridge University Press, Cambridge (1988)
98. Rolfsen, D.: Knots and links, *Mathematics Lecture Series*, vol. 7. Publish or Perish Inc., Houston, TX (1990). Corrected reprint of the 1976 original
99. Seifert, H., Threlfall, W.: Seifert and Threlfall: a textbook of topology, *Pure and Applied Mathematics*, vol. 89. Academic Press Inc. [Harcourt Brace Jovanovich Publishers], New York (1980). Translated from the German edition of 1934 by Michael A. Goldman, With a preface by Joan S. Birman, With "Topology of 3-dimensional fibered spaces" by Seifert, Translated from the German by Wolfgang Heil
100. Serre, J.P.: Faisceaux algébriques cohérents. Ann. of Math. (2) **61**, 197–278 (1955)
101. Serre, J.P.: Géométrie algébrique et géométrie analytique. Ann. Inst. Fourier, Grenoble **6**, 1–42 (1955–1956)
102. Serre, J.P.: Analogues kählériens de certaines conjectures de Weil. Ann. of Math. (2) **71**, 392–394 (1960)
103. Serre, J.P.: A course in arithmetic. Springer-Verlag, New York (1973). Translated from the French, Graduate Texts in Mathematics, No. 7
104. Shafarevich, I.R.: Basic algebraic geometry, study edn. Springer-Verlag, Berlin (1977). Translated from the Russian by K. A. Hirsch, Revised printing of Grundlehren der mathematischen Wissenschaften, Vol. 213, 1974
105. Silverman, J.H.: The arithmetic of elliptic curves, *Graduate Texts in Mathematics*, vol. 106. Springer-Verlag, New York (1992). Corrected reprint of the 1986 original

106. Silverman, J.H.: Advanced topics in the arithmetic of elliptic curves, *Graduate Texts in Mathematics*, vol. 151. Springer-Verlag, New York (1994)
107. Siu, Y.T.: Invariance of plurigenera. Invent. Math. **134**(3), 661–673 (1998). DOI 10.1007/s002220050276. URL http://dx.doi.org/10.1007/s002220050276
108. Spanier, E.H.: Algebraic topology. Springer-Verlag, New York (1981). Corrected reprint
109. Spivak, M.: Calculus on manifolds. A modern approach to classical theorems of advanced calculus. W. A. Benjamin, Inc., New York-Amsterdam (1965)
110. Spivak, M.: A comprehensive introduction to differential geometry. Vol. One. Published by M. Spivak, Brandeis Univ., Waltham, Mass. (1970)
111. Sternberg, S.: Lectures on differential geometry, second edn. Chelsea Publishing Co., New York (1983). With an appendix by Sternberg and Victor W. Guillemin
112. Stone, A.H.: Paracompactness and product spaces. Bull. Amer. Math. Soc. **54**, 977–982 (1948)
113. Tate, J.T.: Algebraic cycles and poles of zeta functions. In: Arithmetical Algebraic Geometry (Proc. Conf. Purdue Univ., 1963), pp. 93–110. Harper & Row, New York (1965)
114. Voisin, C.: A counterexample to the Hodge conjecture extended to Kähler varieties. Int. Math. Res. Not. (20), 1057–1075 (2002). DOI 10.1155/S1073792802111135. URL http://dx.doi.org/10.1155/S1073792802111135
115. Voisin, C.: Hodge theory and complex algebraic geometry. I, *Cambridge Studies in Advanced Mathematics*, vol. 76. Cambridge University Press, Cambridge (2002). DOI 10.1017/CBO9780511615344. URL http://dx.doi.org/10.1017/CBO9780511615344. Translated from the French original by Leila Schneps
116. Voisin, C.: Hodge theory and complex algebraic geometry. II, *Cambridge Studies in Advanced Mathematics*, vol. 77. Cambridge University Press, Cambridge (2003). Translated from the French by Leila Schneps
117. Warner, F.W.: Foundations of differentiable manifolds and Lie groups, *Graduate Texts in Mathematics*, vol. 94. Springer-Verlag, New York (1983). Corrected reprint of the 1971 edition
118. Weibel, C.A.: An introduction to homological algebra, *Cambridge Studies in Advanced Mathematics*, vol. 38. Cambridge University Press, Cambridge (1994)
119. Weil, A.: Numbers of solutions of equations in finite fields. Bull. Amer. Math. Soc. **55**, 497–508 (1949)
120. Wells Jr., R.O.: Differential analysis on complex manifolds, *Graduate Texts in Mathematics*, vol. 65, third edn. Springer, New York (2008). With a new appendix by Oscar Garcia-Prada
121. Weyl, H.: On Hodge's theory of harmonic integrals. Ann. of Math. (2) **44**, 1–6 (1943)
122. Weyl, H.: The concept of a Riemann surface. Translated from the third German edition by Gerald R. MacLane. ADIWES International Series in Mathematics. Addison-Wesley Publishing Co., Inc., Reading, Mass.-London (1964)

Index

Printed by Publishers' Graphics LLC
BT20130401.09.07.36